The Interlopers

THE INTERLOPERS

*Early Stuart Projects and
the Undisciplining of Knowledge*

VERA KELLER

Johns Hopkins University Press
Baltimore

To Toby Takusagawa

© 2023 Johns Hopkins University Press
All rights reserved. Published 2023
Printed in the United States of America on acid-free paper

2 4 6 8 9 7 5 3 1

Johns Hopkins University Press
2715 North Charles Street
Baltimore, Maryland 21218
www.press.jhu.edu

Library of Congress Cataloging-in-Publication Data is available.

ISBN 978-1-4214-4592-2 (hardcover)
ISBN 9978-1-4214-4593-9 (ebook)

A catalog record for this book is available from the British Library.

Special discounts are available for bulk purchases of this book. For more information,
please contact Special Sales at specialsales@jh.edu.

CONTENTS

The impetus for this book stems from a question I could not answer in my 2008 PhD dissertation research on Cornelis Drebbel (1572–1633) (Princeton, 2008) under the guidance of Tony Grafton and with committee members Thomas DaCosta Kaufmann and Peter Lake (and external examiner Tara Nummedal). How could it be, I wondered, that someone aggressively anarchical like Drebbel could not just survive but seemingly flourish in a hierarchical setting like the early Stuart court? I came to understand that this was due not to Drebbel but to the entire intellectual setting he joined. That realization launched the research that became this book.

During this long transformation and development of my research interests, I accumulated debts to innumerable individuals, only a few of whom I can acknowledge here. These include the many with whom I have held conversations or from whom I received advice and encouragement concerning this project, who read parts of it in draft, or who invited me to present on this material: Catherine Abou-Nemeh, Eric Ash, Tina Asmussen, Ann Blair, Arianna Borrelli, Klaas van Berkel, Paola Bertucci, Simon Brown, William Cavert, Stephen Clucas, Fokko Jan Dijksterhuis, Mordechai Feingold, Paula Findlen, Anne Goldgar, Matt Jones, Helge Jordheim, Lisa Kattenberg, Marika Keblusek, Ted McCormick, Elaine Leong, Pamela Long, Peter Mancall, Alex Marr, Elizabeth McHenry, Michelle DiMeo, Martin Mulsow, Carol Pal, Cesare Pastorino, Denise Phillips, Ulinka Rublack, Richard Serjeantson, Margaret Schotte, Gary Shaw, Lisa Skogh, Pamela Smith, Caroline Surman, Rienk Vermij, Carl Wennerlind, Kelly Whitmer, Jessica Wolfe, Elizabeth Yale, Koji Yamamoto, and Anya Zilberstein.

I also owe a vast debt to the many archivists and librarians in the collections and libraries in Europe and the United States where I have done research that has fed into this book, from my days as a graduate student until now. I made especially heavy use of the University of Oregon's Interlibrary Loan program. I completed this project during the COVID-19 pandemic, when travel was often not possible, and I relied upon many digital resources, some open access—such as the *Hartlib Papers* and the Internet Archive, for which I am very grateful—and some through subscription by the University of Oregon—such as Early English Books Online and JSTOR. For a limited time, I had access to Virginia Company Archives Online, State Papers Online, and East India Company Archives Online, through which I was able to supplement research previously done on site in the British Library and at The National Archives, Kew. Thanks to Michele Pflug for checking on a source for me. I am also grateful to the Rijksmuseum for its generous image licensing policy. This book has also benefited from a year of leave enabled by the Charles A. Ryskamp Research Fellowship from the American Council of Learned Societies in 2015–16.

Parts of the book draw upon work I previously published in earlier form as articles and book chapters. These include "Air-Conditioning Jahangir: The 1622 English Great Design, Climate and the Nature of Global Projects," *Configurations* 21 (2013): 331–67; "Deprogramming Baconianism: The Meaning of *Desiderata* in the Eighteenth Century," *Notes and Records of the Royal Society* 72, no. 2 (2018): 119–37; "Into the Unknown: Clues, Hints and Projects in the History of Knowledge," *History and Theory* 59, no. 4 (2020): 86–110; and "A 'Wild Swing to Phantsy': The Philosophical Gardener and Emergent Experimental Philosophy in the Seventeenth-Century Atlantic World," *Isis* 112, no. 3 (2021): 507–30.

Ted McCormick has been my long-term coprojector on the history of projects and deserves to be singled out. Co-organizing with him a 2012 international conference at the Early Modern Studies Institute at the Huntington Library on projects and coediting the special issue in *Early Science and Medicine* that emerged from it was a formative experience.

I dedicate this book to my daughter, Toby Takusagawa, whose generation will have to fashion a more limited place for humans in the world than that claimed by the interlopers.

The Interlopers

Figure 1. Science and *Curiosity.* Cesare Ripa, *Iconologia* (Venice: Pezzana, 1669), 554 and 179. Courtesy of University of Oregon Libraries Special Collections and University Archives

Introduction

The Undisciplining of Knowledge

... they do invade
All as their prize, turn pirates here at land,
Have their Bermudas, and their straits in the Strand.

—*Ben Jonson, "An Epistle to Sir Edward Sackville,
now Earl of Dorset"*

Up for Grabs

In Cesare Ripa's book of personifications, two figures embody knowledge in very different fashions (figure 1). Well-groomed Science stands serenely. She embodies certainty, order, and agreement. Her neatly draped, blue and gold robes indicate her heavenly status. Two petite wings sprouting from her head raise her thoughts above the mortal coil. In one hand she balances an orb and a triangle. In the other she holds a mirror at which she gazes, because Science has no need to look at anything but herself. In the mirror, she considers abstractions from sense perceptions, which she comes to understand not through external evidence but by considering "their Essence" with her mind. Her smooth orb signifies the "Uniformity of Opinion." Her triangle indicates the almost mechanical way in which certain knowledge is produced from logical procedures: just as three even sides make up a triangle the three universal and unchanging terms deployed in logical propositions together produce science.[1]

In contrast to Science, Curiosity stands in desperate need of a good hairdresser. Her hair springs wildly upward with the energy of her thoughts, exceeding the frame of the print. She throws up her hands and "thrusts out her Head in a prying Posture." Her humongous wings attach to her body rather than to her head, since the world in which she flies about, hungry for knowledge, is mired in material and bodily needs and desires. She runs "up and down to hear, and to see, as some do after News." Ears, symbolizing "the Itch of knowing more than concerns her," cover her dress. So do frogs as "Emblems of Inquisitiveness, by reason of their goggle-Eyes."[2] Curiosity is desiring, unstable, always prying into uncertain, changing forms of knowledge. She finds secrets of all sorts—of art, nature, and empire—irresistible.

Science holds knowledge securely in place; with Curiosity, knowledge is up for grabs. Science represents knowledge combed by the severe structures of logical reasoning; Curiosity lets it all hang loose. This book tells the story of what happened to knowledge when Curiosity's style became the fashion. Many accounts of the changes of knowledge in the early modern period stress increasing order, discipline, community-building, collaboration, consensus, and etiquette. In this volume, by contrast, I argue that when we put our knowledge in the mirror and take a good look, Curiosity is still what we see, not Science's logical abstractions and rigorous methods. We have been hanging loose for the past four hundred years.

For good reasons, Science had installed safeguards on knowledge to keep Curiosity in her place. She restrained humans from situating themselves and their fallible senses as the arbiters of what knowledge mattered. These guardrails restricted rationality to highly formal and artificial modes of reckoning. They delineated careful distinctions between forms of knowledge, with appropriate levels of certainty and modes of reasoning pertaining to each.[3] Although precise social and epistemic divisions varied from place to place and were continually being renegotiated, these broadly similar assumptions pertained across Europe. The most elevated were those who prayed (and studied), the clerks. They represented the mind, using reason, like the petite wings on Science's head, to reach up to the heavens and free themselves from the errors of the senses. At the heart were those who fought, the nobility, animated by spirited passions and ambition. Below, in the region of the gut and sexual organs, were those who desired, that is, everybody else, embroiled in an insatiable, mercenary hunt to satisfy their hungers.

Like tendrils intertwining in the stony lacework of a cathedral, these orders, levels, and hierarchies knit together soaring architectures of society, knowledge, and the cosmos. There was a place in the cosmos for everything. Interlocking structures resonated, from the stars in the night sky, to the understanding of the human body, the seasons, parts of the world, colors, music, architectural forms, household ornament and practice, types of knowledge, and levels of society.

Together they created a magnificent symphony—both stunningly beautiful and horribly constraining.[4]

These intersecting hierarchies also wove the texture of everyday life. Consider the image of abundance and domestic stability in one cushion cover (figure 2). Hierarchy, this image suggests, offers safety and predictability. At the top of the hierarchy, we see the four elements: here, the light elements Fire and Air at the top, the heavy elements Water and Earth at the bottom. Their differing qualities determine distinct climates and their inhabitants, with hot, exotic animals clustering around dry Earth, fish swimming at Water's feet, and the English countryside enjoying a temperate climate in the middle. The parrot and monkey disturb this order somewhat; their presence next to the human figures serves to represent the nature of humans as mere imitators of divinely authored nature, since monkeys and parrots are emblems of imitation. All forms of human industry, from the windmill beneath Air to the ship at sail next to Water, imitate and rely upon the structures of the four elements.

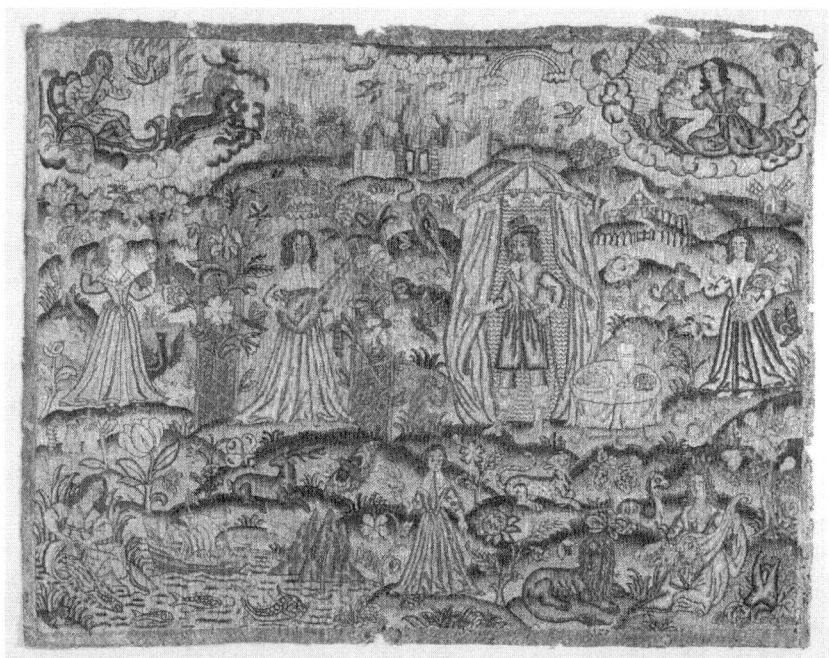

Figure 2. The Five Senses and the Four Elements. British, second quarter of the seventeenth century. Gift of Irwin Untermyer, 1964. Metropolitan Museum of Art, 64.101.1315. Public domain

The two central human figures double as representations of male and female, as well as of two of the five senses. Even among the human senses there is a hierarchy, with Sight (the least material and thus most noble sense), Sound, Taste and Smell all arrayed above Touch, the most ignoble sense. Hierarchies are also represented in the human world with a large, brick manor situated above a humbler dwelling with a thatched roof. This is a very full world. There are no empty spaces and little room to maneuver. Intersecting hierarchies interlace into a thick intellectual, cultural, and political fabric. This resilient network could handle shocks and changes, springing back from pressures or undergoing a repair from a minor tear. But thoroughly pulling out a central thread of the warp or weft meant unraveling the rest.

Order pertained in this world when everyone pursued the forms of knowledge appropriate to their place in the cosmic hierarchy. Those at the top should enjoy access to the most certain and most perfect forms of knowledge, freed as they were from the passionate, bodily entanglements of merchants and craftsmen. One Middle Scots play, *The Three Estates*, written in 1552 and published in 1603, identified deceit with merchants and falsehood with craftsmen. Yet, it also criticized the "mervelous" "misordour" in the realm by remarking upon how this world order had been turned upside down; tailors were "far mair expert / In thair pure craft and in their handie art" than were prelates "in thair vocatioun."[5] This contravened the nature of things. Purity and certainty should belong to the clerks, adulteration and doubt to the craftsmen. Such perceptions of disorder shows that these social and epistemic hierarchies were always normative but also that these norms mattered. Disorder was a symptom of disease in the body politic. For this reason, a 1533 Henrician statute demanded that things be what they seemed to be. Society ruled by a just and majestic king should be clearly legible, in everything from social status marked by sumptuary laws, to correct weights and measures, to coinage based on truly valuable metals.[6]

The collapse of such a secure ideal of order sent knowledge flying off the rails, crashing through sources of authority and security in every realm of knowledge and society. Senses, desires, passions, and ambition all challenged the sovereignty of reason. Earth, Air, Fire, and Water all came to appear fungible, with climate manipulable. Space was unbounded, and many spaces that were not in fact empty seemed empty and available for invasion and transformation. The ideal monarch as the upholder of an ordered realm where everything was what it appeared to be was dissolved by new concepts of the prince as a crafty manipulator behind the scenes. With the etymology of "king" related to "cunning," the interpretation of public affairs as deliberate dissemblance became a common practice of prudent or politic understandings of the world.[7] While order came apart at the seams, wild Curiosity came striding forth, eager to claim the hold over knowledge that Science once possessed. The unraveling of one thread affected many other intertwined domains, posing questions for scholars about changing order as much for political, social, economic, and cultural history as for intellectual history and the history of science.

Each generation has considered the wider effects of this shift through the lens of its own time and has offered differing accounts of where and when the collapse of widely accepted natural, social, and epistemic structures occurred and what it meant. For the nineteenth-century Swiss art historian Jacob Burckhardt, throwing off the weight of tradition in the early Italian Renaissance city state opened up visions of individualism and the dignity of man, embodied in polymath geniuses like Leonardo da Vinci and Michelangelo.[8] For the early twentieth-century Russian philosopher Alexandre Koyré, the opening up of the cosmic structure freed the human mind for a new understanding of nature by other geniuses like Galileo and Newton, a view of purely intellectual, apolitical origins of science that received great political support in mid-twentieth-century America.[9] On the contrary, Marxist British historians like Christopher Hill and Charles Webster integrated labor, craft, and social phenomena into stories of changing understandings of the natural world. They understood the antimonarchical social rebellion of the English Civil War as part of the effort to further human power over nature and benefit society by following what these historians termed the "Baconian program."[10]

Such accounts replaced the unraveled understandings of the basic structures of the universe by rushing to fill the void with new powers and structures, such as, in the history of science, the scientific method. The research of the last four decades has challenged these narratives, offering a much messier picture of change in which no single body held either the cultural or scientific authority to immediately calm the chaos, install a new dispensation, and provide avenues to certain knowledge. Instead, sociologists of science looked to social factors to explain the changing shapes that knowledge was taking. Most recently, historians of science have been emphasizing the roles of artisans, women, and extra-European populations in order to offer more complex accounts of the period that involve many more agents and forms of knowledge than anachronistic accounts of individual genius could.[11]

This volume aims to address a few issues that, perhaps unconsciously, persist in current efforts to hear manifold voices in the chorus of early modern knowledge, an effort that can sometimes play down the question of their dissonances. Recent work often treats the advancement of knowledge as a good for which credit ought to be shared more widely, reinforcing an idealized and normative view of what science is and thus the very ideology of modern progress that these accounts aim to subvert. The forms of knowledge included in stories of emerging science often conform to standards of objectivity, trustworthiness, certainty, and empiricism. Rather than identifying science with the characteristics we might expect to see—increased certainty, objectivity, rationality—and then seeking a wider cast of characters creating knowledge with those values, we might reject wholesale "the positivist view of the 'New Science' as the inevitable outcome of an early modern crisis of knowledge."[12] Instead, we can sit with early modern discomfort. We can try to understand the meanings that a sense of unraveling structure bore for early modern people and explore the diverse ways in which they responded to living in a time when knowledge seemed rudderless. We might offer, as María Portuondo does in the case of late sixteenth-century Spain, or as Anne Goldgar has for the early seventeenth-century Netherlands, accounts of the various knowledge cultures that responded to the breakdowns of social, moral, and epistemic safeguards.[13] Both their works are excellent examples of histories of knowledge as Martin Mulsow has urged us to write them: through thick descriptions, based in actor's categories, drawing on a wide range of endeavors, emotions, epistemologies, distal political and economic structures, and particular material and environmental affordances.[14] Emergence and entanglement also suggest correctives to the teleology that drove older styles of master narratives.[15] The response to what was a more general sense of immense change and uncertainty will thus differ widely based on particular cultures of knowledge, the commitments and concerns of local settings, and varied social positions within those settings.

In this volume, I look at what happened to knowledge in early Stuart England when order was removed in ways that not only challenged the structures of society, but even the concepts and moral valences of structure itself. The antiheroes of this story did not create the crisis of knowledge, which had much larger, long-term, and well-known causes, such as the Reformation and the challenges posed to European understandings of nature by encounters with far different epistemologies. They were opportunists, looting from the rubble that these larger changes created. Yet their actions not only further destabilized knowledge but also turned destabilization itself into a distinctive and influential ideal and a practice. While Portuondo compares the *Magnum Opus* of Benito Arias Montano (1527–1598) to the later and better known *Novum Organum* of Francis

Bacon (1561–1626) as two similar attempts to offer overarching new structures of knowledge, in fact the two differ sharply. Arias Montano's response to doubt was to chart a newly authoritative, comprehensive epistemology grounded on undeniable evidence of the Bible and nature, once and for all. Bacon, by contrast, never finalized such a conclusive new epistemology. Rather, informed by the probabilistic, risk-taking, and politicized setting in which he operated, Bacon leaned into destabilization, deploying it to impel the continued further advancement of knowledge.

I argue that what typified shifts in early Stuart knowledge was not greater certainty, order, and method, but doubt, chaos, and a loosening of knowledge structures. Moreover, such stretching and loosening across knowledge domains was encouraged in some socially elite quarters. Initiatives and enterprises that overcame limits offered ways to augment the power of the state. Elites disdained social divisions in types of knowledge. They arrogated even the traditionally lowest forms of knowledge for themselves, including the mechanical arts that involved physical labor. This arrogation of craft knowledge went beyond the engagement with some sophisticated and luxurious manual arts, such as ivory turning and clockmaking, that European princes had long justified as a means to develop their ingenuity. Early Stuart social elites involved themselves with toxic, polluting, large-scale, and money-making industries that had no such redeeming qualities, such as soap, alum, and saltpeter manufactures. Meanwhile, figures near the bottom of the social hierarchy also claimed ownership of knowledge that was traditionally not theirs to handle. The destruction of widely accepted views of natural order offered opportunities, from heads of state to inventors, to remake the world into a space of infinite possibility.

By focusing on knowledge, I hope to avoid some of the normative associations of the term "science." This means engaging not only with a wide variety of forms of knowledge but also with failure and ignorance. Knowledge, especially in its more destabilized, high-risk forms, moves through the world suffused with half-truths, silences, and emotions running the gamut from arrogance, greed, ambition, and fear, to pleasure, wonder, and delight. Knowledge entangles with power dynamics and can be deployed as much to enslave as to liberate. A portrait of knowledge on the loose engages with both emotion and with forms of power and knowledge suppression that Robert Proctor and London Schiebinger have dubbed "agnotology."[16] I am not arguing that that looseness, social injustice, or those agnotological practices meant that scientific and technical knowledge did not grow over the course of this period. In fact, if the advancement of science and technology could be neatly separated from the power, profit, and corruption that this book explores, then that would weaken my argument concerning the importance and agency of the undisciplining of knowledge. Rather, it is precisely because undisciplining knowledge was also intellectually productive that it challenges us to consider the ways in which unequal power relations and plunder were inseparable from the growth of knowledge and rooted within the very idea of its advancement.

When the safeguards of science were removed, Curiosity could really let her hair down. The period of Curiosity's ascension over Science generated innumerable techniques for further loosening of knowledge and abandoning restraint, from essays, to projects, to wish lists, to hints, and sketches. Rather than systematic, methodical accounts, these were the signature forms of early modern natural knowledge, and especially of those forms of knowledge that engaged experimentation and innovation. As William Watts, an early commentator on experimental philosophy, told Samuel Hartlib in 1635, method would not allow for the increase of learning, since "Systematical Method is like a bag or sack w*h*ich come[s] bound vp."[17] Curiosity let the cat out of the bag, and an array of undisciplined forms spilled forth: incipient suggestions, poorly supported inferences, incoherent snippets of knowledge, futuristic imagining, and jumbles of many types of evidence, argument, and explanation. The extreme intellectual looseness and ambition of the early Stuart period was, I argue, particularly provocative for knowledge innovation and growth. However, if we accept that knowledge is "never pure," to borrow Steven Shapin's phrase, we might consider knowledge in all time periods as a mixture, that is, neither completely pure nor completely worthless.

Knowledge Monsters

In the terminology of the period, which so valued stable nomenclature, these unstable knowledge hybrids would be considered monsters. Since at least the time when Pliny described the waters of the Nile as always generating new, grotesque hybrid beings, monsters were seen as contraventions of natural, stable orders, the product of unrestrained mixing. They were collected in curiosity cabinets for their shock value, as they jarred and questioned systematic understandings of nature.[18] They showed how many phenomena in this world could not be predicted by the abstract reasonings of science. The most remarkable part of this story is how Curiosity, attended by pullulating hordes of knowledge monsters, managed to put on the robes of Science and assume the authoritative mantle of philosophical science for herself and her companions without in essence ever really changing who she was.

Itchy ears, goggle-eyes, and all, insatiable Curiosity flies forth still today in the guise of science, although all our scientific knowledge remains merely probable—at most falsifiable but never provable—as well as often hybrid and metamorphic, continually pushing up against the boundaries of established disciplines and generating new interdisciplinary knowledge clusters. In a pendant book project, *Curating the Enlightenment*, I will explore how the undisciplined forms of knowledge I investigate in the current volume, which were largely developed outside the formal structure of the academy, were later imported back into academic disciplines. The merger of these two contrary tendencies of bounded structures and boundary-abrogating innovation endures in research today, which continually generates new knowledge amalgams.[19]

Despite this ongoing hybridity and metamorphosis, Curiosity has been able to assume the authority of Science in part due to a later belief that her passions have been restrained, quelled, and yoked by new techniques. These techniques supposedly disciplined knowledge by tying down every step of knowledge production to a series of narrow, seemingly mechanical steps, thus excluding leaps, assumptions, biases, and other forms of human error. Scholars, and periodically, members of the popular press, have pointed out repeatedly the mythical nature of scientific method.[20] Nevertheless, an idea of increasing restraint, certainty, and method still lurks in accounts of the period.

One of the main questions in the recent history of science has been how knowledge blends and hybridizes, given the now accepted diverse sources and characteristics of early modern knowledge-making. Rather than chaotic knowledge monsters, these blended forms of knowledge have often been described in mercantile terms as the results of mutually respectful trades, premised upon dispassionate, self-regulated, rational behavior in the intellectual marketplace. The blending of knowledge is sometimes described as "predisciplinary," hybrid, or mixed knowledges, brought together via "go-betweens" or other agents of communication, in a "trading zone," through the "circulation of knowledge," or according to a flow of "cognitive goods."[21] Intellectual trade is characterized as substantive and mutually beneficial exchange.[22] "Cognitive goods" are defined as "shared epistemic tools of knowledge-making disciplines that can be transferred across disciplinary boundaries," such as "methods, concepts, models, metaphors, formalisms, principles, modes of representation, argumentative and demonstrative techniques, technical instruments, institutional arrangements, and intellectual, theoretical, and epistemic virtues."[23]

Yet, what if such tools are not only shared, but stolen, if tools function as weapons, or if vices are as easily communicated as virtues? What if disciplinary boundaries are not well-constructed fences, over which good neighbors trade mutually needed tools, but rough borderlands of violent aggression or savvy dissimulation? As Lukas Rieppel, Eugenia Lean, and William Deringer have argued, "histories of the way knowledge circulates often belie . . . the arduous labor, coercion, expropriation, and at times even violence" that occurs within the circulation of knowledge.[24] When knowledge circulates, it does not do so as hygienically packaged epistemic units from which all other aspects of humanity have been sterilized. It moves through the world enmeshed with ignorance, emotions, desires, and sometimes violent interventions into the lives and knowledges of others. When knowledge monsters are unleashed, they are not easily reined in again.

This volume aims to go beyond pointing out that coercion can accompany the circulation of knowledge, in order to analyze how such coercion functioned as an agent in knowledge's movement across realms and was constitutive of new knowledge mergers. Equitable exchange, free flow, or the voluntary crossing of boundaries are not the only ways in which the blending of knowledge can occur. Theft itself holds agency by constituting a mechanism for intellectual hybridizing and innovation. For this reason, this book explores a period that has often been underemphasized in the history of science due to its corruption. However, the corruption of a society, in and of itself, does not mean that that particular society was not a knowledge-producing one, unless we conceptualize knowledge in highly normative ways. Rather, corruption can itself be seen as a powerful vehicle of knowledge mergers, in the ways that it disregarded property and propriety.

This book presses into the canker of corruption in that society: projects proposed and pursued in Europe, Asia, Africa, the Americas, and the Arctic. Projects—from *projicere,* to throw forward—were understood as "purpose, designment, forecast, intent."[25] They once held meanings now obsolete, including the sense of a project not just as a mental plan, but as a specific type of written document with its own characteristics as a textual genre. This book concentrates on the early Stuart period, not as the very first instance of the appearance of projecting as a practice or of the project as a written text, but as a period when projects were not only extremely politically contentious, but also particularly unhinged in their imaginative visions and loose forms of reasoning. The term "projector," as the figure who devised and led a project, was new to the early Stuart period, and long remained unique to English, with terms such as *Projektemacher*, for instance, imported into German only in the eighteenth century.[26]

Early modern projects were engines of bricolage, mixing a hodgepodge of arguments, interests, political connections, and skills in a competitive, for-profit marketplace of ideas. They imposed upon and did damage to those sources from which they seized intellectual, material, and social resources. They also incited sometimes violent backlashes. The projects this book studies were often seen as a central cause of popular unrest, foreboding violence against the Crown that was ultimately realized. As the Earl of Northumberland, Lord High Admiral of England, wrote in 1638, at a time of Scottish defiance of the English monarch's authority, "The People thorough all England are generally so discontented, by reason of the Multitude of Projects daily imposed upon them, as I think there is reason to fear that a great Part of them will be readier to join with the Scots, than to draw their Swords in the King's Service."[27]

Projects, and even more so, the "projectors" who framed them, were the subject of immense opprobrium, and often the sharpest critiques of them were voiced by fellow projectors. Thus, the retired admiral Sir William Monson (1569–1643), dedicated one volume of his naval tracts, "containing diverse projects and strategems," to "the Projectors of this Age." "If I could think of a more proper Word than Project, to entitle this ensuing Book," wrote Monson, "I would do it; for the name of Projects, and the Inventors of them, are grown so hateful, and contemptible, that all honest Men abhor and detest them."[28]

Projects are so ubiquitous in our lives today that it is easy to overlook their relatively recent genesis in the early modern period and the extreme social opprobrium they often once evoked, although several works on the history of projects are beginning to make these points far better known. Until recently scholars have largely concentrated on the history of projects in the eighteenth century, particularly since Daniel Defoe (1660–1731) dubbed the very end of the seventeenth century a "projecting age."[29] The eighteenth century did witness many ingenious essays, satires, and defenses on the topic of projects, such as Samuel Johnson's definition of a project as "Whatever is attempted without previous certainty of success."[30] However, such reflections only emerged after two hundred years of the existence of projecting as a practice and the genre of the project as a written text. Early Stuart projecting has recently received revived attention (following foundational studies by Linda Levy Peck) through the work of Eric Ash, William Cavert, Ted McCormick, Cesare Pastorino, and Koji Yamamoto.[31] This book builds on that excellent research. However, it takes a somewhat different tack by focusing not on the expertise to be found in projects nor in their eventual taming. Rather, it leans into their wild and violent nature, looking at how, in part through corruption and aggression, projects exercised enduring agency upon knowledge by undisciplining it. Furthermore, it connects domestic and global projecting.

As this volume explores, linkages between individual industry and ideas concerning the position of the state in global commerce continually entangled global and domestic enterprise.[32] Projectors hunted out resources and knowledge wherever they could, from Virginian medicaments to English privies, for deployment in ambitious, intrusive, and probabilistic undertakings around the world. By exploring how individuals engaged in intertwined global and domestic enterprise that they conceptualized through theories of the state, the book sheds light on how state and non-state actors interacted in the creation of empire. It follows projects across a wide range of domestic enterprises, entangling such domestic industry in practices of projecting worldwide that encompassed Asia, Africa, and the Americas. Despite wider cultural and sociopolitical upheaval, such attitudes toward domestic and global projecting, and even specific projects, continued through the Interregnum and the Restoration.[33] The interloping culture of the early Stuart period was distinctive, differing from what came before, but influential for what came afterward, and therefore deserves its own study, both to understand the period on its own terms and to better understand what came out of it.

This linkage of domestic and colonial projects nuances the conclusion of Yamamoto's argument in *Taming Capitalism before Its Triumph*. Yamamoto beautifully portrays the socially coercive and divisive projects of the early Stuart period. He argues, however, that projects were tamed in the Restoration and thereafter by relying on more "benign" and less distrusted motivations, such as financial profit. However, as Yamamoto himself points out, such a shift occurred in tandem with the development of colonial and slave-based projects, and so one must ask for whom such later seventeenth-century projects could be considered benign. Capitalism, I would argue, was never tamed. The undisciplined emotional, moral, and intellectual positions of interlopers contributed to its further global unleashing. Rather than capitalism being tamed, the rest of society joined the interlopers in abandoning restraint, coming to reframe constant societal innovation as a positive value; to believe in the existence and importance of private interest in the structure of society; to see their own careers as risky ventures to be navigated through cunning strategies; to view profits as something always to be maximized; to consider nature as a resource which should be used to the uttermost; and to view the world and the lives and knowledges of others as theirs for intervening in.

Abandoning the assumption that early modern knowledge will be restrained, mutually beneficial, or morally sound opens up new questions. How did knowledge blend when the sources of knowledge disagreed with one another, were socially opposed, and were culturally heterogeneous? What happens when the advancement of one form of knowledge obliterates another? What have been the enduring consequences of knowledge that both creates and destroys?

Interlopers

This volume explores these questions by studying a particular stance toward knowledge that I characterize as "interloping." Interloping is an early seventeenth-century English coinage from the Dutch, "to run or leap between" (*inter* + *lopen*).[34] In trade the term referred to small, illicit traders making rapid, profitable forays into the realms of larger powers and authorized trades.[35] Interloper also denoted a free agent in the world of espionage. Interlopers in intelligence might be untrustworthy, but they were also valued for their ability to open "all passages" and to "find out any thing," as the colorful Sir Balthazar Gerbier (1592–1663), himself both a double agent and a projector, wrote.[36] As an identity located between trade and intelligence, the interloper is perfectly situated for this study, which aims to revisit models of intellectual commerce, and often on the seamier side of business.

One could study interloping from a variety of angles; here I do so through the practices of projecting. This focus might seem to render the adoption of the term "interloping" redundant, in place of simply using the term "projecting." I have chosen the lens of interloping for several reasons. First, projecting comes loaded with associations that would be helpful to keep at bay. Second, the term "projector" used to apply primarily to the head of a project and not to its entire personnel. Third, the term "project," refers to the charting out of a future action. While the futurity of the project is very important, it can also distract from other moves, such as the lateral appropriation and combination of many ideas and resources. Fourth, in ways that "projecting" does not, "interloping" implies taking action in contravention of something or someone else. Projectors often claimed their endeavors harmed nobody, utilized neglected resources, and maximized advantages for the good of all. Interloping highlights the ways unbounded actions always intrude into the sphere of others and come at often severe costs. There is no blank space into which action can be projected. Finally, while I am using projects as a means to explore interloping, I will situate other endeavors that were not strictly projects in the broader field of interloping, such as the plans for knowledge of Francis Bacon discussed in the final chapter, and the ways that interloping practices of thought continued in later experimental philosophy.

According to the metaphor of interloping, the "authorized" intellectual commerce that interlopers invaded would be the knowledge safeguarded and divided according to traditional social and epistemic disciplines. These included the divide between art and philosophy, the latter understood as offering causal understanding. Interlopers also undermined the further divide between the liberal (free) and mechanical (servile and profit-oriented) arts. In fact, they often reversed these traditional knowledge hierarchies, elevating the mechanical above the liberal arts, and the arts in general above rational philosophy.

As discussed above, such social and epistemic divides had been long entangled with various other categories and schemata for organizing the world. Interlopers also overran these. The nature of various social types and professionals, as well as of national types and inhabitants of different climes, were understood to be divided in classificatory frameworks such as hot, dry, cold, and wet; air, fire, earth, and water; and sanguine, choleric, melancholic, and phlegmatic. According to what is called the Hippocratic view of climate, certain areas of the world were believed to be specialized in different skills and forms of knowledge supported by the local climate. It was these divisions, according to Queen Elizabeth, that created complementary needs and resources between peoples and allowed for international trade.[37]

Gleefully transgressing assumed epistemic and natural divides, interlopers flew about the world on the wings of Curiosity. As a matter of course, they blended knowledge from a wide range of sources through theft and appropriation, raiding insights, discoveries, resources, and forms of advantage from wherever they might be found. Sometimes inspired by a dizzying sense that the structure of the world was in fact unknown, and at others galvanized by ideas about the world's infinity or nature's transmutability, interlopers denied elementary and climactic constraints, appropriating skills, resources, and knowledge from around the globe. They attempted to contravene customary rights and liberties. Viewing space as fungible, they transplanted people and nature from one continent to the next. They were particularly enamored with efforts to master and control climate, such as the arts that talented gardeners deployed to nurture up tender global exotics in chilly England, or that projectors wished to deploy to cool the palace of the Mughal emperor in sweltering Southeast Asia.

The linked dissolution in the intertwined structures of knowledge and society is perhaps still most hauntingly conveyed in the poetry of John Donne

(1572–1631). Donne deployed the newly fashionable ancient idea of atomism as a way to valorize relational structures and paint the dissolution of intermeshed hierarchies as immoral and dangerous, for society as much as for knowledge.[38] Most famously, Donne lamented the dissolution of the world into atomistic chaos in his poem on the death of Elizabeth Drury in 1611:

> 'Tis all in peeces, all coherence gone;
> All just supply, and all Relation:
> Prince, Subject, Father, Sonne, are things forgot,
> For every man alone thinkes he hath got
> To be a Phoenix, and that then can bee
> None of that kinde, of which he is, but hee.[39]

With the breakdown of interrelated nature and society, Donne depicts mankind running amok, each individual thinking themselves a peerless being, leaping alone from the ashes of extinguished fire (whose element the new philosophy had "quite put out"). Every sort of man "did in their voyage in this world's sea stray." Nothing remained to reunite them and "fasten sund'red parts in one." Incoherence, Donne concluded dismally, was "the world's condition now."

If Donne's familiar poem on Drury lamented the breakdown of knowledge, his lesser-known elegy on Julia forecast the rudiments of a horrific, new, future-oriented form of knowledge that would be generated within the broken chaos by that breakdown. Julia's mind, "that Orcus," held

> Legions of mischiefe, countless multitudes
> Of formless curses, projects unmade up,
> Abuses yet unfashion'd, thoughts corrupt
> . . . These, like those Atoms swarming in the Sunne,
> Throng in her bosome for creation.[40]

The "jaws of Orcus," according to Virgil, formed the entrance or hellmouth to all of Hades, imprisoning vast hordes of misshapen monsters. Donne's depiction of a mind as hellmouth suggests a pullulating chaos that would soon spew into the world above. "Projects unmade up" represented the ultimate form of broken knowledge, pure potentiality primed to take on new shapes. With all coherence gone, all hell broke loose. Orcs of knowledge were unleashed upon the world, and there was no one ring to rule them all. For some the breakdown of older social, natural, and epistemic structures occasioned bewilderment and distress. For others it meant an exhilarating venture into a world of conjecture and possibility. Donne himself, as we will see later, often promoted colonial projects, which makes one doubt the sincerity of the emotion his verses so gorgeously express.

With so much new doubt cast upon the structure of the world, individuals across the social spectrum seized upon forms of knowledge that, according to the lost coherent social structure described by Donne, had not previously been theirs to handle. In the early Stuart period, merchants and craftsmen integrated forms of philosophizing into their handiwork, while social elites plumbed manual and experimental knowledge. They often overrode boundaries in types and formats of knowledge casually, almost flippantly, dilating into new areas of inquiry in the most unexpected of places. The so-called metaphysical poets like Donne and Sir Frances Kynaston, who explored the latest scientific theories through verse, offer one example of this tendency; the casual appropriation of natural philosophy and causal explanation by practitioners like the gunner Robert Norton or the apothecary and gardener John Parkinson give another.[41]

The merchant and frequent financial projector Gerard Malynes (1586–1641), serves as a case in point of the casual way in which epistemic divisions were transgressed. In a seminal work of economic thought replete with innovative projects, Malynes considered grand ideas about nature. After describing experiments he performed with others to test the nature of magnets, he recounted at length an imaginary conjecture about the structure of the universe he once explored in a wide-ranging conversation. While Malynes was in Yorkshire in 1606 to develop alum and lead mines, Malynes tells us, he and Ralph Eure, 3rd Baron Eure (1558–1617) visited Tobias Matthew (1546–1628) in order to congratulate Matthew on becoming the new Archbishop of York. Later in the same work, Malynes identified Eure as one of the investors who helped him bring seventeen German workmen to the alum works in Yorkshire and to mines elsewhere. This investment came to naught, according to Malynes, because another "great personnage" (likely Robert Cecil, Earl of Salisbury) pursued his own "priuat designs" in this venture while publicly applauding the efforts of Malynes and Eure. The "actions of these two Lords [Eure and Cecil] were like unto Phaetons horses," as they pulled against one another in unbridled frenzy, working at cross-purposes and setting all "into a combustion." As a result, the German workers "went begging homewards," and the investors lost their expected future profits.[42]

The conversation between Malynes, Eure and Matthew likely took place far removed from the noxious setting of the alum works. Malynes depicts how this group, drawn from across social divides (one clergyman, one nobleman, and one merchant), together developed "conjectures" concerning the possible shape of the world. The threesome viewed the cosmic structure as lying open for their imaginative consideration. They invaded all parts of nature and dismissed the doctrinal authority of astronomers ancient and modern. They transported themselves across vast expanses of conceptual space with remarkable ease. First, they imagined themselves at the center of the earth. Then, from this "lowest Center,"

> wee did clime and ascend to the highest Climate by imaginary conceits (for so is all the studie of the Circle of the Zodiacke, and the appropriation of the twelve Signes therein:) and after many reasons of the earths stabilitie, against the Pithagorians and Copernicus doctrine of Mobility . . . we did find all this to be imaginarie, and in that consideration and imagination wee did descend to the lower Center againe; and thereupon conclude, That whereas the Center is taken to be as a point of a great Circle, and so all weightie things falling thereunto, it may as well be a great Circle whereupon all other Orbes runne in circumference circularly.[43]

After all their cosmic traveling, the group concluded that one theory "may as well be" as the other.

In this reported conversation, we might note the palpable sense of exhilaration that this idea of cosmic unknowns afforded Malynes, Matthew, and Eure, in great contrast to the sense of loss that Donne expressed. This group journeyed now up, now down, the whole time in their imaginations, which they considered just as empowered to theorize about the nature of the heavens as those of astronomers whose ideas were all imaginary anyway. Here Malynes betrayed a far different attitude toward astronomical imagination than he had earlier in the same book, when he had criticized imagination in astronomy. Rather than adhering firmly to Aristotelian positions concerning cosmic hierarchy, Malynes made opportunistic use of eclectic natural philosophical ideas in a conjectural manner.[44] Malynes set this conversation in a work whose subject was not the study of nature, but rather the furtherance of industry through risk-taking ventures. He charted a social, emotional, and cosmic setting where one might as well cross traditional boundaries and make mental leaps into the unknown.

Riskscapes

The premodern landscape of knowledge mapped onto other structures conceptualized as natural: the human body, the four corners of the world, the temperaments of peoples. With such structures crushed, knowledge was up for grabs. The incursions of interlopers across climates, practices, and customs further eroded categories. We might consider what replaced the traditional, internally divided structure of knowledge, society, and nature as a riskscape, a term developed to account for how we live, take action, and feel about the world in the Anthropocene. Risk "is constituted in the face of a future that remains undecided: open to anxiety, surprise, struggle and aspiration."[45] This uncertain space offers no traditional guides or handholds. One viewpoint "may as well be" as authoritative as another, and decisions whether to venture one way or another often fall to cost-benefit analyses of what might be lost or gained.

Such unbounded landscapes might seem to offer freedom and individual agency, as in older accounts such as Burckhardt's and Koyré's, which associated the opening of world structures with the liberation of the human mind. However, this view is misleading. Unequal power relations structure varied experiences of risk and make the idea of an unpredictable, open future appear very different to different people. While some, like Malynes, imagined a riskscape as an opportunity from which they might easily extract themselves in order to embark upon new ventures, others might contemplate a riskscape with feelings of fear, subjugation, and impotency.

By the time Malynes published his account, his Yorkshire mining projects, as he notes in a later section of his book, had failed. This failure spelled horrible suffering for the hard-working laborers who went unpaid, and great damage to the environment that was subjected to a noxious process. Malynes moved relatively smoothly on to other ventures (although he would wind up in debtor's prison in 1619 due to another project that floundered). As discussed in a later chapter, after the Yorkshire alum mines failed, one of its elite projectors, Sir John Bourchier, had his enormous debts forgiven by the Crown, and he immediately reinvested in a new project. The laborers in the mines, meanwhile, were paid in paper "tickets," or IOUs. Malynes's ability to recall later his imaginary journeys with such delight speaks to the ways that emotions relate to power. The overriding of constraints upon knowledge that allowed for large-scale, risk-taking ventures might appear very differently to figures like Malynes, than to those workmen who lost their livelihood, to the residents of the area of the mines who were forced to live with its aftermath, or to the free miners whose customary access to mineral rights was curtailed by invasive projectors.[46] For those with the financial or social capital to extract themselves from failure, risk can be exhilarating. For others it can spell destruction.

Discipline, Undisciplining, and Failure

Historically, the purpose of discipline had been to prevent exactly the type of risky interventions into knowledge and denial of authority that Malynes's conversation about cosmic structure exemplified. Discipline had ancient roots in subservience to the teachings of a master, whose doctrinal authority crystallized knowledge in order to transmit it intact to the disciple.[47] Discipline aimed to control knowledge in tight, supervised forms that could be passed down as sources of authority amid chaos, preserving knowledge from the instability, uncertainty, and change that both time and the wayward nature of humans might generate. Thus, disciplines attempted to transmit knowledge in clear, crisp, authoritative, and well-delineated ways. Failure meant the loss of knowledge, a blow to civilization itself.

Schools of thought led by a master, as in Aristotle's Lyceum, might range across many realms of knowledge, but all sanctioned by the authority of its individual leader. The innovation of the medieval university was to set the responsibility for safeguarding knowledge not on the shoulders of individuals, but upon faculties, which sliced and diced knowledge, delineating the means of argument and types of proof appropriate to each, and setting expectations for the content and manner through which each member of the faculty, an expert in his own field, should transmit knowledge to the next generation. The rise of licensing bodies in universities and guilds offered official ways to supervise the transmission of knowledge, ensuring that disciplines passed on knowledge to students in the appropriate manner demanded by the status of their specific areas in the *ordo artium et scientiarum* (order of arts and sciences).[48] We might consider the medieval universities and guilds to represent the highpoint of disciplined knowledge. To achieve mastery in any field meant proving that one had acquired the techniques and knowledge of the teacher, whether those skills were displayed in an academic disputation or in a masterwork presented for guild inspection. The approaches to knowledge of a master weaver or soapmaker would be quite different from that of a master philosopher, but both passed on knowledge according to the master-student relationship in a particular discipline.

Interlopers rebelled against the rule of masters and the concept of distinct disciplines. They invaded forms of knowledge in which they had no training. They mingled approaches and subject matter from all across the knowledge hierarchies. In so doing they were able to rapidly assemble innovative knowledge mixtures, which, as they often boasted, neither traditionally learned scholars nor craftsmen would ever have devised. They also removed quality controls and safeguards for knowledge preservation. Such safeguards were intended to increase certainty and lower risk. Adopting a risk-taking, "nothing ventured, nothing gained" stance, like the "fail fast, fail often" attitude of today's disruptive Silicon Valley start-ups, interlopers brashly invaded areas in which they had no expertise and shook off failure with alarming aplomb.

Such failure was magnified by a massive growth in scale and speed that the abandonment of discipline enabled. Ensuring that work fit the qualifications of the master, with every i dotted and t crossed, meant producing neatly bounded, complete works on a scale that could be inspected. Interlopers, by contrast, hastily spun plans often of national or global proportions on the basis of newly discovered or rapidly cobbled together ideas. Often projectors did not realize the fortunes they dreamed of, especially given the lavish lifestyles that often intersected with the ambitions of elite projectors. For instance, Sir Walter Cope (1553–1614) was a noted collector, garden enthusiast, and investor in and promoter of Virginia, who participated in a "rising metropolitan fashion for subscribing to colonial projects."[49] He died deep in debt, as did Sir John Danvers (ca. 1585–1655), who participated in similar fashions and trends (and served as a witness to Cope's will).[50] Indeed, noting the appointment of Sir Walter Cope and others (including Sir Francis Bacon) to a 1612 commission "to devise projects and means for money," John Chamberlain gossiped that "the world thinks it a strange choice, since most of them are noted for not husbanding and well governing their own estate."[51] Despite the fact that the career of their ventures is shot through with failures, environmental costs, enormous disruptions to people's lives and livelihoods, lost investments, the breakdown of personal relationships, and intense societal antagonism, interlopers trained attention upon the future, continually reinvesting in hopes of their next big break that would dramatically shift the parameters of possibility for themselves and (they claimed) for the common weal.

With so many projects left unrealized or a failure, the undisciplining of knowledge might seem a very vague or broad topic for historical analysis. Indeed, projects did not fall into any category; that was what made them projects, rather than a particular craft or discipline. Overarching though they may be, projects were an actor's category for discussing just such interventions that abandoned any trade or mastery. Furthermore, undisciplining is no more of a broad process than the disciplining of knowledge, which has often been the subject of study.

Undisciplining knowledge bore its own agency, even in failure. The ability to distract from failure with promissory visions of future fulfillment represents another way that Curiosity managed to acquire Science's prestige. As Jürgen Renn points out, scientific knowledge today claims authority through its peculiarly cumulative nature, as each generation sees further by standing on the shoulders of its predecessors. However, this claim is founded on a selectively retrospective gaze, which overlooks the massive amounts of knowledge continually falling into oblivion, those great carcasses of previous scientific knowledge that do not continue on into the future.[52] Modern accounts of change over time in knowledge have tended to focus on a continual chain of new discoveries, charting a triumphant, cumulative progress into an ever-greater future, to the exclusion of what has been lost. They explore cultures of growth without reflecting upon how growth became entangled with failure and loss.[53]

Early moderns argued that as knowledge grew in one area, it did so at the expense of others. Guido Pancirolli (1523–1599) saw new discoveries (*nova reperta*) as balanced against lost things (*deperdita*). In Pancirolli's accounting, the scale tipped far to the side of loss.[54] "Le profit de l'un est dommage de l'autre" (the profit of one is the loss of another), wrote Montaigne.[55] Bacon, a reader of Pancirolli, argued, "Whatsoever is some where gotten is some where lost."[56] John Donne preached in 1628, "here in this world, knowledge is but as the earth, and ignorance as the sea; there is more sea than earth, more ignorance than knowledge; and as if the sea do gain in one place, it loses in another, so it is with knowledge too; if new things be found out, as many, and as good, that were known before, are forgotten and lost."[57]

Unless knowledge is perfect, whenever knowledge is transmitted, it is entangled with various forms of forgetting and ignorance.[58] Undisciplining knowledge unleashed a notion of futuristic, continual growth that mesmerized attention with certain glittering achievements and even brighter hopes and dreams up ahead. Worshippers of this juggernaut willingly threw self-restraint and order beneath its inexorably forward-moving wheels, despite the loss, destruction, and failure in its dust. This undisciplined culture of growth, fueled by projects, has continued to roll right through to the present day. In response to this destruction, a new project for the undoing of projects has arisen. Due to the unsustainability of rising risk levels for the entire planet stemming from a culture of growth, the Degrowth movement seeks to return to a premodern understanding of the limited human place in an interdependent cosmos.[59]

I would never attempt to argue that premodern Science had ever been kept pure of human bias, but that it was a form of knowledge specifically designed to minimize the socially situated human perspectives that Curiosity magnified. The undisciplining of knowledge allowed not only for the mixing of previously segregated forms of knowledge, but also a greater degree of intermingling of knowledge with human aims, biases, and emotions. The ascension of curiosity unbridled the passion for knowledge at unequally distributed expense, destroying some forms of knowledge while creating others. Power shaped emotional responses to epistemic risk-taking, the benefits to be gained from disruptive innovations, and even who was included in calculations concerning knowledge's hazards and benefits.

The Civilizing Process and Social Disciplining

This volume seeks to counter an old and still powerful narrative that equates modernity with discipline and restraint. In twentieth-century accounts of the origins of modernity, man disciplined his (and in such accounts, it was his) mind, abjuring myth, magic, and flights of fancy.[60] Man disciplined his will, tamping down irrational desires in favor of Neostoic right reason. Man conquered nature, chaining her with the ineluctable force and power of a mechanical worldview and a new scientific method.[61] According to these accounts, the disciplined self stood at the origins of both the scientific knower and the economic rational agent.[62] Modernity itself has been seen as premised upon a "personality disposed to regard respectfully, and to derive self-respect from participation in, the endeavor to create knowledge production, knowledge curation, and knowledge application enterprises under the flag of discipline and on the basis of an individual and collective discipline."[63] To this "modern disciplinary personality" has also been attributed "the commercial success and capitalistic accumulation" of early modern European economies.[64] A brief foray here into twentieth-century historiography will trace how the narrative of a disciplined modernity was forged and what its enduring influences have been.

As Peter Miller has written, the outsize role of discipline in the intellectual history of modernity had its roots in a "central problem" facing the German historian Wilhelm Dilthey: "the Reformation liberated the individual without supplying any theoretical structure that could make a community of such radically empowered men." The unleashed "particular passions of individuals" threatened to overturn common bonds and shipwreck societies. The solution Dilthey and many subsequent historians identified was the supreme power of reason and constancy (*constantia*), restored from the ancient Roman Stoa, which might allow man to discipline himself and pacify society.[65] Rather than obeying the master, the individual now had to apply self-discipline.

The early 1980s through the 1990s saw versions of this thesis appear in which the drivers of such supposedly interiorly generated discipline were in fact external: the expectations of social norms or the demands of the emerging absolutist state. Three works were crucial here. The first was Norbert Elias's 1939 *The Civilizing Process*, published in German in Switzerland (while the German-Jewish Elias had fled German academia to Great Britain), which appeared in English translation between 1969 and 1982.[66] It advanced the thesis that increased interconnectedness of European society heightened the need for self-regulation, a detached view of the self, the restraint of the body in public, and a pressure toward the peaceful resolution of disagreements. Quickly, Elias was criticized for his Eurocentric view of civilization. His defenders claim that his point was "not to share in European self-congratulation, but to understand the processes that led to the sense of cultural superiority."[67] Others pointed out that the Third Reich's Final Solution, following soon after the publication of Elias's book, showed where the civilizing process tended (Elias's own mother was murdered in Treblinka in 1942).[68] Nevertheless, self-control and the civilizing process or project play a large role in current accounts of early modernity, including early seventeenth-century England.[69]

A second fundamental work was Gerhard Oestreich's unfinished 1969 *Geist und Gestalt des frühmodernen Staates* (English translation published in 1982), equating discipline and the militarized state with Neostoic philosophy.[70] Oestreich's account of Neostoic discipline was both normative and teleological, smoothly migrating from the influential political views of the Flemish Justus Lipsius (1547–1606), to the development of the police state in Brandenburg-Prussia. A central vector he stressed was the martial discipline of the "Dutch Drill" of Maurice, Prince of Orange, advertised internationally through such texts and printed images as the *Exercise of Arms* (*Wapenhandelinghe* [1608]) of Jacques de Gheyn.[71] The Neostoic individual was likewise a continual soldier, a *miles perpetuus*, subjecting his will to the daily drill of discipline.[72]

Oestreich drew upon Elias, but he set the idea of self-discipline within a new context of the autocratic state, recasting the civilizing process into the social disciplining (*Sozialdisziplinierung*) demanded by absolutism.[73] Thus, Lipsius's *Politica*, which sets out a hardball politics for use by a prince, loomed larger in this interpretation of Neostoicism than did Lipsius's *de Constantia*, which offered subjects mental and moral resources for retreating from society and enduring the violent politics of the age. "Ministers and officials, officers and soldiers, entrepreneurs and artisans, in fact all the subjects of the state were disciplined in their work and their attitudes," Oestreich claimed.[74] This same discipline applied both to man's desires and his approach to nature. "He sought to attain self-control, which was the highest goal. He disciplined nature too, with the artistically clipped trees and hedges of seventeenth-century parks and gardens."[75]

At the end of the day, to Oestreich, *disciplina* appeared less in the form of self-restraint than in the state's powers of *Polizei,* a term that Oestreich defined as the extralegal authority a ruler arrogated to himself to order his state. Oestreich's work contributed to narratives about the rise of the early modern "police state," run as a machine, that have only recently been called into question.[76] He premised his model of the Neostoic disciplined police state on a view of the central European state as a completely obedient and manipulable machine, a view that itself was based on an understanding of early modern science as mechanistic, Hobbesian, and certain.[77] Both these views of the state and of science have now been revised. Mechanical philosophy did not dominate early modern science in general, and certainly not the views of nature held by cameralists, or servants of the exchequer (*Kammer*) in central Europe, who often drew on alchemy for a perfective, and thus improvable and unpredictable view of the future.[78] Heads of central European cameralist states never enjoyed the control that twentieth-century historiography claimed for them.[79] Oestreich thus does not provide a reliable guide to the level of discipline of the central European state. How much less, then, can he be relied upon for accounts of discipline in very different settings? Nevertheless, his model would be applied often to England, and to English science in particular.

Oestreich's Third Reich–inflected fantasy of a highly efficient, militarized state apparatus centrally informed his account.[80] It ran counter to Franz Borkenau's (1900–1957) anti-Fascist views of the social roles of Neostoicism. Rather than self-discipline and the curbing of one's thought within an autocratic state, Borkenau had seen in Stoicism "the self-assertion of rational independence amidst the decline of the feudal world."[81] He also saw signs of the same self-assertion popping up in a variety of locales and forms, in sometimes contradictory ways; libertinism was also a sign of the same self-assertion. Unlike Oestreich's account, Borkenau's was critical rather than normative. What was accepted as natural and rational in the period was not objectively rational.[82] Borkenau's account, however, never gained as much traction in the West as did Oestreich's.

Although Oestreich's account of the police state was a normative one, his emphasis upon discipline was reinforced by the critical picture that Michel Foucault offered of absolute governance as a system of meticulous attention and control, embodied by surveillance in the modern prison. Foucault's vision was also informed by Lipsian Neostoicism, although less centrally so.[83] For Foucault, discipline was more a tool forcefully applied to segments of the population than a choice made by individuals; Foucault's sense of discipline thus better fits the historical realities studied in this book, where enforced and enslaved laborers were subject to discipline, but those meting out that discipline prided themselves on their own unbounded agency. Despite differences in their views, Foucault nevertheless functioned alongside Elias and Oestreich to make discipline a central watchword in seminal works in the field of science, technology, and society that first emerged in the 1980s and early 1990s, and in the history of science more generally. For example, Peter Dear argued with reference to Foucault, "as with the disciplined activity of the ideal modern prison, laboratory procedures too have their authorizing protocols and their sequestered enactments, which lend their knowledge-pronouncements the air of inevitable, bureaucratic truth."[84]

Discipline and the Passions in Studies of Early Modern Science

Unlike Borkenau, Oestreich did not draw any connections between Neostoicism and science. Indeed, his translator, Helmut Georg Koenigsberger, considered "scientists . . . by the nature of their subject, largely indifferent to contact with a wide public," and thus not affected by political change as were the visual arts, political thought, and drama.[85] Although it was Borkenau, and not Oestreich, who drew a connection between Neostoicism and science, Borkenau has been largely overlooked in influential studies of early modern science.[86] By contrast, Oestreich has played an outsized role among founding sociologists of science and "externalist" historians of science. In contrast to internalist accounts that traced the development of scientific theories and discoveries in relation to one another, these historians looked to the wider culture and society to understand how science was conceptualized and practiced differently through history. However, in selecting which aspects of society to relate to science, such accounts notably zeroed in on contexts that could offer explanatory mechanisms for shaping characteristics we associate with modern science, such as dispassion, order, and objectivity.

Sociologists and historians of science turned to political and social forms of discipline to explain how knowledge could remain disciplined in the absence of scientific method or widely agreed upon theory, just as discipline had served Dilthey to explain how the individual liberated by the Reformation could remain dispassionate. Seeking to explain the rise of what was presumed to be a more disciplined way of knowing, these scholars found Oestreich's view of discipline very compelling. Self-regulation and social disciplining were equated with the etiquette of early modern experimental science. Early modern academicians were understood to have set self-imposed "limits" on their inquiries mirroring "the similarly cautious boundaries the neostoic philosophers of the period drew around their political analyses."[87] Such disciplining of the body and the self has been seen as seminal to scientific trust, verification, and thus the ability of empirical knowledge to develop.[88] English gentlemen-philosophers, it has been argued, found a point of overlap in their identities as gentlemen and as philosophers in the ways that "neo-Stoicists" defined gentlemen according to their "*integrity* and *independence.*"[89] According to one summary of this literature, scholars have "concentrated on two aspects of ethos cultivated by the early modern natural philosophers: (1) *civility* or politeness, usually associated with aristocratic cultural modes of verbal exchange . . . and (2) *asceticism* or control over the passions, which confirmed the moral probity of the natural philosopher."[90]

Early Stuart figures, like Sir Francis Bacon, have been recruited into narratives of a disciplined scientific practice often linked to a mechanistic worldview. Stephen Gaukroger, citing Oestreich, compared Bacon's "plan to direct scientific activity by inculcating new habits in scientists" with "subjection" to early modern absolutist state "regimes" requiring a "significant degree of self-regulation." He sees "self-respect, self-control, and internalized moral authority" as particularly central to Bacon's *New Atlantis.*[91] Gaukroger merges a mid-twentieth-century idea of a "Baconian program" with the sociological view from the 1980s and 1990s that Bacon's work exhibited social disciplining.[92] Thus Gaukroger describes Bacon as setting out to chart a "centrally administered natural-philosophical programme."[93] A "crucial ingredient" in this "programme" was his "reform of its practitioners," which Gaukroger situates "as part of a general concern with the reform of behaviour which began outside scientific culture but which was rapidly internalised in English natural philosophy in the seventeenth century." This was "civility."[94] It was connected to how Bacon's "method," according to Gaukroger, provided "a strict regimen which continually curbed the spontaneous tendences of the mind."[95] Much recent scholarship on Bacon has focused on the "medicine of the mind" through which, scholars have argued, he sought to prevent error through Neostoic discipline of the mind.[96]

By focusing on court etiquette and gentlemanly self-regulation, sociologists and historians of science intended to show how social concerns shaped notions of method. They therefore sought to show how defining characteristics of science were epistemologically baseless. For example, Dear argued that by "showing the crucial *groundlessness*" of the authority of the absolutist regime in early modern Europe, "we also see the fundamental groundlessness of [scientific] expertise when understood as a translation of absolutist political authority."[97] However, while removing the epistemological cause for these characteristics of science, this argument still assumes that the science of the time possessed those apparently essentially scientific characteristics of authority, inevitability, and truth. When science, as the object of investigation by historians of science, is identified by its air of impartial undeniability, then forms of knowledge that were openly politicized, motivated by interest, dubious, and contested are excluded from analysis. As a result, studies of science emphasizing civility and etiquette compartmentalized courteous aspects of experimental practice from the latter's frequent engagement with violence and greed. The outcome has been a view presenting science as "never pure," but at the same time, as one whose chief actors, such as founding fellows of the Restoration Royal Society, appear as almost crystalline in their dispassion, restraint, and moral probity.[98]

By making experimental science rely so much on the trust and verification afforded by etiquette and the civilizing processes, these accounts troublingly equate experimental science with civilized European men, suggesting that other populations, including socially inferior Europeans, could not participate openly in the development of experimental philosophy because of their overly embodied and passionate behaviors.[99] Those who did not and could not abide by elite etiquette were forced into "invisible" roles, according to Steven Shapin.[100] This emphasis on gentlemanly behavior has tended to overlook individuals who were voluble, impolite, and far from invisible. They will make themselves heard loudly in the pages of this book.

It becomes difficult to warrant identifying the architects of early modern science as dispassionate and disinterested once attention is drawn to their heavy involvement with colonialism and chattel slavery. Founding members of the Royal Society, such as Prince Rupert, Sir Kenelm Digby, and Sir Nicholas Crispe, were privateers, slave traders, and projectors.[101] Robert Boyle, a member of the Council for Foreign Plantations from 1660 to 1664, praised the slave-based sugar industry as an ideal example of how experimental philosophy benefited the public. Sugar offered "a recent Instance of the transplanting of Arts and Manufactures." When a Brazilian visitor dropped "some Hints" about sugar to a curious English planter in Barbados, the industrious Englishman, according to Boyle, successfully developed the industry in Barbados. Boyle praised the subsequent employment of thirty thousand enslaved Africans in sugar production, which he considered a chief example of how the application of experimental natural philosophy could benefit the public.[102] He attributed credit for this achievement to experimental philosophy, without acknowledging that sugar production relied on the knowledge brought by enslaved populations.[103] Elsewhere, Boyle praised the "great benefits" "not only to single persons, but to whole communities, and sometimes even to nations," afforded by "man's dominion" over creatures, "improved by sagacious and industrious virtuosi," as evidenced by sugar and tobacco.[104] He also treated the sexual violence inflicted upon African women by English men with horrifying indifference.[105] Are these examples of self-restraint and politesse?

Integrity, independence, civility, moral probity, ascetic retreat, pacifism, and self-regulation are normative ideals that do not reflect the actual practices of seventeenth-century naturalists; nor, I argue, were these even their ideals. European men of science did not produce their new forms of knowledge because they had disciplined themselves to a greater extent than had women, people further down the social scale, or peoples from beyond Europe. Rather, they improvised by disregarding disciplines and boundaries and triumphantly seizing both intellectual and material resources from elsewhere.[106] This book, by exploring the modes of knowledge operating in early Stuart projects on four continents, including American colonialism and the African slave trade, shows the violent seizure of knowledge was not only not hidden, but was celebrated as a sign of daring, heroic courage, and greatness of mind.

In his *Passions and the Interests: Political Arguments for Capitalism before its Triumph*, Albert Hirschman influentially described the "demolition of the heroic ideal."[107] He argued that "untamed and vicious passions (such as the pursuit of glory and honor) that in the Middle Ages and the Renaissance had been endorsed as heroic virtues" were replaced in the early modern period by calculations of interest, offering newly pacified forms of moral and commercial reckoning.[108] The idea that early modern knowledge producers were disciplined and dispassionate in part rests upon this larger view of commerce as less violent, less passionate, and more civilized than it was in a time dominated by noble, martial passions and the pursuit of glory. Hirschman described the advent of interest as a Kuhnian paradigm shift that suddenly appeared and was instantaneously accepted by nearly everyone "as a veritable message of salvation."[109]

Against this view, I have previously argued that the idea of interest engendered heated, critical debates in the seventeenth century, that many did not see interest as a new form of rationality distinct from the passions, and that arguments over who gets to decide what best serves the public interest, or even whether individuals are acting in their own best interest, remain contentious today.[110] In the early modern period, the nature of interest was debated across a wide variety of genres, including essays, political speeches, sermons, and very often the texts of the projects that are explored in this book. While I will flag how languages of interest appear and are deployed in projects, my argument here focuses on a different set of responses to Hirschman's thesis.

This volume emphasizes how glory and honor remained central to many intertwined marketplace and epistemic initiatives, how interest remained bound up with emotion, and how, with the founding of overseas colonies, trade and violent conquest entangled more than ever before. Given the structural role that slavery, genocide, and colonialism would come to play in global trade, the argument that eighteenth-century trade offered an alternative of gentle commerce (*doux commerce*) to previously violent martial mores could only be promulgated and maintained through a rather strenuous act of disknowledge, that is, claiming something to be true that deep down we know is not.[111] It rests on the infamous idea that slavery was in the best interests of the enslaved.[112] In essence, accepting the *doux commerce* theory means claiming that the lives lost in the violence of global trade were not lives that mattered and that violence inflicted upon non-European populations and Europeans of lower social orders was not violence.

This falsehood motivated the writing of this book as an attempt to supply a new account of intertwined intellectual and mercantile commerce that does not look away from the centrality and growing scale of violence. Hirschman himself notes that the "persistent use of the term *le doux commerce* strikes us as a strange aberration for an age when the slave trade was at its peak and when trade in general was still a hazardous, adventurous and often violent business."[113] Nevertheless, Hirschman largely accepts the view that "the expansion of commerce and industry" imposed "much needed discipline and constraints on both rulers and ruled,"[114] and that this restraint of passion ultimately produced a "comparatively peaceful, tranquil, and business-minded Europe."[115]

By contrast, I argue that noble passions such as rivalry and pursuit of glory not only endured, but expanded and developed new arenas, including new ways of linking knowledge and wealth to glory. Glory had always related to risk-taking in the form of noble adventure. As wealth became the central measure of the greatness of a state, such risks became unabashedly for-profit. As the pursuit of glory and the defense of honor entangled with the pursuit of profit, traditionally an ignoble affair, noble profiteers were eager to distinguish themselves from social inferiors. They did so by considering trade as adventures in which they were more willing than others to run great hazards and attempt the unknown, inflamed by their noble passion. Steven Connor has recently dubbed as "epistemopathy" the study of emotion in knowledge. His account, however, has not engaged with how unequal power relations constrain who feels which emotions about knowledge.[116] I investigate such emotions in knowledge as they relate to inequalities of power. Passion served as a mark of magnanimity, or greatness of soul, that set interlopers apart and justified their invasions into the more constrained realms of others.

Commercial Piracy, Projects, and Intellectual Depredation

The period's transformation of the military elite into a mercantile one involved a fusion of violence and commerce at the heart of global colonialism, the slave trade, piracy and profiteering, and other projects of attempted domination. This intermingling of martial and mercantile aims and values contributed to appropriative, violent commercial practices that in turn shaped intellectual commerce. As Adrian Johns remarks, the period in which ideas of intellectual piracy emerged was "not coincidentally" that of the corsairs.[117] Projects intertwining the study of nature and the pursuit of profit abounded in coercion of many sorts, from the forcible seizure of private property, to violent colonization in Ireland, to global enforced labor. If our own origin stories of early modern intellectual commerce do not also embrace the violent nature of trade, they serve to justify rather than to explain the emergence of modern forms of knowledge and their entanglement with projects of global domination.[118]

The Grand Remonstrance, a list of grievances presented by Parliament in 1641 to the Crown, depicted the average Englishman caught between pirate and projector, both violently restraining the liberty of others. The seas "were left so utterly unguarded that the Turkish pirates ranged through them uncontrolled . . . consigning to slavery many thousands of English subjects." Meanwhile, "a man's unavoidable daily wants . . . his trade, his employment, his habitation, anything, served as the pretext for some vexatious restraint to his liberty. If he would build near London, he found such building was adjuged a nuisance, and had to beg some projector for permission. . . . If he would trade at sea, he was surprised, even there, by the projector, as by a foreign enemy."[119] Pirates and projectors were paired because they were indeed common threats that made many feel left with little room to navigate to safety. They were also paired because they both lay at the nexus of trade and violence and thus became the lens through which the dissolution of traditional mores and social roles was explored and processed.

As Claire Jowitt has discussed, piracy became a way to talk about social and epistemic disorder, as for-profit behaviors invaded both the realms of knowledge production and of noble action that according to established norms ought to have been segregated from the pursuit of profit. What so confused and sometimes enraged contemporaries was the way violent conquest and commerce became entangled, leading to a "breakdown of the traditional ideological division between mercantile and gentlemanly behaviour."[120] I would argue that the same set of concerns, in related ways, motivated the heated discourse concerning projectors in the period.

Recent scholarship has argued that piracy was a far more prominent affair in everyday society and politics on land than we had previously thought, playing a central role in English politics around the world.[121] English shipping nearly doubled between 1582 and 1629, while Barbary pirates expanded their operations into the Atlantic in the 1620s and were even sailing up the Thames. As a result, piracy, practiced by and upon the English, became a major concern for many under the early Stuarts, with approximately 8,000 English men, women, and children captured and sold into slavery.[122] However, while piracy was certainly a real and life-threatening concern for many, the category of piracy was also highly malleable, with constantly shifting legality, along a "spectrum of piracy."[123] Both James I and Charles I mounted expeditions to quell the Barbary pirates, but the line between pirate and legitimate trader was ambiguous and porous. The admiral of one of Charles's fleets sent to suppress the pirates of Sallee, for instance, was surprised to find a ship of Sir William Courten (1572–1636), the crown financier and a scion of the Flemish mercantile diaspora, already in port, apparently amicably trading with the foe.[124] Piracy was favored, especially by Charles I, as a response to the Habsburg dominance of global trade and to Algerian piracy in the Mediterranean.[125] Several figures of this book were accused of acting as pirates, while they claimed that they were legitimate traders who were being interloped upon by others. Piracy was not a clear-cut category, but an ever-present accusation.

Contrary to the view of Francis Bacon's disciplining of knowledge, he embraced the identities of interloper and pirate in his idealized institution of Salomon's House in *New Atlantis*, his posthumously published and most widely read work. There he described the various institutional roles of knowledge production as though they were differing roles of commercial trade, an analogy that has often provided a touchstone in discussions of intellectual commerce. However, closer inspection of this commerce reveals that it does not offer an aboveboard business model of equitable exchange. The Merchants of Light "Sayl into Foreign Countries under the Names of other Nations (for our own we conceal);" they "bring us the Books, and Abstracts, and Patterns of Experiments of all other Parts," but the rest of the world knew nothing of their society in return. Another group, the "Depredators," that is, pirates, unabashedly plundered "the Experiments which are in all Books." Meanwhile, one might think that a third group, called the "mystery-men," would be devoted to upholding and protecting the knowledge of the traditional guild or "mystery" system. This was not the case. They gathered "experiments of all Mechanical Arts. And also of Liberal Sciences; And also of practices which are not brought into Arts." They were thus called mystery-men not because they possessed their own mystery or guild, but because they invaded the protections raised by the traditional crafts and the gate-keeping of academia and made off with their proprietary knowledge. Moreover, the mystery-men blended these diverse forms of knowledge together and with other forms of knowledge that did not even rise to the level of a recognized discipline. No fewer than eighteen fellows of Bacon's imagined ideal institution worked to gather knowledge from elsewhere, often by means of subterfuge, unequal exchange, and willful disrespect for traditional protections and categories. The process of innovating the most penetrating trials of nature involved first secretly taking material from all over and blending it together, before reconsidering the assemblage anew.[126]

By offering a wider portrait of interloping in early seventeenth-century England, I hope to shed further light on the culture of intellectual appropriation that Bacon here celebrates. I whole-heartedly accept Deborah Harkness's assertions that Bacon stole knowledge from others, however, I would disagree that such theft means that his work, unlike Elizabethan collaborative empiricism, proved negligible.[127] I argue that not only was such theft par for the course, but that his contribution was to teach others how to steal on ever grander scales. Bacon demonstrated how to embed and entangle a wide array of knowledge thefts into dynamics of wishful thinking. Intellectual interloping set ethical and emotional patterns for knowledge. It reframed ambition, human desire, and large-scale gambling as the virtues of epistemic entrepreneurs and conquistadors. Such epistemic risk-taking, I argue, would prove more fundamental to the future scale, ambition, and hybridity of science and technology than the careful Elizabethan empiricism and intimate sociability that Harkness describes.

Many over the subsequent centuries sought to cast themselves as adherents of Bacon because Bacon provided a rhetoric, an aesthetic, an emotional palette, and an array of literary technologies to support, deepen, and extend interloping practices in a heroic, imperial vein. The sociotechnical imaginary that Bacon provided in the *New Atlantis*, and the rich metaphors of epistemic empire he developed in the *Advancement of Learning* and the *Novum Organum*, placed at the center of natural study not empiricism, nor mechanics, nor artisanality, but the pursuit of the seemingly impossible. Bacon magnified the scale of epistemic risk-taking, showcased modes of suggestive and loose reasoning, and justified the publication of incomplete, partial, uncertain, and preliminary findings.

Over forty years ago, against the grain of contemporary interpretations of Bacon as establishing a sensible, grounded program of social benefit for all, Carolyn Merchant offered a controversial view of his works, stressing its patterns of ecological extraction and social domination.[128] This interpretation has been much criticized, most recently for her focus upon Bacon's futurist rhetoric rather than on contemporary, small-scale material practices, such as experimentation in domestic housekeeping and thrift.[129] Although there are some aspects of Merchant's interpretation with which I would disagree (such as her emphasis on mechanical philosophy), overall, the history of projects bears out her larger claims against these criticisms. Projects are the means through which material practice moves out of the small-scale and domestic setting, into a sphere of public action conceptualized in futuristic terms of transformation. Although rhetoric concerning waste-saving strategies certainly is a standard part of these for-profit projects (often by identifying any unused resource as "wasted"), the story of projects recounted in these pages is one of massive expenditure of resources. Intellectual pirates did not so much as spare a glance at the sinking hulks from which they had drawn their treasures.

Content of the Book

The first part of the book introduces us to some key overarching arguments, ideas, and historical settings. This introductory chapter has contested the view of the early modern period, and the aspiring absolutist court in particular, as an agent for disciplining both society and knowledge. Chapter 1 presents the broad structures, social dynamics, and attitudes of projecting in early Stuart England. Chapter 2 introduces us to a representative cast of characters involved in projects. Many of these figures will be central to the following chapters, where we will also meet further characters. The next part of the book dives into interloping practices on the ground, as we look at particular projects proposed for massive interventions around the world. These follow a rough trajectory from projects in locations better known to the interlopers, such as the domestic alum or soap industry, to enterprises in situations of great unknowns. However, we notice a continual back-and-forth between the domestic and the global, as domestic sites of intense experimentation also serve as ventures into great unknowns, generating modes of conceptualizing knowledge, from experimental philosophy to garden philosophy to the philosophy of the voyage (discussed in the last section of the book), that supported leaps between different kinds and sources of knowledge. In chapters 6 and 7 we explore a number of proposals to refashion knowledge in undisciplined ways, from the establishment of a new London academy mixing forms of knowledge drawn from a wide social arena to an institution for researching new inventions, the Vauxhall operatory, founded by Charles I. In the conclusion we reflect upon what the world of the interlopers can tell us about the development of experimental philosophy. We will situate Francis Bacon among the interlopers and point to some enduring legacies of early Stuart undisciplined knowledge in the "epistemology of the hint" that proved popular in the Restoration Royal Society. Finally, we will consider the ways in which knowledge today remains undisciplined. Once all hell broke loose, the gates of Orcus never shut.

CHAPTER I

The Political Economy of Projects

Projectors and the State

Projects and projectors were the object of much widespread social distrust and obloquy because they ran counter to widespread mores and expectations. Whereas "custom" was a central positive value and crux of popular politics, "innovation" was considered a danger.[1] Customary rights were understood to protect areas of common benefit.[2] Projects did not fall into the nomenclature of any customary endeavor—this was why they were called projects rather than something else. The texts of projects, by contrast, often describe common resources and spaces as underutilized and empty, and thus failing to serve the public. Such resources, spaces, and even populations served as the malleable material from which projects could be formed. Yet, those currently using or occupying those resources and spaces saw them not as empty but as already very full, and indeed as central to the commonweal. Thus, the arena of common custom became the zone of contention. Did traditional structures uphold and benefit all society, or did they conceal areas of untapped potential that could enrich the state through the unconventional plans of projectors?

Projectors were seen as undoing the structures of the world in ways that arrogated unwarranted authority. As one satirical character boasted, by taking on the trade of "projector throughout the world," he gained the ultimate power to destroy, to create, and to take away from others.

> doe what thou wilt, command what thou wilt, and take what thou wilt, who is't shall dare for to controule thee? I my selfe which now am called projector throughout the world, when first I took upon me that same trade I was not worth the third part of a groat: but now through my Projects, who of late could have done more in the world than I? what, and who was hee which I commanded not? I have pulled downe and raised up: I have snatched away from some, and given again to others some, but kept the best part to my selfe.[3]

The question "What is a projector?" recurs continually in the period.[4] This was not only because the term itself was new, but because the projector was a destabilizing identity that had no established place in society. The projector's only specific profession was intervening in other people's professions and traditional practices. Václav Hollar's 1641 print showed the patentee, a "Wolfe like devourer of the Common wealth," as a monster with a wolf's head, tenterhooks for hands, screws for feet, and a cloak sewn from a patchwork of patents in many industries, from salt to tobacco pipes to playing cards to coal to soap. The patentee asks, "Who am I, what am I like, what nobody." In Brugis's *Discovery of a Projector*, a courtier lures members of a dozen trades with the promise of rapid profits: "they took up the name Projectors . . . utterly disclaiming all their former Trades, Professions, Arts, Sciences, Mysteries, Crafts and Occupations."[5]

By definition, a projector was somebody who departed traditional crafts and professions, and the social and epistemic positions that those crafts held. Projectors ranged broadly across many forms of enterprise, seeking advantage wherever they could. They argued that such searching out of opportunity enhanced the wealth and strength of the state as a whole. This view was anathema to a conception of social harmony composed of stable parts, each set in defined relationships to each other, and together making up a just society; such professions were seen as vocations and identities and not merely a means to profit. As one Norwich preacher argued in an antiprojector sermon titled *The Projector*:

> Publike justice is that which hee doth exercise to others in the Common-wealth in his particular calling, as he is a Magistrate, Minister, Lawyer, Phisitian, Merchant Mechanicke, or the like. Hee makes a conscience of his calling, and knowes he must giue an accompt for the imployment of his talents; and therefore vseth himselfe in his place, not as if the end of his vocation were onely to gather wealth, and enrich himselfe and his posteritie, but to doe God service, and other men good.[6]

These were bedrock notions of social structure that projects and projectors undermined.

In addition to locating himself in a no-mans-land between professions, the projector also situated himself ambivalently as both an individual agent and as an internal operant in state affairs. As Eric Ash wrote in the case of the drainage of the Fens, a marshy environment that projectors wished to transform into arable land, through "the rhetoric of improvement, projectors asserted that draining the Fens would make the land more rational, reliable and productive; the inhabitants more prosperous, healthy and civilized; the commonwealth larger and richer; and the realm more governable." In short, it would "literally and figuratively expand the purview of the English state."[7]

What was the state? This has long been a contentious question. Recently, English historians such as Steve Hindle and Michael Braddick have sought to identify who precisely makes up the state through attention to a wide range of offices held in local politics. Hindle, noting the "experimental" and "ramshackle" assertions of centralized authority, described how, as a result, historians have concluded that centralized power could not be equated with the state. They therefore sought to identify the state with a broader base of people.[8] Hindle suggests that "a working definition of the early modern English state would have to stretch not only as far as individual magistrates, but beyond: to head constables, petty constables, churchwardens and overseers."[9]

By contrast, I use the term "state" as it was theorized in the Italianate notion of *ragion di stato* (reason of state), a form of political reasoning that explicitly was not identifiable with law or with regular political institutions and offices. The state meant dominion and properly belonged to a prince.[10] However, it was a porous concept that allowed for widespread participation.[11] Such participation was not that of the subjects or the citizenry; rather, participation meant identifying oneself as a knowing insider, with the majority of subjects assumed to be excluded from knowledge of state affairs. Projectors situated themselves as contributing to internal state affairs, in contrast to the subjects of the realm, who were painted as insufficiently politically adept to understand the intricacies and advantages of projects. In a 1609 essay on projectors, Sir John Melton defined them as not only "directers & plotters of state-business" but also as "Pedling informers of the State."[12] A 1620 sermon on projects described projectors as "the stirring braines of a State."[13]

Projectors assumed this role of state insiders even though they often did not hold political office (or their offices did not relate centrally to the vast array of their projects); in fact, many projectors were not English. The reason of state, which sought to drain advantages from rival lands, encouraged, as it were, a magnetic notion of the state that aimed to attract to it ever more knowledge and ability from rival states, and thus grow in power. Grants of special rights in the form of projects served as a key lodestone to attract others to participate in augmenting the powers of the state.

The Secrets of Empire, Global Survey, and the Political Arguments of Projects

The reason of state is a form of contingent, probabilistic, and ingenious thinking that pries into interstate rivalries on a global plane. It was not based upon customary, bottom-up, and widely held notions of the commonweal. In its continual search for the maintenance and expansion of power as a matter of survival in this competitive realm, the reason of state overrode local law, custom, and morals. It explicitly allowed simulation and dissimulation. According to John Donne, who took a dim view of this form of knowledge, *ragion di stato* (the reason of state often appears in Italian, even in English texts), could be defined as "deceit and falsehood" in service to the public good.[14] From the reasoning of state in the first half of the seventeenth century slowly emerged the concept of interests, including a weaponized notion of the public interest that divided a holistic notion of the commonweal or good into a field of competing interests; assertions of the public interest could therefore be deployed to define and override particular interests.[15]

The reason of state was initially developed as advice for princes; according to Donne, a synonym for *ragion di stato* was the *arcana imperii* (secrets of empire.)[16] Both were also equivalent to "police" or "policy." While today we use the term "police" to refer to a disciplinary force executed upon those who break the law, in the early modern period, *police* or policy was itself a form of abrogation of the law. It did not offer a set of predictable rules for society to follow. Even in Gerhard Oestreich's account of the early modern police state, *police* was the special prerogative of the absolute ruler to abrogate law on his own authority. It was *police* that allowed rulers to grant special "private laws" or privileges through letters patent.

Rather than setting out rules for citizens to obey, policy often used the element of surprise and disguise. In early seventeenth-century English, *policie* or *policy* signified an unpredictable, quickly shifting, clever and often depraved political trick or craft. It illustrated qualities of ingenuity, wit, instrumentality, and craft verging on deceit. "Engin" might be translated as, "An engin, toole, instrument; also, vnderstanding, policie, reach of wit; also, suttletie, fraud, craft, wilinesse, deceit."[17] Likewise, "finesse" might be understood as "Craft, subtiltie, guile, deceit, policie, cunning, fraud, sleight, slinesse, wilinesse."[18] "Machiauel-lisme," meanwhile, meant "subtill policie, cunning roguerie."[19] John Bullokar in 1616 defined "Strategeme," as "A policie or subtill deuice in warre, whereby the enemie is often vanquished."[20] A proposer of "Projects, Or New Inventions for Warr" around 1624 described several uses to which a light, easily wielded "smale peece" of ordnance could be put. Besides for the uses that the projector could think of in advance, there would be "other like uses, as the pollicy of war will soone finde out."[21] Policy was a quickly adaptive form of strategic thinking, hunting out new advantages and uses in emergent situations, giving rise to effervescent projects and designs at the center of courtly power.

The reason of state was not a form of certain knowledge, but of contingent strategizing that valued qualities of ingenuity, astuteness, opportunism, and adaptability. Rather than making claims that their knowledge was certain, projectors often criticized forms of knowledge that claimed to be universal and certain. They argued that no certain rules could be followed to ensure success in any untried endeavor. They presented risk as evidence of a greatness of mind and spirit, that if pursued, could win glory. When in 1639 Gabriel Plattes "repeatedly stressed the limitations of his knowledge," stating that no "'certaine rule can be given for so variable a worke' as husbandry," he was not countering the claims of previous projectors so much as echoing them.[22] Thus, for example, Sir Arthur Gorges argued that even when projects do not meet with success "in every point, answerable to hope and expectation," nevertheless, the desire to serve the "Common-wealth" ought to absolve such "honest designs" from any "corrupt aspertions."[23]

As a result of their state role, the structural nature of projects demanded that individuals propose interventions framed in terms of political thought, at the same time that James I and Charles I aimed to shield state matters from view. The early Stuart court is famous for the sharply possessive grip that James and Charles sought to maintain upon the *arcana imperii* (secrets of empire), which comprised discussion of state affairs.[24] If in general, according to Oestreich's theory of the disciplined modern state, aspiring absolute states attempted to keep their subjects each sticking to their last, then this was the hottest button issue for them: preventing subjects from debating public affairs and involving themselves in the business of the crown. The practice of granting privileges for projects created an unintended situation where, with the sanction and even through the demands of the crown itself, individual subjects and groups of subjects not only intervened in what were considered intimate state affairs, but penned political arguments about their interventions. Such arguments discussed and contested public and private interests, freedom, restraint, and advantage. They did so in an arena not typically considered a part of the canon of political thought. Perhaps precisely because of the suppression of more formal voices of political counsel, this undisciplined and uncustomary zone of political discussion could develop nakedly utilitarian views of state power, private profit, and money-making that ran counter to traditional political virtues of justice and restraint.[25]

These arguments engaged the core of political *arcana* and reason of state theory, and they did so in a high stakes, contested competition with rival projects, opening up political secrets to vigorous debate and critique. This was not a zone that was explicitly designed for debate by the citizenry or that had clearly defined rules of engagement. Rather, it was a jostling market of ideas. It created a volatile situation when individual proposers assumed for themselves abilities to comment on and design affairs of state at the same time that monarchs attempted to lock down the secrets of empire.

This incursion into the secrets of empire was merely the first of many lines that projects crossed. The substance of projects also mixed a wide array of forms of knowledge. By interloping into the knowledge and practices traditionally held by disciplinary bodies, such as the "mysteries" or guilds, projects mixed together secrets of art, nature, and empire from across society and around the globe.

In making their cases for how their interventions and innovations served the reason of state, projectors often referred implicitly or explicitly to the economic theories developed by Giovanni Botero (ca. 1543–1617) in works such as *Reason of State* and the *Universal Relations*, a collection of global reports on resources and political structures. Botero defined the reason of state as "knowledge of the means fit for founding, conserving and amplifying a dominion."[26] He argued that money served as the sinews of war and was thus the true moving force behind all power. Surveying the world for industries and importing them into one's own realm, as Botero did in his *Universal Relations*, offered a means for advancing one's territory through internal augmentation of power and wealth rather than through mere conquest of territory. Botero's ideas called for the valorization of industry, especially industry that offered an advantage over neighboring states, often by imitating and outcompeting their arts. This competitive economic dynamic called for continual growth of industry, since other states were likewise always on the hunt for ways to claim and naturalize the resources and abilities of other states. This set up a dynamic of surveying the world for resources, a practice that was both global and domestic, including actual processes of land surveying and enclosure or colonization.

As Linda Levy Peck has discussed, it was Jacobean policy to survey the world for knowledge and resources and transplant foreign industries onto English soil, from the Mortlake tapestry works, utilizing the designs of the German Franz Klein (ca. 1582–1658) and the skills of Netherlandish weavers, to swordsmiths brought from Solingen.[27] In this, James I followed the practices of global survey and appropriation advised in the economically oriented reason of state. Botero, for example, described how Chile "beareth all sortes of fruit brought out of Spaine, and transporteth many Cattle and store of Ostriges," while in Brazil the Portuguese had introduced "all sorts of Europe plants with good successe, and have erected many Ingenios, to try their Sugars."[28] Botero encouraged a competitive analysis of the ways that various states had seized the advantage in the balance of trade by overriding the constraints of climate and claiming parts of global nature for their own. According to one English guide to travel, those in foreign parts were prompted to "survey the best places, where those Arts are to be learned; as in *Germanie* all manner of Formers, Potters or figulists are to be found in perfection. In *Italy* Architecture, Limming, Painting, Engrauing, Imagerie, *Textorie,* and weauing, and Artes ingenious may be learned. So according to euerie Countries seuerall commodities seuerall and peculiar Artes do flourish."[29] The goal of this study of national differences, however, was to erode them by transplanting such arts to English soil.

In a pan-European economic battlefront of knowledge capture in which its honor and glory was at stake, England was far behind many of her competitors. In 1589, an anonymously published work similar to Botero's *Relations, The Political Treasury,* was published in Italian in Cologne. It purported to be a compendium of the secret and highly valued surveys of lands composed by Venetian ambassadors. Its "Discourse on England" took a dim view of England's international commercial standing.[30] *The Political Treasury* would see many editions and translations, and beginning in 1600 the title of the work also pointed out that it offered a "perfect intelligence" of the "reason of state," and the ability to gain "an entire knowledge of interests, aspirations, designs and revenues" of the world's leaders.[31] Here, an "entire knowledge" meant delving into the ways that interests and resources entangled with future plans and aspirations, not only for one region or state, but in interrelated ways around the globe. These interrelationships were ever shifting, which posed both a challenge and an opportunity. Those who could gain a global insight into the constantly morphing political kaleidoscope could use that knowledge to insert themselves to their own advantage into the changing world order. The desire to capture "an entire knowledge" grew from such ambitions as the Spanish crown's quest to gather an entire knowledge (*entera noticia*), or perfect information, of the Americas. As Arndt Brendecke has discussed, such knowledge was not understood as disinterested and objective, but as rather fully interested and politicized. Knowledge of the competing interests of various actors in New Spain could be utilized in order to create an environment of surveillance and reportage that continually conveyed knowledge inwards to the state, where it was then kept among the secrets of empire.[32]

The political essays of Sir John Melton have been cited as offering examples of how in England, also, "the early Stuart regime further developed" injunctions against commenting "on the public business of the realm" or the *arcana imperii*.[33] Yet Melton also described how projectors put forth political arguments in ways that undercut the regime's hold on the secrets of empire. In order to acquire patents for protecting the invention or importation of new trades, projectors had to demonstrate that their projects benefited the crown. They thus situated their projects within political arguments, even if their knowledge of state theory was secondhand and not as astute as they supposed. Projectors were "men that have lapped up from the vommit of other mens wits, some excrements of Court-phrases, and thereupon turne Factors about the court & contrive Proiects and strange devises of tobaco-pipes . . . brown paper, french wares . . . and a thousand of the like conceites."[34] By this Melton refers to the political arguments that perforce were required in making a case for a protective patent on any such industry. Projectors set out those recycled court-phrases and secondhand political argument within the "bundell of papers" that Melton describes the typical projector constantly carrying between "their lodgings" and "the Court."[35] Buoyed by an inflated assessment of their own political cunning, projectors would never concede, according to Melton, that their projects might be rejected through any fault of their political reasoning; rather, as Melton wrote, they "spitefullye fret and fume, and in their secret corners mutter, that fewe or none of the state were able to understand, or sound to the depth of their proiects."[36] Melton should know, since he himself was later complained of in Parliament "as a Projector in Coale at New-Castle."[37]

Such "bundells of papers," although put forward with increasing frequency through the reigns of the early Stuarts, have received little attention for many reasons. They are often archival and unpublished. They are short and informal and have not risen to the level of being considered as a genre. When they are considered, it is usually more from the perspective of the history of technology or economic history than in relation to the political ideas that these projects put forth. However, it was the political economy of projecting itself that encouraged the framing of large-scale projects in purported service to the commonweal that often included some new, newly improved, or new-to-England invention or art that could be patented.

This set up a situation riven with internal contradictions. Even while supporting the royal grip on *arcana imperii*, subjects infringed upon those secrets. For example, Sir Francis Kynaston, in the *True presentation of fore past Parliaments*, argued in favor of absolute royal control of the secrets of empire and strenuously against "popularity," the legislative power of parliaments, and being "too rigid and strict for legality in a business, that the King directs to be done."[38] At the same time, he set forth political arguments concerning the state in the framing of his project for a noble academy, and he did so in a debate with the rival arguments of competing projectors (as discussed in chapter 6).

Likewise, the royal chaplain William Loe vigorously defended the royal control over secrets of state, especially in relation to projects, in a 1623 sermon that aimed "to teach in brief . . . what the carefull condition of the king [is], what the loyall submission of a subject [is], and what proiects are onely to best purpose." Loe related the word "king" to "cunning," marveling at the "great skill" necessary in "Proiestments," that is, projects, which showed the "high pitch of policy they flie, who mannage, and sway the scepter of kingly cunning." The extreme difficulty of "policy of state" and the high rate of failure of projects was itself an argument in favor of submission to the crown, since only divinely appointed kings had access to the heavenly help that might grant projects a greater chance at success. Yet even here, in defending the exclusively divine projecting abilities of royalty, Loe drew upon practices of political argument and legal justification that would appear to infringe upon state policy. Loe set out his own views on particular policies and framed general legal conclusions such as, "All our Proiects, and Purposes are then, and then only legall, and Evangelicall, when God saith Amen unto them, otherwise to Religion, Republique, and private Estate, they prove detriments, and Deiectments."[39] The Bible offered many examples of kings who pursued impious projects and were brought down. Thus, even in a defense of the divine prerogative of kings to project, Loe entered into a zone of debating the legality and political advantages and disadvantages of projects.

Such simultaneous demands for politically savvy arguments and injunctions against making them put projectors in difficult and sometimes dangerous situations. Take the case of the patent for an office of general commerce taken out by Sir Arthur Gorges (1569–1625) and Sir Walter Cope (1553–1614). Gorges and Cope received a twenty-one-year patent to open an office of general commerce at Britain's Burse, or the New Exchange, that would centralize commercial information and offer new tools for credit. However, they could not enact this new plan without buy-in from the public, since the plan relied on private individuals volunteering their commercial information. Gorges thus published not only their patent, but an "overture" describing the political benefits of their plan. It was a project aimed "at the advancement of mutual Commerce, the bond and sinewes of humane society." As was par for the course, Gorges identified his proposed institution with "the publique good," noting that, "according to the Custome of former times and of all well policed States," the king had given "Grace and preferment to diverse honest Projects and profitable Inventions which have truly tended to the publique good without wronging the particular right of any." Preemptively, Gorges identified any potential critic of the office as an "undutifull or malignant spirit" who might doubt the justice of this new office due merely to his "self private gaine."[40] Those who cast aspersions of corruption on his project were themselves "malignant humours," that is, toxic influences within the body politic.[41]

Despite, or perhaps because of, this menacing characterization of their potential critics and their overly successful affiliation with the "well policed State," the general public proved too skittish to participate in this office. In a second pamphlet, Gorges and Cope responded to "some misreports" that had spread that their register "will bee a meanes to enter into Mens estates, and lay them open to euerie enuious Inquisitor."[42] It was also "doubted that in his Register there lurkes some mysterie and policie of State, dragging after it a hidden unconvenience not yet seene."[43] This illustrates the difficult position projectors were in when they simultaneously posed as canny strategists and informants to a "well policed State" and relied upon winning the trust of a larger public. It was precisely the language of political utility through which Gorges and Cope described their register that raised the suspicion that the register would indeed serve the *arcana imperii* and thus in some secret way be used against private individuals who entered their names there. Although arguing for the political utility of their intervention had won Gorges and Cope a patent, it spelled failure for their project.

The repercussions for voicing political views about patents and projects could be still more dire. Several of the individuals discussed in this book spent time in prison due to speech or to the perception of the seditious possibilities of their political engagement. Thomas Russell, Esq., languished in prison for several months due to the accusation that he had spoken uncivil words about King James.[44] Sir Edwin Sandys (1561–1629) was imprisoned for his work on behalf of the Virginia Company. It was suggested that the council of the company, while meeting under the "fair pretence" of developing the colony, in fact served as a "seminary for a seditious Parliament," since council members were "deep politicians, and had farther designs than a Tobacco plantation."[45] The supporter of Sandys, Nicholas Ferrar (1592–1637), suggested that it was the creatures of the Spanish ambassador who "insinuated to the King, that the matter was too high and great for private men to manage: that it was therefore proper for the King to take it into his own hands."[46] It was true that the Virginia Company took a particularly active approach in analyzing projects and how they related to the state, often in public and printed works (as discussed in chapter 5). Ultimately, James revoked the company's patent in 1624, making Virginia the first royal colony directly administered by the crown.

The Patent System

A patent, from "letters patent" or "open letters," were royal grants contained in documents that were open to view and addressed to all subjects. In this sense, unlike in modern law, there was no distinction between a patent, privilege, or what is now called copyright; these were not rights but favors that were exceptions to the law.[47] Grants made by letters patent might be of anything that lay within the royal prerogative, including land, and thus patents were central tools of colonialism. They might also confer protection from competition for an industry or invention, often one introduced from elsewhere. Patents for industry and invention are often studied separately from projects for global invasion by letters patent, but practices of projecting and the legal and political arguments made by petitioners for patents of both these kinds were related.[48] This is not surprising, given that not only was the political means for granting these favors the same, but that many undertakers participated in both sorts of enterprises.

The practice of granting patents for the exclusive exercise of an industry, often an imported one, emerged in England in the mid-sixteenth century in imitation of Italian city-states such as Venice.[49] According to this system, economic protections and rewards for enterprise were granted to individual undertakers, or more often, groups of undertakers, with a cut often going to the crown and to the various highly placed courtiers who helped to secure the patent. Thus, the patent system relied on individual enterprise rather than on formal court servants to initiate a plan and request a patent, but it also relied on the court to grant that request. This was also true of individual initiatives in global conquest.

As proposals for patents passed from the signet office, to the privy seal office, to the office of the Lord Chancellor, and to the office of the great seal, before they were enrolled in the patent rolls sent on to the Office of the Rolls Chapel, they generated large amounts of paper.[50] In the Elizabethan period, the careful William Cecil, Lord Burghley (1520–98), served as a central bureaucrat who could review requests for patents. He especially protected the production of commodities that were often imported, such as "alum, glass, soap and saltpetre";[51] these would remain goals of patentees in the early Stuart period. Burghley investigated potential new industries with an eye to lowering risk and increasing chances of success.[52] Nevertheless, success eluded most of these attempts to establish foreign industries.[53]

Despite this, rewarding patents continued under the early Stuarts, to an increasing popular outcry and at a much greater clip. Whereas Queen Elizabeth had issued fifty-five monopoly patents over forty years between 1561 and 1603, Charles I's administration would reach more than that number in just the first ten years of his reign, eventually granting ninety-five patents of protection for new inventions alone.[54] As Koji Yamamoto has pointed out, the Personal Rule of Charles I coincided with a peak in patents for new inventions that would not be matched until around 1685, during the personal rule of Charles II. It is likely that use of the patent system as a whole (not just confined to new inventions) was "greater in the 1630s than in the 1690s," Yamamoto notes, as the Crown sought alternative sources of revenue in lieu of Parliament.[55]

In 1610 King James declared monopolies illegal, except for those "projects of new invention, so they be not contrary to the Law, nor mischievous to the State, by raising prices of commodities at home, or hurt of trade, or otherwise inconvenient."[56] Thus, acquiring a patent came to require not only court connections, but an ability to situate the desired endeavor within a global economic and social viewpoint, illustrating how the endeavor would support the common good and not produce inconveniences. Only at a later date would the arguments of patent applications be standardized, so that in 1852 in England an applicant for a patent merely had to sign a declaration that an invention would prove "of great public utility."[57] Moreover, patents later became a right that individuals held to intellectual property; if they could prove that their idea was an innovation, they would receive a patent.[58] In early modern monarchies, patents were gifts from the crown made at royal discretion, and thus proposers set out ingenious arguments and claims about how a particular design augmented state powers.

Despite claims that patents would not disturb existing industries, these often proved (excuse the pun) patently false. Projects were supposed to avoid restraining the liberty of others in order to avoid being considered a monopoly. A new or new-to-the-area technique or ingredient, it was argued, should not harm existing industries as it had not existed previously. The same arguments were deployed to justify the legality of global colonial adventures, as it was argued that English invaders were making use of resources not currently in use and thus could not be damaging the interests of residents around the globe. Moreover, actions that might seem like harm or restraint, according to justifications for global violent attacks, could in fact be reinterpreted as benefit and liberation.

For example, in Robert Gray's *A Good Speed to Virginia* (1609), Gray noted the argument that the original inhabitants of the land had no concept of private property and thus "if the whole lande should be taken from there, there is not a man that can complaine of any particular wronge done unto him." Additionally, those "people are vanquished to their unspeakable profite and gaine," since they would be converted and freed from sin. Thus, by establishing that incursions into their area did not harm their interests or restrain their rights but in fact bestowed advantages upon them, Gray concluded, "we might lawfullie make warre uppon the savages of Virginia, our project having the endes aforesaide."[59] Likewise, the 1616 commission for Sir Walter Raleigh's last voyage to the Americas specified its aim as discovering "some commodities and merchandizes in those countries, that be necessary and profitable for the subjects of these our kingdomes and dominions, whereof the inhabitants there make little or no use or estimation." Incidentally, this might have the effect of introducing Christianity to the region, the commission noted, which would be a benefit to the inhabitants.[60]

Under the guise of introducing some beneficial innovation, projectors were able to invade entire regions of the world as well as gain control at times of existing domestic industries, such as glass and soap production. Little to no investigation was undertaken by government offices prior to granting the patent to ensure that these techniques were new or did no harm, as projectors claimed. By contrast, in the Netherlands, especially in politically volatile situations, initial investigations were required before a patent was granted to ensure that a project would not harm others and create social dissension.[61] In England it was largely up to the proposer of the project to make a case for both the novelty of the project and for its social benefits. This encouraged the framing of very ambitious projects concatenating many different interventions. The combining of multiple interventions served to demonstrate that nothing like such a project had existed previously, even if individual elements of the projects were not new. Thus, it could not harm existing interests. Projectors often also argued that their complex, wide-ranging projects conferred multiple benefits.

At the same time that projectors argued that their projects were novel, they also argued that the practices they proposed were *not* political innovations, but age-old, common practices of states. The notion of taking on the role of innovating in the state would be anathema both to the public and to the crown. Thus, Robert Johnson argued that transplanting excess human population was "no new thing, but most profitable for our State."[62] His project of enclosure of land and resettlement of its people, wrote the surveyor John Norden (ca. 1547–1625), might seem a "distastfull novelty," but considered correctly, it could be shown to serve the common good, or *"commune bonum."*[63] Neither claims to novelty nor claims to the lack thereof should thus be taken at face value.

Sometimes, the arguments for defending what were in fact infringements upon the liberty of others took highly convoluted forms. As the merchant (and frequent financial projector) Gerard Malynes reasoned, if a new invention, like a better kiln for drying malt, could be shown to be superior to older ways of doing things, then requiring everybody to use the new invention would not be a restraint of liberty so much as a convenience, even if "the publike libertie seemeth thereby to be restrained." The placement of these kilns everywhere for the required use by all could be considered rather "a common distribution than a restraint whensoever it bringeth a generall good and commoditie to the Commonwealth."[64] Indeed, when Sir Nicholas Halse, Sir Philiberto Vernatti, and John Battalion petitioned to found a Society of Maltsters that would reserve exclusive use of a malt-drying oven that Halse had invented for society members, ninety-two people subscribed.[65] However, the logic that requiring everybody to use such a new oven would be a convenience rather than a restraint was one that could be easily twisted (as in the case of the new soap company discussed in chapter 3).

Another common tactic was framing those individuals who might be negatively affected by a project as pernicious to the commonweal. Thus, the ways projects restrained their rights and enforced their behavior was itself a public benefit, and moreover, it was also to the benefit of the population being infringed upon, who could not understand what was in their own best interest. This argument was especially frequent in controversial projects for the enclosing of what was assumed to be common land and the forced removal of its inhabitants, projects much associated in the period with the "improvement" of lands. John Norden proposed enclosing royal forests, parks, chases, and wastes into farmland because of the ways lawless inhabitants were multiplying in these open areas. Into these places, "infinit poor yet most idle inhabitants have thrust themselves, living covertly, without law or religions, *rudes et refractorii* by nature, among whom are nourished and bred infinite idle frye." It would already be a service to the commonweal if his project merely cleared the land of such spawning rebellious folk, without providing any other "commodity else." However, it could grow the commonweal "abundantly" if the poor were not only "removed" but also "replanted"—to where in particular Norden does not specify—and others put in their place. The spaces that the idle had vacated could serve for "a more full plantation . . . for men of more worth" who could turn "those unprofitable places into farms and dwellings." Undoubtedly the poor that were removed would object, but it could be shown that they could earn more money in this way, and thus replanting was to their benefit.[66]

Language relating a specific intervention to large-scale notions of improvement and benefit to the commonweal became almost a requirement for projects. As Liliane Hilaire-Pérez has discussed, in England the system of patents granted to individuals or corporations encouraged non-state actors to set small interventions into more ambitious projects, integrating "inventions into the form of an economic project." In France, by contrast, inventors came to be seen as servants of the state, without any need to draw such connections to larger structures of the commonweal.[67] Rather than a technology of the state for controlling and delineating forms of commercial enterprise, the practice of projecting in England encouraged constant innovation and bricolage. Projectors set new or transplanted ideas within larger projects that responded to often multiple political aims. This volume explores the intellectual habits and larger implications of this constant extrapolation from limited, private interventions to claimed grand social benefits.

The Social Status of Projects

Surprisingly, given current assumptions concerning the social status of projectors, projectors often stemmed from the social elite, a group that tended to emphasize daring and courage as a sign of status. As Koji Yamamoto has pointed out, the social identity of the projector remained "malleable."[68] Projectors could come from the lowest social ranks. In a later chapter, for instance, we will meet John Williams, elephant showman and theater projector. More often than not, however, in the early Stuart period the term "projector" applied to somebody who was at least of the social status of a merchant. Moreover, projects were ordinarily not the work of a single individual, but were structured according to an approximately pyramidal model, with the Crown at the top, followed by courtiers, projectors, investors, agents, inventors, skilled craftspeople, unskilled labor, and finally, enforced skilled and unskilled labor.

The frequently elite nature of the projectors who helmed such teams of personnel bears far-ranging implications for the emotional, intellectual, and political aspects of projects, since elite projectors conceptualized their initiatives in terms of chivalric honor and conquest. Yet, although the personnel of the project was often arranged hierarchically, the destabilizing effects of interloping meant that this was not a well-regulated system of top-down control, at least until we reach the level of poorly paid or enforced labor. The distinctions between ranks at several points along the project's hierarchy could overlap or be fuzzy. Courtiers could also be projectors, investors, agents, and inventors. Inventors might also be skilled craftspeople, work as agents, or invest financially. There were projects that populated many levels of this pyramid with distinct individuals, and there were others that compressed many roles into the hands of a few. Such fuzziness and lack of discipline among the ranks was where the culture of projecting differed from militarized notions of ranks.

While this distinction between a projector and the rest of the project's personnel might seem a fine one, it raises many questions about our current picture of early Stuart projecting. We have often pinned the seeming eccentricity, improbability, ostentation, and ambition of early Stuart projecting on socially marginalized figures, when those qualities were in fact flaunted by the highest echelons of society. Socially elite figures exploited those beneath them and often attempted to appropriate their knowledge and even engage in their crafts. Meanwhile, socially inferior figures proved impudent and undisciplined, claiming authority from below. The culture of triumphant seizure of intellectual and material resources from elsewhere encouraged attempts to interlope in many directions between these various roles on the project. Interlopers seized knowledge from below and from above.

Practices of interloping belie attempts to pigeonhole the personnel on projects too neatly into social roles. They also call into question various historical lenses through which early Stuart science and technology have been previously interpreted, from the model of a "Baconian program" to that of court patronage. Aurélien Ruellet's *Maison de Salomon*, for instance, compares science and technology in French and English courts in the early seventeenth century.[69] Perhaps colored by the French case, however, Ruellet sees the world of early Stuart science and technology through traditional practices of patronage according to which technicians and scientific innovators were patronized by their social betters. The patronage model underemphasizes the ways that early Stuart elite projectors upended such expectations, sometimes working as their own experimenters or inventors or exploiting those in their employ.

By interloping downward, social elites invaded trades in ways that did not fit their status. Conceptions of various estates or "sorts" of society were often fuzzy around the edges and underwent change and modification, as well as vigorous discussion.[70] Both Sir Thomas Smith and Henry Peacham, for instance, purposefully conflated nobility with gentry in their discussion of English social order. When it came to practitioners of the mechanical arts, however, Smith and Peacham were very clear. Craftsmen could never join these ranks. According to Smith in *De republica Anglorum* (1583), nobility was divided into "major nobility (dukes, marquises, earls, viscounts and barons)" and "minor nobility (knights, esquires and gentlemen)." They certainly did not include "yeomen, craftsmen and labourers."[71] Henry Peacham in *The Compleat Gentleman* (1622) argued that nobility could be won through effort, "either out of knowledge, culture of the mind, or by some glorious Action performed" that was "usefull and beneficiall" to the commonweal. However, "mechanical arts and artists" who labored for their livelihood, "have no share at all in Nobilitie or Gentry."[72]

In the projects discussed in this book, however, we will observe behaviors that belie this distinction, such as gentlemen becoming soapmakers. The arguments about benefits to the public that projectors advanced, as well as the general undermining of disciplinary boundaries within the culture of interloping, also served to query what had seemed a clear threshold dividing gentry from craftsmen. In an anonymously published 1629 work, Edmund Bolton (1575–1633) proposed that the mechanical arts could be noble, based on their extreme usefulness to the commonweal. Since "civill Arts" and "flourishing Industrie" were the "sinewes, and life itself of Common-weale," it would not derogate from nobility even if a nobleman chose to serve as an apprentice in a city art.[73] Governance itself was a "civil art," and all "wise Masters in the most noble civill Art of government" have sought to "quicken, & inflame affections" for industry.[74] Since industry, not war, empowered the sinews of the polity, then craftsmen, not martial heroes, could warrant a claim to nobility. Given that interlopers were regularly inserting themselves into arguments concerning the public good and matters of state, it was more possible than ever to make such claims for the elevated social position of the mechanical arts.

Lawrence Stone long ago argued for the presence of aristocrats among the projectors, especially in the Elizabethan period but continuing under the early Stuarts. He offered the example of eight early Stuart earls active both in industry and overseas colonization. It would be, he contended, "impossible to draw up a list of City merchants or country gentry with such a wide range of interests."[75] Stone's view was undercut by John Wiedhofft Gough, who traced a far wider social spread of projects in his *Rise of the Entrepreneur*. Gough, however, concentrated on the Elizabethan period.[76] In contrast to Stone, Joan Thirsk and others have unfavorably contrasted early Stuart ostentatious projects with the more sensible ones of the Elizabethan age, which Thirsk called the "constructive phase of projects."[77] John Cramsie has even doubted whether early Stuart schemes deserve the name of projects.[78] Paul Slack has argued that in England after 1580, "most projectors and patentees were no longer inventors and skilled craftsmen but courtiers and speculators, the 'blood-suckers of the commonwealth.'"[79] In the recent history of aristocracy, projects appear only in so far as they relate to the sale of titles and the "inflation of honours."[80]

Elizabethan projects are often accorded a major role in developing science, technology, and practical expertise.[81] Rules developed for patenting have been related to the development of scientific method itself.[82] However, as a result of the view of early Stuart projects as corrupt, their contributions to science and technology have been seen as minimal. According to such accounts, noble projectors accomplished little in the realm of science, technology, and society, focused as they were on extracting as much profit for themselves as they could.[83] I argue that corruption alone cannot serve as the grounds for arguing that an endeavor played no constructive role in science and technology. Rather, corruption was thoroughly interwoven into a suite of behaviors, practices, values, and emotions concerning knowledge that would inform emergent experimental sciences and their relationship to technology.

Noble projectors of the early Stuart era did not serve a formal state bureaucracy and did not exhibit the habits of a scholar. As Stone has written, noble naval rovers had long sent their ships "out over the uncharted oceans in pursuit of more spectacular, and sometimes chimerical, objectives."[84] After "the outbreak of open war with Spain in 1585, aristocratic interest concentrated upon the more exciting, but sometimes equally unprofitable, pursuit of oceanic privateering."[85] Early Stuart projectors were often members of the recently remilitarized nobility of the sword, much of which by the 1620s and 1630s had seen military service abroad.[86] They emphasized attitudes of conquest and personal glory in their projects, bringing to them a very different flavor than those promoted by more risk-averse merchants or approved by the circumspect William Cecil, Lord Burghley in the reign of Elizabeth.

In contrast to Albert Hirschman's thesis that rational calculations of interest replaced the violent pursuit of glory in the early modern period, we find allusions to greatness and glory often in the texts of socially elite commercial projects. Goals of demonstrating greatness and glory by achieving the seemingly impossible help explain the period's penchant for bizarre and unlikely projects, which were less programs to be followed for an assured route to wealth, than distant targets shot at through an uncertain future. Attitudes of rivalry, honor, and conquest influentially intermingled commerce and colonialism. These attitudes undergirded the framing of knowledge acquisition as a chivalric advancement of epistemic empire. Such triumphant appropriation helped create the knowledge mergers of the period and fueled probabilistic thinking and risk-taking action.

This view of the persistent ideal of adventure in noble-led projects supports Michael Nerlich's account of the "modern" (for Nerlich, that means twelfth century) sense of adventure. Nerlich contrasts this idea with the ancient concept of surviving through an ordeal dealt by the fates. One proved one's constancy in the classical ordeal by lasting through it unchanged.[87] The modern knight, by contrast, sought out an adventure as an opportunity for personal transformation in ways that justified a claim to noble status. From their medieval beginnings, such adventures were to be found in contact with the marvelous—magic, alchemy, and incredible automatons.[88] The fusion of wonderful technologies and adventure in seventeenth-century projects had deep roots in chivalric culture.

Troubling this chivalric view of the project, however, were the ultimately bourgeois money-making goals of projects in the early modern period. Early Stuart noble projectors often came into conflict with merchant adventurers who sought to limit risk rather than to prove their greatness of soul. The figures who controlled access to court patents and privileges in the early Stuart period often favored ambitious, risky projects that proposed a high rate of financial return and also might win honor. Such views could be highly irritating to figures such as Gerard Malynes, a major financial projector from 1600 to 1620.[89] One of his many proposals concerned his plan for silver ore discovered in Scotland in 1601 and brought to the Tower of London, where the Royal Mint was located, and where it was subjected to several competing refining attempts between 1608 and 1610.[90] Malynes sent the same ore overseas to an "expert Mint-master" who could refine far more silver out of it than anyone in England could. He hoped to purchase some of the ore and export it to Holland to have it refined there, but in the end "was put off with this consideration, That it was a dishonour to England, not to have men of as good experience as any were beyond the seas, whereby the Kings losse was 2000 lb."[91]

The industry of a people as a whole, and its ability to compete economically with other nations, came to seem a measure of the honor of a state. Thus, King James I viewed his efforts to establish silkmaking in England as a competition for honor with King Henri IV of France, who had succeeded in this endeavor. In his instructions sent around England ordering everyone who could afford to do so to plant mulberry trees (upon which silkworms fed), James wrote that it was a "principall part" of "soveraigntie" "to increase among their people the knowledge and practise of all Artes and Trades." This was not only to offer each a livelihood, but also "in regard of the honor of their native country, whose commendations is no waie more set foorth then in the peoples actiuness and industrie." Those who sought to further "plantations, increase of Science, and works of industrie" were also advancing "our honour and contentment," just as the French king Henry IV had already "wonne to himself honour" in "the making of Silkes in his countrie."[92]

Aliens and the Transformation of Labor

The introduction of a trade entailed the remaking of the nation into a more productive, industrious force for the Crown. Due to the valorization of transplanting knowledge and resources into one's own lands as a means for drawing advantages away from alien states, projects often recruited foreign talent or retrained local labor in foreign skills. Foreign projectors, who were understood to have easier access to foreign talent and new ideas, received a warm welcome in the early Stuart court. For example, the ancestors of Sir Balthasar Gerbier moved from Spain, to France, to Flanders, and to Zeeland, where Gerbier was born, before Gerbier moved to London, where he became a diplomatic agent and eventually, Master of Ceremonies; the bill for his naturalization passed the Lords in 1628 but failed in the Commons, due to dislike of Gerbier's master, George Villiers, the royal favorite.[93] Prolific foreign projectors such as Johann Caspar von Wulffen of Brandenburg (a.k.a. Wulfe, Woulfe, or Wolffen) and Sir Philiberto Vernatti, originally from Savoy via the Netherlands, were made Gentlemen of the Privy Chamber of Kings James and Charles respectively. Von Wulffen was repeatedly given block grants for the denization and naturalization of foreigners, including a hatmaker, tailor, mariner, cutler, weaver, brewer's clerk, goldsmith, shoemaker, and clergyman.[94] In 1631 Vernatti was granted a block grant of denization to fourteen Dutchmen who were investors and land-holders in the draining of Hatfield Chase in the Fens; this included Vernatti and his nephew, Abraham.[95] Occasionally, as in the case of the Moravian noble-man Johann Christoph Berger von Berg (discussed in chapter 6), foreign projec-tors were fly-by-night figures who took advantage and left. However, contrary to current stereotypes of projectors, this was not the norm. Both Gerbier's children and multiple members of the Vernatti family, for instance, settled in England.

Gender

While women (as well as children) not infrequently supplied skilled and un-skilled labor for projects, they rarely figured in the role of the designer of the project. This is a topic certainly deserving of further scrutiny, and historians have offered powerful suggestions for how to recover women's entrepreneurship in ways that do not appear in the archive.[96] Lady Arabella Stuart (1575–1615) was one figure who promoted several projects, such as one to lower the price of oats to benefit travelers and another to sell wine in Ireland. In 1608 Francis Bacon made a note to himself to be prepared with arguments that might support her in her endeavors.[97] Stuart's projects for oats and wine appeared in a collection of many other texts of projects assembled by Sir Julius Caesar.[98] Another figure who appears in the projects discussed in this book is Lady Mansell, née Eliza-beth Roper, who defended the Mansell monopoly on glass production against all-comers in 1621.

Such figures, it seems, were exceptions in the early Stuart period. Women often took on public roles in projecting as widows or, as in Mansell's case, while her husband was abroad leading an anti-pirate expedition to Algiers. Thus, rather than providing evidence of a more widespread phenomenon, the figure of the "Lady Projectress" in Jonson's 1616 *The Devil is an Ass* was meant to be comedic.[99] Projectors were more likely to be men who appropriated the labor and knowledge of women; doubtless, many recipes and processes forming core technologies of projects developed in the context of heterosocial household ex-perimentation.[100] They do not, however, often leave traces of this domestic zone within the texts of projects.

Women might be called upon to test projects, as in the case of laundresses testing soap discussed in chapter 3. Opportunities for women to receive public credit for their work and serve openly as models seem to have become more common during the Interregnum, especially in collaboration with Samuel Hartlib.[101] Hartlib and John Ferrar, for instance, promoted the sericultural experiments of John's daughter Virginia as a model to be imitated in her namesake colony, Virginia.[102] However, John Ferrar also offered a highly gendered view of colonial experimentation in the work *Virginia Triumphans*, as discussed in chapter 5.

Arguments on behalf of women taking on public roles, such as those of the inveterate projector Sir Balthasar Gerbier, also indicate that women were generally not occupying these positions on projects. In a 1652 pamphlet by Gerbier depicting a conversation between an English captain, a Dutch merchant, and a mysterious Frisian, the latter (the savviest commentator of the bunch) advised that women "ought to be of publik use in a well setled Government." After they had finished child-rearing or if they had "such under them as are capable to look unto their house-keeping," women should be employed as leaders in public affairs, such as the women "directours" of "several Establishments" in Holland, or women who "had in former ages the mannaging of great undertakings." The English captain, by contrast, felt that women were fit only to "order dances," a view that Gerbier by no means shared, but here seems to suggest was particularly English.[103]

Intersections and Ambiguities

While projects were built along a pyramidal social structure, it was an unstable one, with many gray areas and insurrection from below. Individuals throughout the project's structure seized upon forms of knowledge that did not traditionally lie in their purview. Angry, undisciplined fights often broke out among participants in a project. Projectors stole ideas from each other and from those beneath them on the social hierarchy. This could not differ more from Gerhard Oestreich's vision of the absolutist police state, in which everyone knew their place and stuck to their last.[104] The entire identity of the projector is premised on not adhering to a single profession; the term projector applies when no other identity, such as physician, lawyer, engineer, or craftsman, fits.

Thomas Howard, the Earl of Berkshire (1587–1669), an art collector and frequent performer in court masques, offers an example of the wide-ranging roles assumed by elite projectors and the ways they often interloped upon technical abilities, claiming such abilities as their own.[105] Howard served Charles while still a prince, as Master of Horse (1614–1625) and member of his council (1617–1625). Howard held a glass monopoly in 1615, and, by warrant of the Privy Seal on April 11, 1618, 500 pounds per annum out of a gold thread monopoly; a mere eleven days later, Howard's own father was put on a commission together with the Lord Chancellor, Francis Bacon, "to examine, find out, and punish abuses in the making or importing of gold and silver thread." An investigation of Villiers's relative Giles Mompesson, "commissioner for the manufacture of gold and silver thread" ensued, eventually leading to Mompesson's arrest and trial, which in turn implicated Bacon and led to his impeachment by the 1621 Parliament.[106] The revelation of Howard's own glass and gold thread monopolies also precipitated a call for his expulsion from Parliament, which was unsuccessful. He became a farmer of the greenwax fine from 1625, founded his own colonial project, the "Earl of Berkshire's Guiana Company" in 1631–1632, and was a member of the Fisheries Society by 1632.[107]

The earl also claimed to be an inventor. In February 1637 Charles I granted him a patent for kilns for drying malt and foodstuffs, supposedly invented by the earl himself (although one might find this claim doubtful and it was later contested).[108] The earl approached figures, Bulstrode Whitelocke among them, who might help him set up the kilns in various areas. Whitelocke later described how the Earl of Berkshire, "hoping to repair his indigent fortune by setting on foot and engaging in new projects and monopolies," had offered him "a share in his project," which Whitelocke rejected, for he "looked upon such projects as dishonourable and illegal and little better than cheats."[109] In print, Whitelocke did not refrain from noting in his *Memorials of the English Affairs* for 1635, "The Earl of Berkshire whose fortune was lower than his mind, betook himself to some new Projects."[110]

Berkshire's claim to having invented a kiln intersected with another gentleman inventor, Sir Nicholas Halse. Halse had previously been granted a 1635 patent based on a model he presented of his own invention of a kiln for drying malt using iron plates, which he claimed would save in London alone "40,000*l.* per annum in wood and fuel" and reduce smoke.[111] Using the wide-ranging, suggestive thinking typical of many projects, Halse theorized many benefits that could accrue from his invention. For instance, by sparing straw, his invention would allow more cattle to breed, which in turn would produce more compost, "for the better manuring and inriching of land."[112] According to a petition of those who assumed Halse's patent after his death, four other individuals with knowledge of his ovens had falsely claimed the invention as their own.[113] One of them, Nicholas Page, clerk, countered in another petition that he had been "the first publisher, and has the first grant" of the invention, even if it wasn't his idea. Furthermore, the delay in setting the invention in practice was "a great inconvenience to the commonwealth and hindrance to your Majesty's revenue," and reason enough to revoke Halse's patent. He asked at least to be able to build an oven without iron plates.[114] A commission was ordered to investigate who had been the first inventor.[115] Page also entangled with the Earl of Berkshire's claim.[116]

Overtures: Projects as Keys

Projects took the form of short, small texts, or collections of texts proffered to the prince, an elite courtier, or another power, such as a trading company, that could finance and protect a large-scale enterprise. These were not audiences who were expected to devote much time to perusing a text. Like executive summa-

Figure 3. The binding of British Library Egerton 1140, *Great Britain's Treasure*, 1636. Album / British Library / Alamy Stock Photo

ries of reports today, the texts of projects supplied knowledge as a "thin description."[117] They rapidly presented numbers, calculations and arguments in broad strokes, without wasting space on contextualization, evidence, or further discussion. What they lacked in length, projects made up for in bravado.

After Sir Nicholas Halse's death, a collection of his projects was assembled by Francis Stewart (1584–1640), son of the Earl of Bothwell, and presented in a small octavo manuscript to Charles under the grand title, *Great Britain's Treasure: Environing this famous Ile with brazen walls invincible, maintained with great gaine by forces invisible* (figure 3). Stewart cast Halse's projects as a militarized, if invisible force, defending and empowering the nation through powers of transformation on a glorious scale. The "invincible" walls that these projects were said to construct around the nation were maintained, wonderfully, not through great expense, as material defenses would be, but with "great gaine" afforded by the seemingly cost-free ideas that Halse proposed. Halse's projects in this manuscript included his ovens, ironworks "in your new Plantations," a plan for transforming "100,000 sturdie vagabonds and idle beggars" into "laborious and industrious tradesmen in the fishing craft," and much else.[118]

This manuscript addressed the king directly, bound with Charles's coat of arms and the inscription, "Tibi soli, O Rex charissime!" (For you alone, O most beloved King). Held closed with silver clasps engraved with Justice, Prudence, Temperance, and Fortitude, the book presented itself as a key to unlocking treasure for personal rule. As in the etymology of the word *overture* (a term sometimes used for projects) from the French *ouverture* (opening), these designs posed as a small portal accessible to the king alone. Kept secure among the secrets of state, such keys could open up areas for intervention in the world that promised to unlock immense possibility. Thus, for example, Gerard Malynes entitled one 1618 project of his (concerning the valuation of coins), "The Locke & Key of all Projects for Preservacon & Augmentation of the Wealth of the Realm of England."[119] Yet, such rhetoric of a select, simple portal to power was undercut by the chaos of competing projectors, each bombarding the crown with their own set of keys.

CHAPTER 2

Cast of Characters

In early Stuart London, projectors appeared on the stage of the world, and upon actual stages, in dramatic fashion. In a poetic epistle to Sir Henry Wotton, John Donne described how the devious schemes of court projectors far outstripped any depictions of them in the theater. Being "like the courts was a player's praise," in times when "Plays were not so like courts, as courts like plays." The "deepest projects" found in the "mimic antics" of the theater were dull in comparison to the real dramas enacted in the court. In tune with the general theatricality of the early Stuart court, bravado and performance were central to the practice of projecting, as was skulduggery and the element of surprise. Projectors charged the stage of the world with sword outstretched in one hand and a dagger hidden in the other. They proved highly alluring figures for period dramatists and poets who devised many satires upon the projector.[1] Those anti-projector satires, which often located such figures on the fringes of society, have informed our view of who a projector was. However, the identity of the projector in the period was more ambivalent than such fictions would suggest. As we began to explore in chapter 1, projects and projectors not only received support and protection from the court. Figures from the highest echelons of society were also key participants in projects in ways that resituate the outsize ambitions of projects as pursuits of honor and as central contributions to a state's greatness.

In the spirit of the self-conscious performance through which projectors proposed their designs, this chapter offers a cast of characters who made up the personnel of period projects. This cast is meant to be representative: from royalty through elite courtiers, recruiters, financial schemers, figures of exceptional technical ability, subcontracted tradesmen, and unskilled laborers. These represent a small fraction of those in the period who engaged in similar endeavors, and we will meet quite a few more even over the course of this book. The general picture of the various social roles and relations in projects presented in this preliminary chapter, however, will be handy to keep in mind as we continue to dive further into this undisciplined world.

Setting the Stage: The Projector of Masques

Ben Jonson took aim at the theatrical antics of projectors in many of his plays, including his 1622 *Masque of the Augurs*, performed in the new Banqueting House on Twelfth Night by Prince Charles and gentlemen of the court.[2] Masques, danced by members of the court, offered a means to imagine and then quickly quell challenges to royal authority. Typically, a comic interlude, or antimasque, represented a world turned upside down that briefly threatened cosmic harmony and order. The forces of chaos and misrule displayed in the antimasque were then expelled, often with dramatic stage effects of heavenly rulers floating down from clouds above to re-establish order. In the *Masque of the Augurs*, as he often did, Jonson placed projectors in the antimasque as personifications of disruption and disorder. With the return of the main masque, the "College of Augurs," a rationally ordered neoclassical structure descended from the heavens. To harmonious strains of music, the Augurs, Prince Charles among them, expunged the erratic and unreliable projectors from court and foretold enduring greatness for King James and his progeny. As the inauguration of the Banqueting House, this masque served as a showcase of the Italian classicism, based on a Palladian sense of order and reason, that court architect and set designer Inigo Jones was attempting to introduce to England.

Through its assertion of intertwined social and natural hierarchies over forces of misrule, the early Stuart masque has been seen as an embodiment of Gerhard Oestreich's and Norbert Elias's theories concerning the disciplining and civilizing of modern society.[3] Yet, while Neostoic discipline might have been the aim of the court playwright Ben Jonson, it was belied by the historical realities in which the masque performers participated. Off the stage, life proved to be a continual antimasque. In fact, the very characters Jonson hoped to ridicule in his antimasque were supported by the court, including by King James and Prince Charles.

The antimasque of the *Masque of Augurs* featured Van Goose, "a projector of masques" and a "rare artist." This figure offers Jonson's fullest portrayal of a character based upon the court artist, inventor, and philosopher Cornelis Drebbel (1572–1633), whom Jonson frequently satirized over the course of two decades, beginning with his 1609 *Entertainment at Britain's Burse.*[4] Projectors are often seen as lone individuals stemming from the bottom of the social scale. Drebbel is precisely the sort of individual, proposing ambitious ideas and ranging from one enterprise to the next, who in current historiographic accounts would be considered a projector. Indeed, Drebbel has been described as a typical projector full of "crack-brained" schemes.[5] However, Jonson's derisive portrayal of Van Goose as a "projector of masques" was the closest that Drebbel ever came to being called a projector in his lifetime. He worked on projects, but he was not a projector. As in other theatrical portrayals of projectors, Jonson's casting of Drebbel in this role represents wishful thinking on Jonson's part as a critic of projects; he wished to suggest that the culture of interloping in the early Stuart court could be entirely pinned upon a few foreign and socially marginal figures such as Drebbel, rather than being endemic to the entire court, including among its highest-ranking members.

As representatives of classical order and reason, Jonson represents himself and Inigo Jones as maintainers of royal authority in the face of incursions by projectors. Van Goose, having heard that the king's "Poet" and "Architect" (Jonson and Jones) were barren of invention, came to court to offer a witty entertainment. He brimmed with outlandish ideas for odd forms of entertainment, from optical displays of Ottomans and Mughals fighting in the air to Welsh pilgrims wandering in erratic ways all over the stage. The Groom of the Revels wished to "try" Van Goose, the projector of masques, just as other projects were routinely subjected to trial. However, he objected that Van Goose's plans for an antimasque had nothing to do with the themes of the main masque.

In his response, the projector turned this accusation on its head. Van Goose celebrated absurdity. Only by departing from the expected could art really demonstrate its power, as the projector promised to show. In this account of the transformative powers of art, delivered in a heavy Dutch accent, Van Goose tellingly confused "antimasque" with an "antic" masque:

> O Sir, all de better, vor an antick-mask, de more absurd it be, and vrom de purpose, it be ever all de better. If it go from de nature of de ting, it is de more art: for dere is Art, and dere is Nature; you sall see.[6]

Stemming from *antique*, antics suggested frenzied Bacchanalia and the fanciful figures found in the late Roman wall painting that so inspired the airy, imaginative aberrations of the "grotesque" style (Figure 4).[7] This style did not adhere to the assumed structures of nature. Monsters, strange amalgams of humans, plants, and animals, and irrational structures seemed to defy weight

Figure 4. Lucas Kilian, *Newes Gradesca Büchlein*, Augsburg, 1607, etching on paper. Rijksmuseum, Amsterdam, RP-P-OB-8232. Public domain

by posing heavy bodies atop light ones. "Anticks" were also another term at the time for automatons.[8] Thus, an "antick-mask" suggested a machined display and a preference for purposefully bizarre forms of artificiality in contrast to the neoclassical aesthetic that claimed to locate stability, predictability, and rational order upon natural structures.[9]

While Jonson hoped to ridicule Drebbel through Van Goose and his "antick-mask," we might, like Donne, smile at Jonson's "mimic antics." In real life, Drebbel regularly supplied the court with machined displays and other spectacular technologies surpassing nature. The projects upon which Drebbel worked, often for the Crown itself (as discussed later in this chapter), far outstripped Jonson's dramatization of Van Goose's antics in their outrageous ambitions and confidence in art's ability to surpass nature.

There were some ways that Jonson's portrayal of Van Goose's "antick-mask" did reflect reality, in a distorted, world-turned-upside-down way. For one, he shows that Van Goose did not work alone. The group with whom he collaborated represents a comedic version of the real social dynamics at play in early Stuart projects. The "projector of masques" was accompanied by "Notch a Brewers Clarke, Slug a Lighterman," and a "Lady Alewife" and her barmaids, each of whom had a clearly delineated role in the group endeavor. Notch the clerk had first learned of the lack of talent at court. It was Van Goose's "project" to come to court from the tavern, where he recruited the alewife and her two barmaids to accompany them "in person here Errant, to fill up the adventure."[10] The company also needed to gain a point of entry, or some means of access to favor that could get their foot in the door at the court. They relied on Slug the lighterman, who had "credit" at the back door of the buttery, as "one that hat had the honour sometimes to lay in the king's beer there." Jonson thus also ridiculed how the language of honor, knight errantry, and adventure intertwined in period projects with base, commercial industries, such as brewing.

Projects relied on some new or newly imported idea, but needed far more than just such intellectual labor. They required intelligence about opportunities, needs, and desires, to which a project might respond. Here Notch supplied that information. Projects often intertwined several phenomena, often subcontracting out workers with particular expertise. The alewife had brought along a bear trainer and three dancing bears, fed with a special "diet-bread" to be extra docile. Especially crucial was the role of the courtier, represented here by Slug, who knew some back-door entry to royal favor that might help the group acquire sources of power and funding. The remainder of this chapter stages the historical cast of characters who in fact collaborated on projects, beginning at the top of the social hierarchy, with the prince himself. It aims to serve as a corrective to many of our assumptions concerning projectors that stem from popular satires of the period.

Charles, Prince of Projectors

As Pauline Gregg has noted, "Charles himself was genuinely interested in projects, and his eclectic mind enjoyed ranging over the schemes brought before him."[11] Yet, Charles's scientific and technological activities, as well as his personal engagement with projects, have generally been underemphasized.[12] Prince Henry, rather than his younger brother Charles, has typically attracted attention for scientific and technical interests. Deborah Harkness has written lyrically of the urban, mercantile, and global networks of natural study in Elizabethan London. As Harkness put it, the "halcyon days of Elizabeth's reign were thought to be returning when the young Prince of Wales, Henry Stuart, began to take an interest in the natural sciences and technological innovation, but his untimely death in 1612 dashed those hopes."[13]

Scientific patronage supposedly went from bad to worse once Charles I assumed the throne in 1625, a narrative supported by accounts linking Puritans and revolutionaries to a resurgence of invention, improvement, and experimental science during the Interregnum. Based on the theory of Robert Merton associating Protestants and experimental science, Christopher Hill and Charles Webster attempted to account for social reform via scientific reform.[14] As a corrective, this volume therefore pays special attention to the attitudes toward science and technology of Charles I and his courtiers.

Despite my emphasis here, I would argue that practices of projecting, or attempts to develop new, experimental forms of knowledge, were not the property of any particular group, either religious or political. Crypto-Catholics, Anglicans, and radical proponents of further Christian reform all engaged in similar practices of projecting. Of the projectors discussed in this volume, it is often not predictable who would adopt which side in the Civil War, as in the cases of the regicide Sir John Danvers and the royalist Sir Nicholas Crispe. Rather, projecting can be understood as a force further increasing social disruption in general.

Suddenly thrust onto the center stage after the death of his sporty and martial elder brother, the infirm Charles has a bit of a "deer-in-the-headlights" look in his portrait by Simon van de Passe a year later (figure 5). Charles eventually became an art connoisseur and great collector, the interests for which he is best known today. However, he was also keenly engaged in science and technology.

Figure 5. Simon van de Passe, *Charles, Prince of Wales*, 1613. Rijksmuseum. RP-P-1909-412. Public domain

Like his brother, Charles was deeply interested in military invention and training. Prince Henry had established an academy for training the nobility in warlike skills and mathematical abilities from 1607 to his death around 1612.[15] After Henry's death, Prince Charles took an interest in this initiative and resurrected it, as discussed in chapter 6. In their 1624 interview with the French polymath Nicolas Claude Fabri de Peiresc, the Küfflers, who were associates of Drebbel, mentioned several devices that Drebbel was working on for Prince Charles, including a telescope for seeing at night and a new version of his perpetual motion.[16] In 1622 Charles would prove personally very invested in utilizing Drebbel's submarine in a global project of commercial interloping and conquest, as chapter 4 explores. He also pushed hard for a new trade in Persian raw silk. After the death of Sir William Heydon, who was to have led Charles's global project, Charles worked closely with Heydon's brother and replacement, Sir John Heydon, to recruit military engineering talent, especially from abroad, culminating in the establishment of a well-furnished center for invention at Vauxhall in Lambeth. One inventor whom he patronized, Richard Delamain (discussed in chapter 7), even claimed that Charles personally invented several instruments in conjunction with the launch of the warship the *Sovereign of the Seas*, a material embodiment of Charles's hopes for unbounded royal authority. Robert Hubert, alias de Forges, who described himself as a servant of the king, claimed that he drew his Interregnum-era traveling collection (which would later serve as the basis for the Royal Society's collection in the Restoration) largely from the royal collection of natural rarities.[17] He listed King Charles first on his catalog of donors and described in one instance Charles's personal experimentation with one of his curiosities.[18]

George Villiers, Peerless Performer

In the Elizabethan court, the cautious William Cecil, Lord Burghley (1520–1598), served as a central bureaucrat who could assess projects. Merchants, inventors, and craftspeople could approach him directly with mercantile and technological schemes. As Deborah Harkness has described, while inventors and projectors enjoyed risk and showcased versatility, William Cecil increasingly sought certainty.[19] Eric Ash has explored Elizabethan projects as a way to study the testing and supervision of expertise.[20] During the reigns of James and Charles, especially after the death of Robert Cecil, 1st Earl of Salisbury, in 1612, there was no bureaucrat advocating for certainty and expertise in projects. The subcommittee that was briefly established after Salisbury's death in order to review projects and make recommendations indicates some concern about this lack of oversight; to it were appointed Sir Julius Caesar, Sir Henry Hobart, Sir Francis Bacon, Sir Thomas Parry, Sir Walter Cope, Sir George More, Sir George Carew, and others, who were all, as Cesare Pastorino points out, already serving on the "commission for suits" established in 1608.[21] Carew (1555–1629), appointed Master of the Ordnance in 1608, remained for some time available as an adviser on projects. Until he fell out of favor, the younger Lionel Cranfield (1575–1645), Earl of Middlesex, was also in a position to promote projects. Thomas Howard, Earl of Arundel (1586–1646) promoted some projects, but as he was the head of a faction opposing the royal favorite George Villiers, he seems to have been somewhat marginalized in the world of projecting.[22] It was the glamorous and contentious royal favorite George Villiers (1592–1628), who not only monopolized the position of arbiter of projects, but came to embody the culture of ambitious interloping.[23] He was no Burghley.

In 1614 Villiers (figure 6) began his meteoric rise from the son of an obscure Leicestershire gentleman through a succession of positions, becoming Lord Admiral in 1619 and then Duke of Buckingham, the highest position in the land outside the royal family, in 1623. As one song played by three fiddlers (who were prosecuted) put it, "he came to the courte and grewe Cupp bearer, / unto the Kinge he still grew nearer, / In his eye he semed a Pearle / sate downe a viscount, and rose upp an Earle."[24] Villiers monopolized access to power to such an extent that before his assassination in 1628, those who hoped to pursue projects often had to go through him to receive patents and privileges. Rather than care, Villiers aimed to embody carelessness, and he encouraged projects that undermined conventional restraints and boundaries.

Villiers parlayed brilliant physical performances in tourneys, dancing in masques, and diplomatic negotiations into political power. The outfit he wore when he acted as proxy for Charles I in his marriage to the French princess Henrietta Maria in 1625 in Paris illustrates his gorgeously negligent style. Among the twenty-seven suits of clothes that he had his traveling companion, the gardener John Tradescant the Elder, bring with him, was one gray velvet outfit with pearls purposefully attached loosely to it. With "every step he took past the assembled court of France a few dropped off and rolled across the floor." Paying no attention at all, the Duke "made an elegant bow in front of the queen mother before cheerfully exiting, scattering pearls as he went."[25] This performance illustrated the noble quality of *sprezzatura*, that is, disregard or disdain, differentiating the courtier from the thrifty merchant or the nitpicky scholar.

Villiers used his beauty, athletic ability, and daring willingness to traverse boundaries in the court masque, which mixed dancing, music, acting, special

Figure 6. Willem Jacobsz Delff after Michiel Jansz van Mierevelt, *George Villiers, Duke of Buckingham*, 1626. Rijksmuseum, RP-P-1884-A-7727. Public domain

effects, and elaborate costume and set design. In the early Stuart court, such performances, acted out by members of the court, functioned as a vehicle of political expression and maneuvering. As one critic of Villers observed, King James "loved such Representations, and Disguises in their Maskaradoes, as were witty, and sudden; the more ridiculous, the more pleasant. And his new Favourite, being an excellent Dancer, brought that Pastime into the greater Request. . . . No man dances better, no man runs, or jumps better; and indeed he jumpt higher than ever Englishman did in so short a time, from a private Gentleman to a Dukedom."[26]

Beginning in 1618, and until the end of King James's reign, Villiers and Prince Charles put on most masques together, performing as a duo. As "Prince Charles's masquing companion/competitor," Villiers groomed the prince on the importance of theatricality in the seizure of power.[27] Signally for practices of interloping at court, however, Villiers taught Charles to reach not only higher, but also lower, playing alongside him in a variety of rakishly transgressive roles. Like the pearl-scattering outfit, these also represented a noble lack of care.

Villiers commissioned and performed in a number of masques that thrillingly crossed the masque's traditional divides. In the *Running Masque* he gave himself a speaking part, usually reserved for the lowly actors, in a "clear transgression of this performative idiom."[28] In *The Gypsies Metamorphosed* Villiers presented himself "as the captain of a band of gypsies who attempt to pick the pockets of the assembled spectators," and performed palmistry on the king's hand. By appearing as a figure on the margins of society, Villiers "candidly" addressed the criticisms concerning his meteoric rise at court, as well as his troublingly physical relationship with James.[29] While on the sidelines his critics griped about his sudden social rise, Villiers brought his transgressions into the limelight as the ultimate expression of his glamorous, confident power.

Dance in the early modern period was understood as a martial exercise, preparing the masculine body, as with other sports such as riding the great horse, for war. Yet, when Villiers brought the same attributes of theatricality, swagger, and nonchalance that he regularly deployed in his bodily performances to his deployments in the actual theater of war, the results were tragic, as in the battle of the Isle of Ré off the coast of France. According to a logic that often played out disastrously in early Stuart England, Villiers relied on his ability to extend to an enormous scale those skills he had developed on a smaller stage in courtly settings. As one libeler wrote, 'To o're-run Spaine, winne Cales, and conquer France, / Requires a Soldier's March, noe Courtiers daunce."[30] A French pamphlet accused Buckingham of coming to the disastrous battle at the Isle of Ré to play at being at being a knight of the Round Table, as in a "pretty romance."[31]

By attempting to expand small-scale successes on the courtly stage to the large-scale theater of war, Villiers participated in an intellectual habit of extrapolation. The structural nature of patents demanded that projectors spin out large-scale effects and benefits to the commonweal from the often small-scale trials of their technologies. Claims to enormous effects for the patented invention or industry relied on the belief of art's ability to overcome nature and widen parameters for action. Villiers demonstrated such a belief in new-fangled technology in the great reliance he placed on Cornelis Drebbel and the latter's underwater bombs during the naval expedition to the Isle of Ré. While these, like Drebbel's other court performances, may have functioned spectacularly in small-scale trials at home, they failed miserably in the stormy waters off the coast of France.

The new technologies that Villiers attempted to deploy in the theater of international war were of a scale and ambition with other period projects he patronized that overturned natural order. Describing how Villiers promoted "Proiectors, state cheaters," the Earl of Clare criticized the cosmic scale of Villiers's appetites: "all in earth, in sea, in heav'n that dwells, / All is too little for his continent."[32] Villiers, wrote Clare, "diggs through land and sea, he clyms the skies / Colleagues with divells, sells both earth and heaven."[33] Metaphors of new inventions subverting the structure of nature were not mere analogies. In the name of endeavors to improve upon nature, projectors literally upended the natural and built environment. Key examples included the New River Company chartered by King James in 1619 to bring water to London through Sir Hugh Myddelton's system of waterways, into which project, Sir Giles Mompesson, Villiers' relative by marriage, intruded himself as surveyor of profits solely through the power of his connections.[34] The saltpeter project in which Villiers participated, discussed in the next chapter, entailed digging through the privies and henhouses of thousands of ordinary subjects.

Villiers embodied the culture of early Stuart interloping in his venality, rivalry, risky ventures, and great ambitions of scale. His intellectual practices showcased patterns of predation upon others as a form of one-upmanship. For example, he continually emulated previous naval heroes like Sir Walter Raleigh (1552–1618), and it was in part his ambition to upstage these figures that led to some ill-advised engagements. He constructed a private fleet of ten fast and flexible warships, the *Lion's Whelps*, that he employed as privateers to prey on French shipping.[35] These ships pointed to a storied past. The original *Lion's Whelp* had sailed in Sir Walter Raleigh's expedition of four ships searching for El Dorado in Guiana in 1595, and then had been captained by Sir Walter Raleigh in the Anglo-Dutch attack on Cádiz in 1596. Although the hunt for El Dorado failed, the attack on Cádiz spelled an embarrassing defeat for Spain, even if it fell short of the main objective of capturing the Spanish silver fleet from the Americas. Raleigh had been imprisoned in the Tower in 1603, and James I briefly allowed him to leave in 1616 to redeem himself through a new expedition to Guiana. After Raleigh returned without success, but having wakened the ire of the Spanish for predations on their settlements and ships, he was beheaded in 1618. Meanwhile, Villiers seized upon Raleigh's intelligence and project as his own. As the Swedish agent Michel le Blon wrote, "the deceased Duke of Buckingham [Villiers] preserved among his rarities a trustworthy description of a very rich gold mine . . . which the famous sea hero Sir Walther Raleigh had discovered."[36] Raleigh's ship was later given by King James I to Villiers, who then had his whole fleet built on its model, all called *Lion's Whelp*, numbered one through ten. In 1625 he attempted to repeat Raleigh's raid on Cádiz in a new expedition that failed miserably.

Sir Endymion Porter and Sir Balthazar Gerbier, Errant Knights

Two of Villiers's closest clients were Endymion Porter (1587–1649) and Balthazar Gerbier, both better known today as patrons and collectors of art than as industrial and colonial projectors. Both were notorious figures, Porter for his crypto-Catholicism and Gerbier for his general untrustworthiness. Their roles grew in importance after Villiers was assassinated in 1628, leaving behind a power vacuum. Porter became Gentleman of the Bedchamber, and Gerbier was at long last appointed Master of Ceremonies in 1641.

Porter, educated as a page to Count Olivares in Spain, first served in the household of young George Villiers, whose niece he married in 1619, before he entered the service of Prince Charles.[37] As the next chapter explores, one of Porter's early activities in Charles's service was presenting the English East India Company with the prince's ambitious project for using Drebbel's recently invented submarine in the Indian Ocean. This project came to naught. However, in 1636 Porter helmed plans to invade Madagascar and establish Prince Rupert in power there, an endeavor King Charles considered "so hopeful a project," while Rupert's mother Elizabeth criticized the "Romance some would putt into Ruperts head of Conquering Madagascar where Porter they say is to be his squire when [they] shall Don Quixotte-like conquer that famous island."[38] When Porter's friend Charles Davenant later described the scheme in a dream poem entitled "Madagascar," "Endimion" served as the English champion of this new overseas realm, defending it against attacks by "ambitious Wanderers" seeking to interlope into the English conquest.[39]

Such chivalric mores should not blind us to Porter's participation in political and economic projects. Although Porter is better known as an artistic patron than as an investor in manufactures, he regularly operated across a broad range of activities, from the most courtly to the most utilitarian. Porter had a finger in many of the financial schemes of the period, from Fens drainage to the manufacture of soap, iron, salt, and white writing paper.[40] He was put in charge of employing convicts reprieved of the death penalty "in works or trades as his Majesty shall direct."[41] He petitioned the Admiralty for permission to build a harbor and raise a lighthouse.[42] Like his friend Sir Balthazar Gerbier, Porter utilized aesthetic interests and art collecting as a means to forge friendships and obtain political ends. Both worked as agents in Brussels to secure an Anglo-Spanish peace, while befriending artists such as Rubens and van Dyck.[43]

Already in 1616 the gifted and unscrupulous Gerbier, who had come to England in service to the Dutch ambassador, had hitched his wagon to Villiers's star. Gerbier was evidently proud of identifying such a speedily ascending meteor, describing how after he "applyed myself to George Villiers, newly become favorite to King James, the said George Villiers, being immediatly after Baron, Viscount, Earle, and afterwards created Marquis and Duke of Buckingham."[44] Gerbier did not commit his loyalty to others easily, but he always remained enamored of Villiers. The 1634 portrait of Gerbier by his friend van Dyck shows him holding out a folded letter, indicating his role as a spy and agent. It is inscribed, "may the memory of Buckingham live" (*vivat memoria Buckinghamii*) (figure 7).

Gerbier stemmed from a global mercantile family. He claimed distant descent from aristocracy and styled himself "Baron D'ouvilly." His closer relative and namesake was Balthazar de Moucheron (1552–ca. 1630), a founder of the Dutch East India Company. Gerbier criticized English merchants for what he saw as their lack of courage and imagination in refusing to undertake the scale and risk of Dutch global traders such as Moucheron or the related Courten dynasty, from which stemmed Sir William Courten (1572–1636) (who played a major role in English overseas trade).[45]

Over the course of his lengthy career, Gerbier would blend activities from seemingly every conceivable area, including painting, calligraphy, cartography, military engineering, pedagogy, masque design, architecture, cryptography, espionage, forestry, horticulture, art connoisseurship and collection, alchemy,

Figure 7. Paulus Pontius, after Anthony van Dyck, *Portrait of Balthazar Gerbier*, 1634, engraving. Rijksmuseum, RP-P-OB-33.307. Public domain

experimental philosophy, travel writing, financial and banking schemes, colonialism, and the slave trade. Gerbier's description of experimental philosophy in the prospectus for the academy he briefly operated in the Interregnum was one of the first descriptions of such an approach in print.[46]

As an artist and architect Gerbier developed an ephemeral, ornamental style in contradiction to the symmetrical classicism of Inigo Jones (1573–1652), Surveyor-General of the King's Works, with whom Gerbier engaged in a sharp dispute over style. Gerbier's style referenced ancient triumphal architecture, which incorporated the *spolia*, or booty, taken from the conquered.[47] The idea of the *translatio studii* (movement of learning) that accompanied the *translatio imperii* (movement of empire) made the appropriation and movement of knowledge through conquest a well-established dynamic. In aesthetics, these ideas were embodied through the use of *spolia* as well as through *mimesis* (the emulative rivalry of opponents). Gerbier translated these ancient ideas to his many interests and forms of knowledge appropriation in ways that supported global commercial empire.[48]

Just as Gerbier's interests were hard to confine to one area, his social status was also difficult to pin down. This range and ambiguity have often attracted criticism. Hugh Ross Williamson described his "enquiring, if superficial, mind, and his multifarious, if short-lived, enthusiasms," which Jeremy Wood characterized as "preposterous schemes."[49] Gerbier, on the strength of his promotion of the African slave trade alone, was undoubtedly an unsavory character. However, descriptions of his endeavors as superficial, short lived, and preposterous view them out of the cultural context of interloping. The multifarious and outrageously ambitious nature of his enterprises was part of a larger culture of knowledge appropriation as conquest. So too was his racism and casual cruelty. His projects were not short lived, as he continually revived many of them over the course of an extraordinarily long career. In the early 1660s for the benefit of Charles II, for example, Gerbier drew a map of the Gambia river that contained secret information concerning the sources for various riches, from slaves to gold. The information on this map, he claimed, had been given by King James I to George Villiers, who passed it on to him.[50]

The Heydon Brothers, Experienced Soldiers and Recruiters

The career of the Heydon brothers exemplified the ways that erstwhile military nobility sought to forge new paths as projectors. William and John Heydon were the sons of the astrologer Sir Christopher Heydon (1561–1623), friend to mathematicians and astronomers such as Henry Briggs and John Bainbridge.[51] The

Heydons, an aristocratic Norfolk family, fell from power due to their involvement in the Essex rebellion of 1601. William Heydon began the "long climb back to favour and respectability" that his brother John would continue.[52]

Both William and John gained military experience defending Protestant lands abroad. William Heydon (?–1627) served as "Junior Paymaster" for Prince Henry (1611–1612), was knighted in 1620, when he was in Prince Charles's service, and became paymaster for the forces in the Palatinate and later for the English regiments in the Netherlands.[53] Edward Norgate and Edward Stevens (who would later serve Heydon's brother John) worked as his deputies in his role as paymaster for the Palatinate.[54] In 1613 and in 1616, James granted William Heydon patents to raise 16,000 pounds from entailed estates, with 10 percent to go to the Crown. Upon his accession, Charles appointed the battle-tested Sir William Heydon as the Gentleman Usher of his privy chamber. Two months later he made Heydon a senior commissioner of the Navy, and a year later, Lieutenant-General of the Ordnance.[55] His brother John had served in the Dutch army in 1608–10 and then returned to England, where in 1613 he became storekeeper of Sandown Castle, before returning to the Netherlands to captain his own company in the service of the Prince of Orange from 1618. He was then called to help serve in Villiers's campaign in France in 1627 upon the death of his brother. He did not go to France, but he assumed his brother's position as Lieutenant-General of the Ordnance Office, a position that he maintained until the Royalist defeat in the Civil War.[56]

Both their experience abroad and their own interests in pursuing the secrets of nature made the Heydon brothers valuable advisers. One of Sir William Heydon's first tasks as Lieutenant-General was to oversee Cornelis Drebbel's research into submarines and underwater bombs. Heydon received a warrant from King Charles for bombs, cases, gunpowder, and "boates to goe under water" as would be "necessarie in the tryall of these and other practises." The supplies and spaces required for these trials were left to his discretion.[57] William Heydon's various memoranda on innovations in ship design suggest that "he was far more competent on the technical side of shipbuilding" than many of his fellow naval commissioners in 1626.[58] He joined with a group, including Gresham College professors, to devise a new means of weighing ships in 1626.[59] Charles ordered a commission to examine it. A new dimension of a ship that could be "nimble and forcible" to act against pirates was delivered to the commission by the master shipwrights.[60]

Heydon was often consulted for advice on projects. In a letter to Villiers, Sir Francis Crane reported on his consultation with Sir William Heydon about a mining "dessein" Villiers had proposed. Heydon's views had been sobering. He believed that "the nature of Mynes and the state of those partes" meant that a "present return cannot be hoped for."[61] Nevertheless, Crane claimed that his desire to "adde to the glory or greatness of my Master [Villiers] made me ready to putt on the greatest hopes."[62] Even though he had been advised against the project by Heydon, Crane cast the pursuit of the project as a means to pursue glory and greatness.

John Heydon was also a knowledgeable and trusted naturalist. In 1653 he would recall that he "first began to practice Spagyrically in London" in 1615, when he was twenty-seven.[63] John Heydon experimented chymically, often with the help of his brother-in-law, Sir Christopher Gardiner, more famous for his adventures in early New England. Gardiner was celebrated as "a Knight, that had bin a traveller, both by Sea and Land; a good judicious gentleman in the Mathematticke, and other Sciences usefull for Plantations, Kimistry, &c." and also for being "a practicall Enginer."[64] Heydon, however, was the leading light in their chymical investigations.[65]

Always the experimentalist, Heydon utilized Drebbel's self-regulating ovens to hatch chicks in the Ordnance Office and investigate their development at different stages. In a published work of 1644, Sir Kenelm Digby recalled experimenting with the hatching of chicken eggs alongside Heydon, whom he called a "generous and knowing Gentleman; and consummate souldier both in theory and practise."[66] When a commander for the royalist armed forces was sought, an enciphered 1645 letter to Edward Walsingham (secretary of George Lord Digby) suggested Heydon, mentioning Digby's published praise. The informant doubted "if there be 3 in Europe greater masters of the art of war and fitter for a General Commander, or at least to moderate a Council of War, in matters of action I mean, and the manner of execution." Heydon was not the one to devise stratagems, however; Walsingham's informant conceded that "as for design I do believe him not so excellent."[67]

If not the most politick of heads, both Sir William Heydon and Sir John Heydon were experienced soldiers who took a personal interest in inventions, alchemy, and other secrets of art and nature. They owed their position in the informal chain of command of large-scale projects not only to their own practical experience, but to their ability to judge and recruit talent from elsewhere.

As successive Lieutenants-General of the Ordnance Office, it was also their role to test new military inventions. For instance, Gerard Dalby, a master gunner on the *Adventure*, approached Captain Pett in 1634 to demonstrate "an engine of firework" on the water, but the Lords of the Admiralty redirected him to "attend Sir John Heydon, and demonstrate his engine and the use thereof to him, and that Sir John certify his opinion."[68] When, as discussed in the next chapter, Thomas Russell devised a massive new project for producing saltpeter, Heydon, Sir Robert Killigrew, and John Evelyn, Esq. (relative of the better known John Evelyn, FRS) were appointed to test it.[69] Heydon, Evelyn, Killigrew, and Sir Kenelm Digby also partnered on an attempted reform of the Ordnance Office.[70]

The centrality of Sir John Heydon (who is regularly confused with his nephew, the astrologer John Heydon) in early Stuart technology has often been overlooked.[71] Aurélien Ruellet, for instance, has traced a network of German and Dutch practitioners in early Stuart technology by pointing out connections between Hildebrandt Prusen, Arnold Rotispen, Cornelis Drebbel, and Howard Strach. Prusen partnered with the philosopher, inventor, and engineer Drebbel in a Fens drainage project, and after Drebbel's death Prusen and Howard Strach received the privilege for Drebbel's novel furnaces.[72] Missing from Ruellet's network, however, is Heydon. Heydon was the linchpin of this group. His years in the army in the Netherlands equipped him to recruit Dutch talent. Sir John Heydon was also a partner in the Fens drainage project, and Howard Strach (or Strachey) was Heydon's clerk. Prusen served as a witness to the wills of both Cornelis Drebbel and Sir William Heydon.[73] Prusen also served as executor for the will of Henry Briggs, a Gresham College professor and then Savilian Professor of Geometry at Oxford, who was a partner in the Fens drainage project of Prusen and Drebbel, and who had been a friend of Christopher Heydon.[74]

In the terminology developed by Eric Ash for projects undertaken in the Elizabethan court, the Heydon brothers would perhaps fit into the role of the "expert mediator" located between craftspeople and political patrons.[75] However, among the interlopers of the early Stuart period, such clearly defined roles become elusive. Sir John Heydon was also a projector, hands-on experimenter, and philosopher himself, even at the height of his very busy military career in the Civil War. After his enforced retirement, he composed philosophical treatises in which he cited (and disagreed with) Drebbel's writings on the nature of fire, showing how he not only utilized Drebbel's devices but investigated his philosophical views.[76]

Heydon also rubbed shoulders with Drebbel socially. Both William and John Heydon were friends and travel companions of Constantijn Huygens Sr., Drebbel's friend.[77] Huygens described how in England he got to know Sir Robert Killigrew (1580–1633), who would later become ambassador to the United Provinces and Henrietta Maria's vice-chamberlain. Killigrew, well known for his medicaments, was also a projector who partnered with Sir John Heydon on testing saltpeter projects, on reforming the Ordnance Office, and on draining the Fens.[78] Killigrew's circle purposefully aimed to reach across various interests, social positions, and roles. Huygens praised the company Killigrew hosted of "all those who excelled in their art;" these included the poet John Donne, Elizabeth Raleigh (née Throckmorton), the politician and philosopher Francis Bacon (the uncle of Lady Mary Killigrew), the musician and artist Nicholas Lanier, the lutenist Jacques Gaultier, and Cornelis Drebbel.[79]

Hildebrant Prusen, Cooker of Books

Hildebrant Prusen, a.k.a. Pruson, Spruson, Prazen, or even Preiwsen (1576–1637), offers another example of a socially ambiguous stage in the pyramid of projects. Although classified by Ruellet as a craftsperson, Prusen was more of an investor as well as a liaison for gentlemen and craftspeople. His greatest expertise seemed to lie in cooking the books. Evidently, Prusen was a figure Heydon's circle found capable and dependable. The merchant community found him entirely reprehensible.

Prusen was of apparently Netherlandish descent, but his family had been in England for at least two generations. He was a resident of Tower Wharf, and had served as royal sailmaker to the navy since at least 1598, with his father holding the same office for almost as long.[80] An elevated region of Wapping Marsh began to be called Pruson's or Spruson's Island in 1628.[81] Prusen was related by marriage to the corrupt naval official Sir John Trevor.[82] Trevor was implicated in investigations into the navy in which Prusen figured prominently. In the Jacobean commissions of enquiry of 1608 and 1618, Prusen was described as a thirty-two-year-old "salter of London," or member of the Worshipful Company of Salters, a livery company for those dealing in salt and other chemicals. He was criticized as "Sir John Trevor's favorite" and one who "hath married one of Sir John Trevor's kinswomen."[83] For his part, Prusen claimed that Sir Robert Mansell and Sir John Trevor had forced him to be corrupt and to split the profits with them.[84] When a commission of enquiry was formed again in 1626 at the urging of George Villiers, Trevor, hardly an impartial figure, served as one of the naval commissioners, alongside Sir William Heydon.[85]

Due to his venality, his ability to network with social elites, as well as various profiteering ventures, Prusen became a thorn in the side of both the East India Company and the Eastland Company.[86] He was a literal interloper, disrupting the trade of the Eastland Company. On November 3, 1610, he received the grant of the office of "merchant in the East for furnishing the King's ships' for life."[87] The Eastland Company complained that "Pruson, being a sailmaker by this trade, hath procured of late to be free of the Company of Eastland merchants neither by service nor by redemption, but by favour of letters procured in his behalf."[88]

Prusen, promising to save the East India Company money lost by previous abuses, later also became that company's shipping supplier. Suspicions began to develop that he was lining his pockets. On March 11, 1621, Prusen was asked to account for the use of his supplies for the last three ships sent abroad. Prusen replied that he had no time for this, and he would "rather be acting than projecting."[89] The Prusen affair would continue to dominate the Minutes of the Court of the East India Company for 1623, stretching on into the next year. At a meeting of 1624 there was a "long discussion concerning the business of Hildebrand Pruson." It was suggested that Prusen had been questioned "for some falsehood in his Majesty's service," leading him to burn his books. It was also said that "Pruson had joined with Sir James Cunningham and Sir Thomas Dorrington for the overthrow of the Company," and "his abuses were as foul as any man's."[90] Sir James Cunningham, a Scottish courtier, had received a patent from James I in 1617 to found a Scottish East India Company, posing an existential threat to the English company. At some expense, the English East India Company succeeded in having Cunningham's patent withdrawn in 1618.[91]

Besides working for two joint-stock companies and the navy, Prusen was an original colonial investor in Virginia and Bermuda and also personally financed several other voyages.[92] He had his books audited by the Virginia Company due to an alleged financial impropriety concerning a fishing voyage to the north coast of Virginia.[93] In another venture together with Abraham de Baker (originally from Ghent and identified as a merchant), Prusen invested in the *Flying Joan*, one of the ships Sir Walter Raleigh took up the Orinoco River in Guiana.[94] The pair also outfitted three ships "to sail to the West Indies in the Peregrine, the Grissell, and the Hopewell to cut wood there and to discover an island called Green Island or, if they failed, to go to Virginia and trade for goods there." As Prusen complained in the High Court of the Admiralty, suing the legate of Spain, Gondomar, "the ships were taken and robbed by the Spaniards."[95] This suit details a sorry story of Prusen's business methods, a story that includes his rather proud appropriation of plans from others (the ships they sent out were searching for "discovery of an Island formerly attempted, first by the Hollanders, secondly by Sir Thomas Smyth, and lastly by us"), mutiny, piracy, a falling out between business partners, and finally his imprisonment.

Heydon supported Prusen when Prusen's relationships with others soured. Prusen, Henry Briggs, and Cornelis Drebbel partnered with Sir Anthony Thomas and John Worsopp on a Fens drainage project in which Heydon participated.[96] Their Fens drainage plan opposed an alternative plan by John Heydon's enemy, William Burrell (ca. 1570–1630), master shipwright to the East India Company.[97] For the Fens project, Burrell partnered with the appropriately named Sir Thomas Crooke (1574–1630), a suspected pirate. In a printed broadsheet of 1629/30, Thomas and his associates attacked Crooke and Burrell who had "interposed" on their plans for Fens drainage.[98] Eventually, Prusen and Thomas had a falling out, and Heydon took Prusen's side. Thomas complained that the work of draining had been almost finished when Prusen and Sir John Heydon attempted to wrest control from him. Such "is Prusons impudence . . . that he doth report about the Cittie that there are 8000 acres underwater which his clamour doth disharten many Adventurers to this day." Thomas also complained that Prusen had manipulated many gentlemen "of qualitie," including Sir John Heydon, to get them to attend a meeting where Prusen pretended that "his desire was to set the business at right, in truth intending to perplexe it the more as the sequel proved."[99] Much as Prusen had taken advantage of the East India Company while promising to clean up its former corruption, Prusen argued that a commission should be established to examine the proceedings in their area of the Fens, overriding the protocols already in place. Prusen's attack worked, and his group successfully drained their area of the Fens, which was, however, reflooded by protestors in 1642.[100]

Cornelis Drebbel, Interloper from Below

Interloping did not only apply to the way that social superiors drew upon the knowledge of their social inferiors or intruded themselves into arenas from which they were traditionally excluded, such as the mechanical arts. It also applied to the ways that figures lower down the social scale claimed epistemic abilities. A case in point was the Dutch-born philosopher and court engineer Cornelis Drebbel (1572–1633) (figure 8), whom Jonson mocked in the figure of Van Goose as a force of misrule in the *Masque of the Augurs*. Yet, even as the undisciplined nature of Drebbel's investigations troubled social, epistemic, and natural categories, it was applauded from above as a form of knowledge that social superiors might utilize. What attracted socially superior projectors to such individuals was not their thorough experience in a particular craft, but rather the ambition to override boundaries in ways that might push the envelope on what was possible. They were interlopers, not experts.

Figure 8. Christoffel van Sichem I, *Cornelis Drebbel*, 1604. Rijksmuseum, RP-P-1913-2557. Public domain

A freethinking, social maverick, Drebbel repudiated hierarchies of all kinds—natural, social, and intellectual. He rejected discipline and formal learning. As Drebbel's associates the Küfflers reported in 1624, "as he grew in age, he was always growing in inventions, which came from the vivacity of his spirit, without the help of the reading of books, which he always despised, holding it for a maxim that the truth and excellence of the sciences consists in the knowledge of the secrets of nature of which they [the sciences] are entirely composed."[101] He lived "as a philosopher, concerned only with his observations and disdaining all the things of this world and the nobles and would more readily salute a poor man than a great lord. He lived according to the laws of nature and believed in nothing."[102] In 1639 Samuel Hartlib noted that the "binding ones-selfe to any Rule whatsoever dose hinder mightily a Mans free-Invention," and for this reason "Drebbel would not suffer his children to bee taught in schooles."[103] The elegant Constantijn Huygens Sr. characterized Drebbel's face as that of a Dutch farmer but his speech as that of the ancient philosophers of Samos and Sicily.[104] A Bohemian nobleman, Jindřich Michael Hýzrle z Chodů, who witnessed Drebbel's first appearance in the court of King James I in 1607, described him as "a very rude fellow in appearance" (na hledění velmi sprostnej člověk).[105]

Drebbel's many wondrous inventions also transgressed limits, from courtly optical displays that appeared to transform himself instantaneously from a king into a beggar,[106] to his microcosmic perpetual motion that, as he wrote in his letter of dedication of the device to King James in 1607, could either display the natural order of the elements in miniature or reverse them at will.[107] Drebbel's invention of the submarine, underwater torpedoes, or technique of artificially cooling the air, all performed before public audiences, likewise demonstrated a mastery of nature in their ability to transgress elementary divides of water, fire, heat, and cold.

His most famous device, the perpetual motion, incorporated many simultaneous motions moving at differing tempos and communicating a variety of forms of knowledge, including liquid moving in a ring (as though in accordance with the ocean tides), an astronomical clock, and a display of lunar phases. Drebbel claimed that he had discovered his knowledge of nature through the work of his own hands and that his devices offered an easy route for others to do the same.[108] After Drebbel's device was held for observation in Whitehall for several months, James assigned Drebbel to Prince Henry's household and had him install the device and other automatons at Eltham Palace outside London, where this display was visited by many.[109] Drebbel's perpetual motion was also designed to be interactive; it could be set in motion through the heat of a hand placed upon it. As one letter attributed to Drebbel boasted, not only King James but a "thousand" other witnesses were able to test the device by making it move through the heat of their hands.[110] Indeed, as Drebbel complained in another letter to James I, Queen Anne investigated the device with such a "curious hand" that she broke it.[111] Drebbel made many versions of his perpetual motion that were sold across Europe and entered various princely collections.[112] One of the most often described was held in the Saxon electoral *Kunstkammer* in Dresden, where it was witnessed in 1623, and, still working, in 1654, although the keeper of the collections was loath to allow visitors to touch it.[113]

Drebbel's perpetual motion used subtle changes in temperature to keep the device moving; he thus used macrocosmic motions to power his microcosm, which was an acceptable definition of perpetual motion in the period.[114] He argued that his working device served as a demonstration of his wider knowledge of weather patterns and changes in heat and cold that he explored in his published text of natural philosophy, *On the Nature of the Elements.* In his 1607 letter of dedication to King James, Drebbel claimed that he could reorder the elements, making light things heavy and heavy things light. He extrapolated from his own ability to transform nature to James's ability to transform the political realm. "Justice," Drebbel told the king, "does not wish the punishment to outweigh the crime, but rather that the punishment should be lightened through moving compassion, so that all men may taste the pleasing fruit of wise rulers, and in place of cruel war, shall enjoy the sweetness of the arts."[115]

James and other crowned heads cared little that for Drebbel, his perpetual motion might signify how the knowledge of nature lay within the reach of all to investigate. To them, Drebbel's displays of the cosmos served as symbols of imperial power, like other preternatural wonders they had in their collections. His inventions promised control over the elements to the crown. In their portability, navigability, and ability to overturn seemingly natural order and to convey intelligence across great distances, Drebbel's objects seemed to place nature that was otherwise out of reach within the grasp of the monarch. According to the courtier Sir Francis Kynaston, Drebbel invented "a glasse wherin hee could raise both lightning and thunder with another weather-glasse which both were presented to the King."[116]

The court was also intrigued by Drebbel's various microscopes and telescopes, which his associates, the Küffler brothers of Cologne, also sold for him around Europe. Drebbel informed James that he had made an "instrument through which in time of need or of war, letters could be read at a distance of one or two miles, also so that we might extend our vision at night one or two miles."[117] He was working on lengthening that distance even further. According to a letter to King James written after the death of Prince Henry, he was already able to make a telescope that could read letters a mile off, and with the help of the king, he was working on one that could read regular-sized letters at distances of "5, 6, or 7 miles." Through this instrument, the king could also observe what was being done around him at distances of eight or nine miles as comfortably as though it were in his own bedchamber.[118] According to another source in 1622, Drebbel was aiming to construct a telescope that could read writing at a distance of seven leagues (about twenty miles) and to "multiply the light of a star" so that one could read a letter at night and illumine a space thirty paces (about 150 feet) in diameter."[119] Several observers noted how Drebbel used his telescope to observe the moon, sometimes alongside fellow artists. They believed they saw lunar fields, woods, and buildings much like Earth's, claims that, predictably, Ben Jonson would ridicule in his 1620 masque, *Newes from the New World Discover'd in the Moone.*[120]

Drebbel rarely sought to develop his inventions himself into large-scale industry and projects, as his associates the Küffler brothers would after his death in the case of his ovens and scarlet dye. Instead, others quickly saw the potential of his inventions and made use of them. Samuel Hartlib noted that, according to Drebbel's son-in-law, Dr. Küffler, Sir John Heydon, who had already made use of Drebbel's ovens to experiment with hatching chicken eggs, "brought in a Calculation how by Kufflers new- baking ovens 40. thousand lb. might bee saved to the k[ing's] Army."[121] As discussed in chapter 4, both King James and Prince Charles wished to deploy Drebbel's arts of submarine navigation, weight-lifting, and artificial cooling of the air in ambitious projects around the globe. Drebbel also joined the talented cast employed by Villiers, assisting John Tradescant the Elder in outfitting the grounds of his property, New Hall.[122] Between October 1624 and July 1626, Villiers paid Drebbel for "contriving of some works at New Hall," "for his paines about the waterworks," and for delivering "divers models."[123] In 1627 Villiers would call upon both Tradescant and Drebbel to travel to France and assist him as his engineers on campaign there.[124] Villiers leaned heavily on the abilities of figures like Tradescant and Drebbel to extrapolate from their courtly, symbolic, and aesthetic displays of the mastery of nature to applications on a grand scale.

A criticism of Drebbel in the period was that his models, while functioning beautifully as curiosity objects, failed to scale up to useful dimensions. Drebbel's friend, Constantijn Huygens Sr., acknowledged that some "have criticized King James, saying that this perpetual inventor [Drebbel] has never produced works whose costs were repaid by their usefulness." Huygens contended that this was not the case and that most penetrating minds realized that James's patronage of Drebbel redounded to royal benefit.[125] Yet, his comment speaks to the tension at court concerning the royal patronage of Drebbel and doubts about whether his wonderful inventions could ever be deployed to useful ends. As discussed further in chapter 6, the Moravian noble inventor Johann Christoph Berger von Berg, a.k.a. Jan Kryštof Pergar z Pergu, partnered with fellow Moravian Jan Amos Comenius on developing a perpetual motion in competition with Drebbel. In 1635 Berg reported that Sir Francis Kynaston said, "Many things are Mathematically true which are not Mechanically et contra . . . Also that Models faile often as Drebbeli perpetuum Mobile would move or bee true in smal Models but not in great."[126]

Below Drebbel on the pyramid of projects worked his assistants, although their efforts are even harder to trace than his own. When he presented himself before King James at court in 1607, he was described as accompanied by a servant.[127] When Drebbel briefly went to Prague to work for Habsburg emperor Rudolf II, he was assisted by Roger Cock, John Dee's former assistant in Prague, who would eventually leave Prague together with Drebbel.[128] The Küffler brothers from Cologne (Gilles, Abraham, Jacob, and Johann Sibbert), sought Drebbel out, and he took them on as assistants. Jacob, Gilles, and Abraham peddled his microscopes and telescopes in France, Italy, and German-speaking lands. Abraham Küffler accompanied Drebbel on Villiers's naval campaign to France in 1627 and helped him there. The clockmaker Ahaseurus Fromanteel (1607–1693) also used to work for Drebbel, making cases for his telescopes and microscopes.[129]

Disciplined Labor

If discipline appeared in the world of projects, it did so at the lowest ranks of a project's labor force. Many projects framed the disciplining of unruly populations through labor as one of the public benefits of the project. Thus, projects could be experienced very differently by different people. For their undertakers, projects were vehicles for expanding ambition and undisciplining knowledge. For the populations at which projects took aim, they could be tools of restraint and enclosure.

Likewise, many of Drebbel's inventions automated tedious labor in ways that could be freeing for some and constraining for others. Hartlib described Drebbel's self-regulating oven as one "that one needs not looke often vnto it and may goe abroad etc."[130] He also included Küffler's art of hatching eggs on this stove on a list of low-skilled and large-scale employment for the poor.[131] Similarly, Abraham and Gilles Küffler described in 1624 how Drebbel kept a mill for grinding the glasses he needed for lenses "in his house near London"; after putting the glass on the mill, a "young boy" could easily keep it running for him. One could leave and "go for a walk for three or four hours," and the grinding would not have gone astray at all.[132]

Labor was also enforced in various projects on which Drebbel worked, such as in the naval engineering he performed for George Villiers. In 1626 Villiers ordered that "seamen or fishermen or that are practesed in sea fareing" be pressed into the navy's service. By the next year, navy captains were also granted the power to impress men directly from merchantmen at sea to ensure that they were obtaining skilled mariners.[133] Minimally paid, enforced or enslaved skilled and unskilled labor played a role in many of the other projects discussed in this book. As discussed in the next chapter, these included the noxious work of alum miners, who went unpaid or were paid merely with paper "tickets," or IOUs. As discussed in chapter 5, projects included the shipping of children and convicts to labor in Virginia. They also included the use of children as young as seven years old and residents (likely women) of the Bridewell houses of correction to string beads for Sir Nicholas Crispe in order to trade for slaves and gold in Ghana. Other projectors devised machinery, for spinning yarn for example, that could employ up to sixty children as young as ten years old at a time "with double of the gaine that they now have."[134]

Conclusion: The Antimasque That Never Ends

Dramatic portrayals of projectors have shaped our assumptions that these figures were lone individuals, laughed at by mainstream society for their ridiculous ideas. In fact, projects functioned with large groups of personnel and were often helmed by figures with significant political entrée and clout. Satirical portrayals of projectors often served as both implicit criticism of the court and as a means of indirectly petitioning the court against its culture of projecting.

The jurist John Selden (1584–1654), who bitterly opposed projectors, assisted in putting on a 1634 masque, *The Triumph of Peace*, performed by members of the Inns of Court before the king and queen in Whitehall. An antimasque of six projectors paraded various supposedly ridiculous inventions such as: a hollow bridle conveying a vapor to "cool and refresh a horse" so "he shall ne'er tire"; a perpetual motion for threshing corn; a lamp which could "boil beef so thoroughly, that the very steam of the first vessel shall alone be able to make another pot above seethe over"; "a new project" of a case for walking "all day under water" with the air inside kept cooled with "artificial bellows" (the case could be used to "fetch up Gold or whatever jewels have been lost, In any river o' the world"); a "new way to fatten poultry with scraping of a carrot, a great benefit to the commonwealth"; and finally, "a ship to sail against the winds."[135]

The authors of the antimasque carefully delineated the social status of the projectors: a "Jocky," a "country fellow," a "Scholasticke" projector with a "face Philosophicall and beard," and "a grave physician." Drawn from both learned and unlearned ranks of society, they shared little in common other than the fact that none of them were noble nor connected to the court in any way. Bulstrode Whitelocke, who composed music for the performance, described the antimasque as a means of bringing the corrupt culture of projecting to royal attention. Through the antimasque, "an intimation was covertly given to the King of the unfitness and ridiculousness of these projects against the law."[136]

At first glance, then, the antimasque appears to be an effort to inform the court of abuses performed outside of it. Yet the story is not as simple as it seems. As Koji Yamamoto has discussed, one of the chief authors of the antimasque was himself a monopolist involved in the new soap company, one of the most contentious projects of the period, and to the "extent that the antimasque downplayed the centrality of projecting to the early Stuart royal policy, and to the extent that hopeful observers found mild criticism in the caricature and others the comforting triumph of peace over trifling projectors—to that extent the royal entertainment and its participants were *complicit* in the grievous state of affairs."[137]

As in the case of Jonson's *Masque of the Augurs*, the depiction of the projectors in *The Triumph of Peace* functioned more as a petition to the court to change its ways than as a representation of reality. As the Earl of Northumberland grumbled in 1637, "Here hath been lately so much Favor and Countenance shewed to Projectors, that there are few in Court that have not at this Time a Suit either granted or referred."[138] Many of the courtiers in the audience were themselves projectors with projects not dissimilar to those gently mocked in the antimasque.

The seemingly irrational and improbable ambition of many projects of the period can be explained not as the unhinged imaginings of eccentric individuals on the social margins, as in critical antimasques staged by Jonson or Selden. Rather, such ambition often stemmed from elite projectors who brought attitudes of lionhearted valor, aggressive rivalry, and conquest to their grandiose plans. They emphasized their adventure, rather than their mitigation of risk, since the more adventurous and ambitious a project was, the more glorious it would be.

Many early Stuart projects were not a business success. Yet, if projects often did not offer immediate financial returns, they did bear other proceeds. As Lindsay Levy Peck has contended, the graft and misuse of powers that flourished among early Stuart projectors has "obscured the long-term importance of these projects for trade, industry, profit-seeking and changing views of the economy."[139] Projects offered a form of speculative thinking about the intersections of politics, technology, and profit. Projectors were the robber barons of the early modern period. Even as the persona of the elite projector morphed into new, more bourgeois forms as speculator and entrepreneur, many of the relations earlier elite projectors had adumbrated between science, industry, and power endured.

CHAPTER 3

"Projectors are commonly the best Naturalists"

Knowledge Practices

The previous chapter surveyed the varying roles encountered in early Stuart projects. Projects often brought together individuals of differing talents and resources. Yet, that chapter also made the case that different roles on a project were fuzzy. Figures who could interlope across many domains proved valuable in projects. This chapter focuses on a single interloper, tracing his work across several projects. It aims to analyze practices of projecting that appeared across the many fields in which projectors engaged. These included techniques for currying favor and attracting investors and collaborators, framing of projects in ways that ingeniously maximized advantage and entangled various goals together, political and numerical arguments and calculations on behalf of projects, methods of trial for testing feasibility, and ways to profit from failure by reassembling pieces of projects into new endeavors.

The projector we will follow is difficult to trace in the archives, yet worth the effort given the great esteem (and sometimes opprobrium) he garnered from contemporaries. Noting in 1634 that "Projectors are commonly the best Naturalists," Samuel Hartlib, who had moved from Prussia to London six years previously, recorded the view of one of his interlocutors concerning

one Mr Rossel the father of all projectors et now the only Man for the New
sope besides many other Experiments which hee hase found out to turne iron into

steele et divers others about Copras Saltpeeter and the like. To whom all others resort. Hee is not only Mechanical but very Rational being able to give a reason for all hee dose.[1]

This "Mr Rossell spent above a hundred thousand lib. [pounds] or by his Meanes in Experiments," Hartlib also noted. In the course of the same conversation, a few lines down in Hartlib's notes, the same interlocutor discussed his efforts to acquaint others with "Experimental Philosophy."[2] According to Mordechai Feingold, this may have been the first usage of the term "experimental philosophy," which would become the approach with which the later Restoration Royal Society identified itself and which would also become the academic discipline for experimental sciences established at universities beginning in the late seventeenth century.[3]

This interlocutor was the clergyman William Watts (ca. 1590–1649) and "Mr. Rossell" can be identified as Thomas Russell, gentleman, a resident of London whom Francis Bacon reminded himself to seek out in 1608.[4] This chapter offers a close reading of the texts of some of Russell's many projects. It examines the practices of trial to which Russell's projects were submitted, in the context of this early development of notions of experimental philosophy. Russell's practices illustrate how he brought a wide array of forms of knowledge together in his projects. He was, as Hartlib noted, both "mechanical" and "rational," that is, he possessed both practical experience and could reason theoretically about his ideas. Beyond this, he was also politically calculating and ingenious in the ways he entangled diverse phenomena in the design of his projects.

Although Thomas Russell might have been a more skilled projector than most, his projects represent wider practices. Such practices illustrate skills, strategies, and attitudes that could be applied to the experimental study of nature. As Hartlib noted, it wasn't just Russell who excelled as a naturalist, but projectors "commonly" did. Russell was also far from alone in developing the social and political (and not just technical) skills of projects. Seventeenth-century observers often appreciated the various ways projects might be presented to entice participation. For instance, John Aubrey FRS, in writing the life of Thomas Bushell (1594–1674), described how Bushell knew no Latin but "had a good wit" and a "working and contemplative head." His "Genius lay most towards natural philosophy," but he also possessed other skills, such as "so delicate a way of making his Projects alluring and feazible, profitable, that he drewe to his Baits not only rich men of no designe, but also the craftiest Knaves in the countrey."

Using "the strangest bewitching way to drawe-in people," Bushell's "tongue was a Charme." According to Aubrey, Bushell was no technician, but "a handsome proper Gentleman." Bacon "much loved him."[5]

Bushell's "working and contemplative head" recalls Russell's fusion of the rational and the mechanical. Likewise, the gentleman Simon Sturtevant boasted that his own "diverse proiects and new dewises" were both "Litteral as Mechanicall." He cut across such categories by producing a "Scholastical engin" or "Aucomaton" [sic], that is, an intellectual device for speedily teaching Hebrew.[6] He brought his literary accoutrements to bear upon mechanical practices through the rather bombastic, neologism-filled theorizing he applied to practices of invention or "Heuretica."[7] Russell, Bushell, and Sturtevant represent social types that have been retrospectively classified as technicians or miners due to their hands-on experimentation and invention. However, in their lifetimes, these figures were considered gentlemen. Their fusion of traditionally socially divided forms of knowledge, such as book-learning and craft, embodied a period ideal of interloping across social and epistemic realms.

Thomas Russell, gent.: Russell, Bacon, and the Courting of Courtiers

Thomas Russell has featured in the recent history of early Stuart technology due largely to Bacon's references to him. As Cesare Pastorino has noted, Bacon identified Russell in 1608 as a key figure to cultivate, both for his own knowledge and for the networks to which he had access. Pastorino explored the relationship between Russell, Bacon, and mining as a paradigmatic example of a partnership between practical expertise and political patrons. Russell had been one of the figures refining the ore in the Tower of London mentioned by Gerard Malynes (discussed in chapter 1). An official report fully described the "trial" of Russell's technique. Pastorino understands this trial in the sense of a metallurgical assay, but with interesting relations to later experimental reports.[8] Pastorino sees Russell as "most likely an assayer and entrepreneur," and possibly also a medical "empiric" and "iatrochemist."[9] In general, Pastorino situates Russell as a technical expert in a field in which Bacon the gentleman had entrepreneurial interests.

Russell thus fulfills Eric Ash's model of the Elizabethan "expert mediator," that is, "a figure bridging the gap between a particular field of expertise and the patrons and central administrators supervising it." According to Pastorino, "Russell's interactions with Bacon confirm this picture"; Russell would be the technician who could serve as bridge between mining and the forms of political knowledge and privilege enjoyed by administrators such as Bacon.[10] Yet, if Russell was himself a "gentleman," as he is consistently described, and if he engaged in a variety of projects for which he did not in fact possess particular expertise, his relationship with the gentleman Bacon becomes much less dichotomous.[11] As this account of Russell's many activities and relationships will show, rather than providing technical expertise, his role in projects was supplying some new ingenious insight or underutilized resource, or often, a combination of several, which Russell then framed into large-scale projects. The trial of his metallurgical technique belonged to the same practices of testing that applied to all projects, from submarines to saltpeter to soap.

In the passage in which Bacon reminds himself to cultivate Russell, Bacon shows how familiar he was with the pyramidal arrangements of projects. On his to-do list he noted, "Makeing much of Russell that depends upon Sr. Dav[id]. Murry and by that means drawing Sr. Dav. and by him Sr. Th[omas]. Chal[oner]. in tyme the prince."[12] Bacon's eventual aim was gaining access to Prince Henry, and Russell was a carefully chosen means to this end. Ever strategic, Bacon aimed three levels below his target on the hierarchy, at a level that he believed he could access considering his own status in 1608. Bacon evidently considered Russell enough beneath him that "Makeing much" of him would work to win him over.

Bacon also saw Russell as a portal to knowledge, but not, however, restricted to a particular field of expertise. He reminds himself to get "from Russell a collection of phainomena, of surgery, destillations, Minerall tryalls."[13] Russell excelled at making trials. In his discussion of attempts to assay the Scottish silver ore between 1608 and 1610, Malynes discussed several challenges presented by trials, including variation in results based on differing samples, as well as the expense and labor of trials. Initial reports of the quality of the mine in Scotland varied greatly, "according as they found the Ore of severall veynes." Trials of the ore from various parts of the mine varied by vast amounts, from eighty pounds of silver per hundredweight to a few ounces.[14] The Privy Council ordered an elaborate and costly trial to determine the average content of silver in the ore. Ten tons of ore were brought to the Tower of London, of which one ton was "indifferently taken and calcined or grinded together," to which two tons of lead were added and "molten by a continuall fire and hand-blast of foure men." This expensive trial (as Malynes noted) revealed twenty-two ounces of silver per hundredweight of ore, the same proportion found by four others who tried the ore through cheaper means, including "Master Russell."[15]

Russell's involvement in mineral trials does not make him an expert in this field. Many other figures who were not particularly experienced in mining also hoped to test out the silver ore in the Tower. Cornelis Drebbel, for instance, petitioned King James to allow him to try a new process for refining the silver ore taught him by an unnamed friend.[16] In fact, Malynes negatively contrasted all those who attempted to refine the ore in the Tower, including Russell, with an overseas "expert Mint-master" to whom he hoped to export the ore, since this master, he claimed, could refine greater amounts out of the ore than anyone in England could.

Pastorino discusses Russell again in another article, "The Philosopher and the Craftsman," which examines the privileges for new inventions whose texts Francis Bacon was charged with drafting from 1607 to 1617.[17] These twenty privileges included three granted to Thomas Russell, for brimstone and copperas (with Sir David Murray, 1608), for "making of copper by dissolution (with Sir David Murray, 1610), and for "making copper" (with Lord Edmund Sheffield and Sir John Bourchier, 1614).[18] Pastorino notes that many patentees were elite "courtiers and high-ranking officers of state," who "joined the craftsmen for a form of financial investment and speculation." Among the latter, Pastorino identifies patentees "with more technical expertise: projectors or inventors," such as "the metallurgist and mining projector Thomas Russell." However, as the analysis of Russell's career here indicates, he was also a gentleman and an investor and did not possess a particular expertise in a single field.[19] Rather, he interloped among many, including forms of financial and political argumentation central to drafting projects and acquiring patents. It was this position on the project, rather than any particular technical expertise, that made him into a "projector." This was also true of other aspiring patentees whom Bacon reviewed, such as Simon Sturtevant. Sturtevant, for example, was identified as a gentleman in the official patents that he proudly republished, and he had received a bachelor's and master's degree at Cambridge.[20] Thus, the interaction of Russell and Bacon, or Sturtevant and Bacon points less to the exchanges of a craftsman and a scholar, and more of two gentlemen, both of whom interloped into many areas of knowledge.

In fact, Bacon entitled the section where he reminded himself to make much of Russell, "Cast.," apparently an abbreviation for "casts," the term he deployed for projects. Others also interpreted "project" to mean "forecast."[21] Elsewhere in the same text, Bacon delineated the exact mode he should deploy when sketching such schemes. When corresponding with Robert Cecil, 1st Earl of Salisbury in particular, Bacon reminded himself to adopt "a habite of naturall but nowayes perilous boldness, and in vivacity, invention, care to cast and enterprise," a style of projecting that he thought would most please Salisbury and make himself appear the most useful.[22] Thus, not only were both Russell and Bacon gentlemen, they were also both reaching out at this time to the same patron (Salisbury) with their projects.[23] In Bacon's case, one of his projects or "casts" included making much of Russell.

Further blurring clear dichotomies between technical experts and social superiors, those above "Thomas Russell, gent" on the social scale that Bacon outlined (Murray, then Chaloner, then Prince Henry) also brought their natural knowledge and practices of trial to the table. Prince Henry asked Sir Thomas Chaloner II (1564–1615) and Sir William Godolphin (1567–1613) (upon whose mining knowledge Russell also relied) to test proposals for refining silver from lead ore.[24] King James had also requested that Chaloner test Sturtevant's new method of casting earthen pipes, which Chaloner duly subjected to a "full tryal" over the course of a "half yeer."[25]

Innovations

According to William Camden in 1610, Chaloner, "a learned sercher into natures workes," had first discovered deposits of alum and copperas in Yorkshire. Chaloner noticed a strange color of tree leaves in the area, an unusual structure of the trees, and earth of diverse colors that never froze but "glittered in the pathes like unto glasse" at night—all strange phenomena that led him to hunt for minerals in the area.[26] The glittering soil and strangely colored and shaped trees gave rise to an alum industry in 1607 in which Russell also played a role.

Alum, a mordant essential to the dyeing industry, was normally imported into England from either the Ottoman Empire or Rome. A local source would not only save on those imports, but increase profits from finishing textiles, and also damage England's enemies. This was a large-scale, environmentally destructive project that relied upon bringing huge amounts of human urine to the mines; ammonia from the urine was used to precipitate alum out of an aluminous liquor drawn off from the burning of shale deposits. Its effects on the landscape were not pretty. When two other alum projectors tried to establish new works near the Tower of London, neighbors protested about the "insufferable and contagious annoyance" of the "loathsome vapor" these would produce, and the College of Physicians advised that those works be suppressed for reasons of health.[27] The argument that 40,000 pounds would be saved annually in money and commodities, and that this project would be to the detriment of the pope and his alum exports, succeeded in winning Lord Sheffield, Sir Thomas Chaloner, Sir David Fowles, and Mr. (later Sir) John Bourchier (ca. 1560–1625) of Yorkshire a patent for alum works in 1607.[28]

Huge sums were quickly lost on this project: the alum works consisted of six buildings, with each building spending 300 pounds a year on urine alone (at 10 tons weekly).[29] A new bid to run the works in 1615 by a group of projectors— Sir Arthur Ingram, Martin Freeman, and George Lowe, esquires—received a grant and a subsidy of 10,000 pounds in cash. After failing, this group then spent 2,000 pounds to subcontract the works to Sir John Brooke and Thomas Russell.[30] Brooke (1575–1660) was Esquire of the Body and Gentleman of the Privy Chamber, a client of Villiers, and an important figure in Virginian colonization who had journeyed to Guiana in 1610.[31]

Russell's contributions to the alum project consisted of a number of innovations, such as the use of kelp rather than urine, and a redesign of the pans in which the alum was boiled. Kelp was an attractive replacement for urine since the latter, in addition to being noxious, had become strangely expensive given the large amounts of it that were required.[32] Although the chemistry of how it worked was poorly understood at the time, Russell was correct in supposing that kelp could replace urine. About 3 percent of kelp is made of potassium chloride, or potash. To aluminum sulfate dissolved out of the calcined shale, either potassium sulfate from kelp or ammonium sulfate from urine could be added to precipitate alum. Although this venture failed in Russell's time, using kelp to produce alum would become a major industry in the eighteenth century.[33]

In 1616 the Privy Council issued an order allowing Ingram, Freeman, and Lowe to collect, burn, and carry away kelp found on the coasts. As was typical of arguments in favor of projects, this order noted both how kelp was an unutilized resource, considered worthless by others, and also how using kelp could be partnered with a labor project employing poor people:

> whereas we are informed that there is a certaine weede, called kelp, seawracke or sea oare, cast up by the sea in many partes of the coast of this kingdome, which beinge of noe generall or knowne use, is altogether neglected by the lordes of the soyle, and soe suffered to ly upon the shoare, without makinge any benifitt or use thereof; forasmuch as the ashes of that seaweede is found to be of speciall use for the making of allome, and that many poore people would be sett on worke in gatheringe together, burninge, barrellinge up, and carryinge the same to the allome workes.[34]

Sir Arthur Ingram, one of the contracting projectors, stored great hopes in the arrival of Brooke and Russell, saying, "I longingly wait for them."[35] Sir John Bourchier was so confident in Russell's abilities that he tried to buy out Brooke.[36] Despite such hopes, Russell also failed, only producing 230 tons of alum between April and July 1618. Moreover, he had melted down the old lead boiling pans to make new ones of his own design, but when these didn't work, the old type had to be constructed again at great expense.[37] Another issue was Russell's idea of moving the whole operation to Newcastle. A petition signed by eighteen people was submitted in August 1619 on behalf of Sir Arthur Ingram against Russell's idea, saying that it would be cheaper to bring urine from Newcastle and also stating that the original large boiling pans were better.[38]

Ingram did not conceal his disappointment with Russell, calling him "an uncertain fellow who never continueth two hours in a mind."[39] Ingram's negotiation to end the subcontract with Russell and Brooke proved very contentious.[40] Ingram's partner, George Lowe, warned him,

> I am undone by this business and will you be if you cannot work yourself out of it quickly . . . or else . . . provide 5,000 l. to repaire, complete and stock the houses with great pans of the old fashion. Mr. Russell's small pans prove altogether useless . . . and twice as expensive. He would now change them for little, shallow stoving pans, of which he is as confident as he was of his small round pans. He has not made 500 tons of alum this year and has wasted a great quantity of coal, lead and iron, and made such havoc of everything.[41]

In addition to 5,000 pounds to remake the old type of pan, Lowe estimated that it would cost 2,700 pounds to undo other changes Russell made at various sites of alum production.[42] Meanwhile, the workers had not been paid for three or four months and worked "daily sore labour for nothing but tickets, which they set before their wives and children to feed them when they ask bread."[43] Lowe accused Russell of keeping his debt to the workmen of 1,500 pounds a secret so that "he might get me to pay it by degrees in expectation of his new device for making alum . . . with stoving pans 6 inches deep, into which form he would have all the works altered. . . . When he saw I would not pay his debts he stirred the men up against me."[44] Russell's accounting often relied on one project or part of a project to fund another, as in the case where he sought to be paid for refining Scottish silver so that he could fund his new brimstone and copperas works.[45] Here, according to Lowe, he deferred paying the workmen with the expectation that his new design for alum boiling vessels would pan out and provide an injection of much-needed funds. Lowe described the workers as "a multitude of poor miserable people that are ready to starve for want of means to buy bread and will not be quieted with any answer but money." He heard that Russell was barely keeping "them from violence, which cannot long be suppressed."[46]

Meanwhile, Russell also wrote to Ingram in his own defense, saying that he was "forced to devise new courses" for making alum since Lowe had "failed me in delivering ashes, liquor and urine." Russell claimed that he had told Lowe concerning his new design that "if he or any one disliked it when they had seen it I would disclaim all hope and interest in the works," but that Lowe didn't have the "patience or grace" to give Russell's new method a chance. Lowe, Russell claimed, "pays the men with big, roaring words and no money and sets them all against me."[47]

In another lengthy letter in his defense to Ingram, Russell displayed some of his rhetorical skills that allowed him to stand fast against accusations of incompetence and impropriety. Russell accused Lowe of acting "like a bear robbed of her whelps . . . professing against all my courses, chafing, railing, swearing, cursing, exclaiming and what not, by which means he hath brought such an uproar in the country that it will not easily be appeased." Russell claimed that Lowe had not been pulling his weight in the alum business. Lowe and others were supposed to act "as stewards and caters of the work, as I was to be the cook." Alum production was "not one man's work," and Russell would like to see Lowe try to "hold the plough himself, as I have done all this while." In the end, Lowe's "great roaring against me will be but like a raging billow against a rock, dissolving in his own froth."[48]

Many sided with Russell against Lowe and Ingram. Lord Sheffield (who along with Bourchier had previously collaborated with Russell on a privilege for copper production in 1614) wrote to Ingram in 1619, warning "if it appear that the malice of those who oppose Mr. Russell cause these contradictions to shuffle him out of the works I hold myself tied in honour to see him righted."[49] Brooke and Bourchier also stuck with Russell. In 1624 Sir John Bourchier proposed restarting the alum works and pairing it with a soap works, promising that the king would enjoy a profit of 20,000 pounds a year from it; Russell would long figure in this project for soap.[50] Brooke also partnered with Russell again for soap in 1624 and on another large-scale project for the production of saltpeter in 1626/27.

Framing Projects

Russell won his moniker as "father of all projectors" not only for his fusion of the "mechanical" and the "rational" in his consideration of nature, but also for his political savviness in the framing of projects and recruiting investors, collaborators, and personnel. The powerful administrators to whom Russell, like all projectors, had to turn to acquire patents and monopolies did not necessarily engage with him in a mutually beneficial partnership that the model of expert mediation implies. Russell, for example, approached Robert Cecil to make sure he was paid for the silver ore he had been refining, as he required the funds to set up a new brimstone works in partnership with Sir David Murray; he promised to be ever ready to serve Cecil in return.[51] Later that year, however, Russell didn't take kindly to how Cecil intervened on a patent for copper production by what Russell claimed was a new method of dissolution in water. Russell had petitioned the king that this patent should be licensed to him only; Cecil inserted two others into the patent at Russell's charge. Russell was willing to buy them out, if only Cecil "will free him from any other charges and from molestation, according to his promise."[52] This was more intrusion and appropriation on Cecil's part than mediation.

In 1614 Russell and Chaloner's old Yorkshire neighbors, Lord Sheffield and Sir John Bourchier, received a patent for making copper due to the "failure" of the previous attempt of Russell and Sir David Murray.[53] In 1617 Sir David Murray sought another grant for "the sole making of brimstone and Danish copperas," claiming that he had been "cozened in the former grant, but the parties who persuaded him to it promise to perform their pledges, if he can get it renewed."[54] Evidently Russell was not a participant in the fraud Murray claimed. Their previous attempt had failed, Murray claimed, because they had apparently relied on subcontracting to unreliable parties, again indicating the extent to which Russell subcontracted out technical work.

Russell was still dealing with the aftermath of Cecil's interloping into his copper patent many years later. In 1618 Russell managed to interest the Merchant Adventurer and alderman Sir Francis Jones (1559–1622) in a project for making copper in Cornwall.[55] As Russell wrote around 1624, however, Cecil's old intrusion upon his patent was still valid (even after Cecil's death). He had paid 160 pounds for his patent for copper production, only to have the Cecils claim a two-thirds stake in it. At that rate, it became difficult to secure investors, and he hoped they might "be free from further charge to the King, and protected from other encumbrances."

These other encumbrances included the local landowners who evidently objected to him mining on their property. Russell worried lest Prince Charles would decide that the Cornish copper mine he intended to work was "no royal mine, and therefore will agree with the owners of the soil." Russell argued that the copper mine could be proven to belong to the Crown given "the quantity of silver" it contained. He referred to a loophole that claimed mines of noble metals, such as silver, for the Crown even if the intent was to mine base metals there, as Russell was proposing, who claimed to be able to extract one ton of copper and two tons of bell metal from forty tons of ore.[56]

Russell himself was not a mining expert, but rather highly skilled in identifying such loopholes and opportunities. Cecil, at least, did not think much of Russell's expertise in mining. Russell had directed him for information on mining to Sir William Godolphin, a Cornish knight and member of parliament whose family had long been involved in mining.[57] Cecil wished instead to be informed concerning the mines from one John Milward.[58] Thomas Russell wrote to Cecil saying that he had "not heard of Milward, sent down to the copper works" but that "Sir Wm. Godolphin can speak of the produce, &c. from the mine."[59]

Figures in Russell's orbit who could be considered mining experts included the Hechstetter family from Augsburg, which was involved in many mining operations in England, including the refining of Scottish silver ore in the Tower and the mining of copper in Cornwall. Daniel Hechstetter proved quite dismissive of Russell's supposed mining knowledge, writing, "it is no wounder that Mr. Russell made so little copper having never gotten any ewer [ore] that I can heare of woorthe the carriadge to the meltinge howses which makes me admire how wise men should thus delude theym selves and thinck to force copper owt of that wher none is."[60] Even in the case of the Hechstetter family, their knowledge was mostly mercantile and consisted of arranging labor and getting products to market as well as computing potential profits. For the technical knowledge of mining itself, they relied "upon the qualifications of their workmen."[61]

The knowledge of Thomas Russell was removed one stage from mercantile knowledge and two stages from technical mining knowledge. His was the knowledge of a projector able to hunt out innovative approaches and frame them into large-scale projects across a wide array of endeavors. Such skills made him attractive not only to Francis Bacon but to many long-term elite collaborators, including Charles I. During the years of his mining in Cornwall, Russell was elected to the 1614 Addled Parliament, representing Truro. Described as a "gentleman of London," Russell had three sons baptized in Truro between 1613 and

1619 and leased several properties around town.[62] Upon his return to London, he won a position among the servants of Prince Charles. In 1624 a Thomas Russell described as the prince's servant and as a "gentleman," was jailed on the allegation that he had spoken "some uncivill words against his majestie."[63] In an undated document, King Charles I appointed "our servant Thomas Russell our principal agent" in mining copper in "Cornwall, Devon, and other places," to be used "for making brass ordnance for our forts and navy, besides manufactures for employment of our people, which copper mines we have now taken into our hands."[64] By taking control of the copper mine, Charles had clarified the issue Russell mentioned previously about whether the mine belonged to the Crown or not.

Russell was also employed in a scheme for royal mines across England, Ireland, and Wales. Sir Hugh Myddelton (1560–1631) proposed a project "to extract silver and lead, to the employment of many poor, and general benefit of the kingdom" in Wales.[65] Through, it seems, the efforts of Russell, Myddelton's mining operation became integrated in a far wider plan. In 1621 Russell received a grant for "the sole exercising of the art of refining and extracting silver out of lead and copper ores, in England, Ireland, and Wales."[66] In July 1624 Solicitor-General Heath was authorized to complete "the contracts begun with Thomas Russell, for employment of his invention and skill in working the mines royal of England, Ireland, and Wales, and with Sir Hugh Middleton and Sir Bevis Thelwall [Clerk of the Great Wardrobe], for working the mines royal in Cardiganshire, with assistance of Russell's invention." The same day, Heath was ordered to draw a commission for Myddelton and Thelwall "to take and employ such miners as they find needful, for digging for ore in the Cardiganshire mines, according to the plan invented by Thos. Russell, giving them reasonable wages."[67] It is unclear from this description whether Russell's part of the plan lay in some metallurgical or mechanical mining invention, or in his ability to plan other aspects of large-scale projects, such as the recruitment of labor. What is clear from Russell's other projects is that the framing, advertising, and presentation of projects was as central to his skill set as any particular technical knowledge.

Calculations

Perhaps it was the experience they gained in transporting large amounts of urine for the alum works that endowed Brooke and Russell with the confidence to tackle the even more noisome problem of saltpeter. Indeed, Lowe complained that one project for producing alum that Russell had purchased proved to be "idle and a cheat" as it only made nitre (that is, saltpeter), not alum.[68] The failed alum project perhaps planted the seeds of Russell's saltpeter plans.

Despite the foul nature of saltpeter production, it was a common project advanced by gentlemen-projectors of the period. In 1626 a "quartet of Essex gentlemen (Sir William Luckin, Sir Gamaliel Capell, Francis Quarles, the poet and William Lyde) proposed to make five hundred tons yearly."[69] Other saltpeter projectors of the period included David Ramsey, a Gentleman of the Privy Chamber and an inveterate inventor and projector, and the equally prolific Sir Philiberto Vernatti, also a member of Charles's Privy Chamber.[70]

Saltpeter production proved an alluring target because the fearsome new technologies of the age such as fireships, petards, and new gun and cannon designs were impotent without sufficient saltpeter to fuel them. Saltpeter (in today's parlance, potassium nitrate) provided the majority of the substance of gunpowder, and recipes for gunpowder kept aiming to increase their potency by upping their proportion of saltpeter.[71] Charles's forces would prove particularly prodigal in their consumption of it.[72]

England relied on imports of saltpeter from Southeast Asia, or from gathering saltpeter that naturally accrued in outhouses, henhouses, and the like. As Russell recounted in a printed broadsheet advertisement of his saltpeter project, the actions of saltpetermen, given license to hunt for the precious chemical wherever it encrusted in chicken coops and on dung heaps in private residences, gave rise to great complaints. The saltpetermen extorted money from householders to prevent their digging around their properties; this extortion became their main revenue stream in lieu of the actual work of producing saltpeter.

According to the plan of Russell and Brooke, instead of saltpetermen, householders would collect their own urine and that of their animals. Transporting the urine from individual homes to the saltpeterworks could employ "common Beggars about *London,* and in other Cities and Townes corporate . . . whereby they may haue a competent meanes for their liuing without begging." As was par for the course in projects, Russell and Brooke aimed for more than one advantage in their project. Not only did they claim to be able to supply saltpeter, to their own and the Crown's profit and the benefit of the defense of the realm, but they could also employ beggars and rid the citizens' houses of the noxious saltpetermen.[73]

Russell claimed that he originally presented the project to King James and it was examined and recommended by George Villiers, by Lionel Cranfield, the Earl of Middlesex, and by George Carew, the Earl of Totnes. In 1607 Edward

Somerset, Earl of Worcester, had obtained a twenty-one-year license for making saltpeter and gunpowder in England and Ireland. He relinquished it in 1620, and in 1621 this monopoly was given to the powerful trio, Villiers, Cranfield, and Carew.[74] This group, according to Russell, "after long and many serious debatings (divers trials being made of the certainty of making Saltpeter in this new course proposed, and the goodness thereof,) were pleased to recommend it back to the said King James . . . and a Priviledge for the sayd work to be granted."[75] Thus, trials, approved by top courtiers, had demonstrated his process. To further make his argument, Russell also included in his printed broadsheet a rather lengthy calculation of how much better and cheaper his product would be in comparison to saltpeter that was currently imported into England. Therefore, he could also show how much cheaper gunpowder would become. His project, he calculated, would save 6,000 pounds a year for every 400 tons of gunpowder consumed.[76]

In addition to printing his proposal, Russell also communicated a terse document for the House of Lords to consider. On March 24, 1626, Russell, having made a "Proffer" to the "Lords Committees for Defence of the Land" a few days beforehand to make saltpeter with urine on earth and lime, brought in a paper containing six points awaiting the response of the House of Lords, as an efficient means of negotiating the terms of this project.[77] This included hammering down the details of the trial that would prove Russell's abilities. In an effort to characterize the literary forms associated with practices of projecting, I cite this memorandum in full. Like most forms of presentation and argument in projects, it is short, forefronts practices of trial, gives no illusions concerning the skepticism with which its audience greeted projects, and deploys many calculations. It recalls the form of "heads" or "topics" with responses often deployed for the efficient review of information in governmental settings, and thus indicates Russell's familiarity with such political formats.

1. First, how much Salt-petre the House will require to see made, for their Satisfaction.

   R[esponse]: "A Thousand Weight is to be made for a Trial, at his own Charges. To begin on *Monday* next, and to be ready by the last of *May*."

2. The Trial shall demonstrate, that there may be sufficient Quantity of Salt-petre made to serve the Kingdom; and that it shall be as good and useful as is undertaken.

   R[esponse]: "If this Second Part appear to be so upon Demonstration, then it is accepted of."

3. The Person that must see and report it to be appointed.

   R[esponse]: "The Triers of this to be named by the Lords Committees by *Tuesday* next."

4. A provisional Act to pass both the Houses, for confirming the King's Grant, and for raising the Stock of Twenty Thousand Pounds, if the Petitioner make good his Undertaking.

   R[esponse]: "If the Premises be made good, then a Provision may be had for Confirmation of the Letters Patents, according to the Statute against Monopolies, and Provision to be also made for the Twenty Thousand Pounds demanded."

5. The Petitioner averreth, That the Urine which may be had from every Two Hundred Houses (each House having Four Persons one with another) will make a Tun of Salt-petre Yearly.

   R[esponse]: "This Part is to be considered of by the Triers."

6. The Stock of Twenty Thousand Pounds being provided, within One Half-year after there may be Works erected, and sufficient Quantity of Earth ripened to make continually Ten Tun of Salt-petre Weekly, which is Five Hundred Tun Yearly.

   R[esponse]: "This is to be undertaken and performed by Mr. *Russell*, upon the Payment of the said Twenty Thousand Pounds unto him."[78]

In this memorandum, Russell asked the House of Lords to specify the details of the trial, including the amount of saltpeter he should produce. Russell also included some rough demographic calculations of how much saltpeter he could produce annually. Assuming that each household held four people, he could make a ton of saltpeter yearly from the urine collected from two hundred houses. The Lords did not take him at his word, but responded that this sort of calculation would fall under the purview of the "triers" of Russell's saltpeter, who would consider it. Russell demanded a stock of 20,000 pounds to begin the work, which would produce five hundred tons yearly. This means he planned to collect urine from 100,000 houses.

The "sole privilege of making saltpeter" was granted to Sir John Brooke and Thomas Russell.[79] Charles issued a royal proclamation ordering all residents of "Cities, Townes, and Villages within this Our Realme of England" to "constantly keepe and preserve, in some convenient vessells, or receptacles fit for that purpose, All the Urine of man during the whole yeere, and all the Stale of Beastes, which they can save, and gather together" for use by "Sir John Brooke knight, and Thomas Russell Esquire." He threatened that "if any person shall be remisse or negligent in the due observance hereof, We shall esteeme of all such as persons contemptuous and ill affected both to Our Person and State." Brooke and Russell, meanwhile were charged with carrying away the urine from every-one's properties once every twenty-four hours in the summer, and once every forty-eight hours in the winter.[80] The project failed, since householders imme-diately refused any entrance to their properties, not only to Brooke and Rus-sell's beggars, but also to the old saltpetermen. All sources of saltpeter came to a standstill.

Russell then presented an alternative plan to the Privy Council in 1627, ac-cording to which urine would not be collected, but soil would. I again cite this in full.

A Declaration by an experimented way howe his Ma[jes]tie may have as much Saltpeter Earth prepared within half a yeare after it is begun to be set at worke, as will make 1000 Tun of Saltpeter w[ith]out any charge to his Ma[jes]tie, for prepareing the Earth, or greevance to the Subiect but much for their ease in tyme to come.

There are w[i]thin his Ma[jes]ties kingdome of England atmost 10000 countrey Villages or Parishes, besides, Cities, Towenes corporate and Market Townes, and each Village or Parish one w[i]th another, may have 40 howses and everie house fower persons.

Now if every house will but lay a loade of Earth in some corner of a Barne, Stable, outhouse, or hovell, and cast the urine that is made in that house upon the Earth for 3 Moneths together, and then let it rest 3 Moneths longer, it will be ripe for Salt Petre.

Thus 400000 loade of Earth will make 10000 Tunn of Liquor, w[hi]ch will make 1000 Tunn of Petre. This 1000 Tonne of Petre will make 1500 Tonne of Powder, w[hi]ch his Maj[es]tie is to have at 70 lb the Tonne and it may now be sold for 100 lb the Tonne (if his Ma[jes]tie had no occasion for use of it) being but 12 s. the pound and so his Ma[jes]tie will gaine by this 1000 Tunne of Salt-petre 45000 lb.

If these be doubted of and that thereis no present necessitie of Powder, it is best to trie it in 40. houses neere London, and see the effect at the half yeares end, and then if it be found true, and held convenient, it may be proceeded w[i]th. If not I have discharged my duty I owe to his Ma[jes]tie and Kindgomes, in that I have made knowne this thing, w[hi]ch upon certaine triall I have seene done.[81]

Sir Robert Killigrew, John Heydon, and John Evelyn, Esq., were ordered to "presently meete together and examine by a triall of the earth, the said Thomas Russell already prepared with urine, whether it will make good Saltpeter, and that Saltpetre good powder or not." They were also asked to submit a "Report" on the trial "in writing."[82]

This project, like many of the time, is vast in scope, projected outward over the entire nation and every single household within it. It proposes to locate enormous financial resources and to manufacture other crucial mainstays of power in the form of gunpowder out of worthless waste. The process through which these grand aims would be achieved is described as a simple matter of depositing urine on earth to ripen into saltpeter ("if every house will but lay a loade"). The project claims to serve both the Crown and the general populace. It would also redound to the profit of the projector and any associates he might have (which would have to be many, given the scale of the project), who would sell the gunpowder to the Crown.

In the name of benefiting the kingdom at large, this project arrogates extreme power to intervene in the most privy business of every Englishman. The claim that this project served the public lay in its unstated contrast with the aggressive and intrusive saltpetermen, as Russell had already described in his printed broadsheet. Collecting aged, urine-soaked earth here is presented as a more desirable alternative to Russell's previous idea of collecting large amounts of liquid urine on an almost daily basis. In August 1627 Charles I issued a general letter "to various cities and towns" encouraging investment in Russell's project "for procuring saltpetre, without the inconveniences then common."[83] While Russell claimed this project served the "ease" of the subject and would occasion no "greevance," the text fails to account for the many ways this plan intruded upon people's lives and properties, from having to store six-month's worth of urine-soaked earth on their property, to dealing with beggars. It also goes unstated that the ways and means by which urine turned into saltpeter had long proven impenetrable to Englishmen who had not cracked the sequence of bacterial and chemical phenomena required for the transformation of urine into saltpeter.[84] Thus, this highly intrusive plan on a national scale was based on a body of very doubtful knowledge, and followed many other failed saltpeter plans.

The project anticipates doubts and objections, and it advances forms of argument and evidence common to period projects in order to assuage them. One was the eyewitness testimony of the projector, who claimed to have seen this working before. He evidently had no expectations that he would be believed, as he proposed a small-scale trial of the project, on the basis of which the project could be extended to the entire kingdom. Both the centrality of experimentation and the dismissal of any problems that might arise from scaling up from 40 to 400,000 houses, were typical of projects at this time. As was also typical, Russell deploys many figures and extrapolations from those figures to make his argument. He makes no pretense about the imprecision of his figures, as he cites no authorities from which he drew these numbers and he makes many generalizations, such as assuming that every household contained four individuals.[85]

Russell's deployment of calculations in support of his proposal in general underscores William Deringer's arguments concerning the widely understood role of calculations as forms of argument in probabilistic, risk-taking settings.[86] It queries, however, Deringer's contention that such forms of calculation only became a regular part of politics after 1688. Calculations as deployed by Russell here (and in even more extensive forms in the many financial projects of the period) were par for the course in projects. As Paul Slack has discussed, Botero's reason of state, which assessed strength in terms of population and wealth, encouraged demographic study.[87] Since projects were regularly framed in terms of contributions to the reason of state, they often included large-scale analyses of costs, profits, materials, trades, and population. Such projects were then continually discussed in governmental settings and elsewhere and remained central to politics from the early Stuart period through the Interregnum and Restoration.

The history of calculation in projects thus also questions Deringer's etiology of this form of argument, which he sharply distinguishes from science and technology and locates as originating in political argumentation alone. However, politics, trade, science, and technology were inseparable in the practices of projecting. It would be artificial to distinguish the form of calculation in Russell's project as political, when it deals with a form of natural knowledge (making saltpeter from urine and earth) and in the period was indistinct from many similar forms of calculation and numerical conjecture voiced by investigators of nature and art.

Deringer is correct, however, that such calculations by and large did not form part of a "civic epistemology" in the period, as Deringer argues that they did after 1688.[88] Sheila Jasanoff coined the term "civic epistemology" for the generally shared practices and styles by which a society judges and debates knowledge.[89] The loose reasoning, massive scale, and improbable nature of projects made them matters of derision at best, fear and loathing at worst, for many in the period. Rather than enjoying the authority of a shared epistemology, the reasoning and calculations found in projects ran counter to widely held epistemic values and ideas of knowledge.

In June 1628 Russell was already back with another petition for saltpeter, read to the House of Commons, where he did not receive as enthusiastic a hearing as he had before Charles and the Lords. It was ordered "that Russell Should bring a certificate that he was able to effect" the making of saltpeter with urine and earth "and then the House would countenance his project."[90] At the same meeting, the patent that had been given to Villiers and Carew for making saltpeter was condemned as "a grievance both in creation and execution."[91]

Scaling Up and Extending Out

The same formats, calculations, and manners of reasoning deployed in domestic projects also appeared in proffers for ventures overseas. A case in point is a 1620 project proposed by "Mr. Russell" to the Virginia Company for "making of Drinck in Virginia being an artificiall wine made of vegetable growinge there naturally."[92] The Virginia Company, like other joint-stock companies, regularly reviewed projects presented to them for deployment in the areas under their management according to the terms of their founding charter (which was itself a patent). This included, for example, a proposal by an unnamed projector for the planting of licorice, which the company reviewed and refused because the proffered terms were too high for something the company considered easy to accomplish.[93]

Given that Sir John Brooke, who would partner with Thomas Russell on his saltpeter, alum, and soap projects, seems to have had insider knowledge of the project for artificial wine (as discussed further below), this Mr. Russell is likely once again our Russell, the "father of all projectors." Mr. Russell began his project with the same demographic calculations we observed in the saltpeter project, noting that there was "like to be shortly 3000 people" in Virginia. Russell presumed that these "3000 people . . . being most labouringe men, and the country very hote, will drink one w[i]th an other a wine pottle a day, w[hi]ch for 3000 people is 23 tunne and 28 od gallons accompting 4 hoggesheads to the Tunne and each hoggeshead to contayne 64 gallons wyne measure." He proposed a drink whose ingredients were so readily found in the colony that it could be made for 6 shillings a ton or 12 pence a barrel. This drink never went sour. It did not make anyone drunk. It functioned as "an excellent preservative against scurvy and other disease and infections" for those on board ship or those subject to "many unholsome fogs and damps." It was ready to drink the first day it was brewed. Finally, it did not cause heartburn.

Russell envisioned how the future supply of this drink might relate to a colonial situation whose population might rapidly escalate: "If there were in the collony ten tymes as many people, they may have sufficient of this drinke at the foresaid rate." Projectors regularly offered estimates and calculations of population, comparisons of rise and fall of population over space and time (including projections into the future), analyses of other phenomena causing or resulting from population change, and proposals for the transplantation and metamorphosis of various populations. Such projects established habits of demographic thinking that would continue to develop later in the century in Sir William Petty's "political arithmetic."[94] Ted McCormick points to the qualitative, if not quantitative roots of political arithmetic in the many discussions among the interlocuters of Samuel Hartlib between 1639 and 1660 concerning the intersections of food, labor, and population on an imperial scale.[95] We can already observe such practices developing earlier in the century in projects such as Russell's.

Russell offered to demonstrate the making of the wine in any amount, "either in Tonne or 10 tonne, or what one pleaseath." He only asked, "upon demonstracion of it here to the company," for 1,000 pounds, besides the benefit of supplying the colony with the drink at the stated rate. The company agreed with Mr. Russell, the "chimist," after "some little varyacion." A note on this document in the papers of John Smyth (1567–1640), states: "Sir John Brooke . . . told mee, that of his c[orrec]t knowledg, this wine was made of sassaphras & licoras boyled in water: he had of the drynk."[96] Thomas Russell's long-term collaborator, Sir John Brooke, was heavily involved in Virginian colonization. Among the incorporators of the Virginia Company, according to the 1609 charter, Brooke was elected to the council in 1620, took out a patent for a Virginian plantation in 1622, and was appointed by Charles I to the royal commission that took over the governance of Virginia in 1631.[97] Likewise, Russell's other long-term collaborator, Sir John Bourchier, became an adventurer in the company in 1620, around the time when his daughter emigrated to Virginia, and took out a patent from the company for a private plantation.[98] Bourchier worked with Sir John Brooke and others on the company's efforts to oversee the Virginian tobacco trade.[99]

Russell's artificial wine was a form of root beer that drew on the medicinal use of sassafras. As English writers noted by 1610, sassafras was used medically by indigenous peoples in Virginia.[100] It became an important ingredient in European medicinal "diet drinks," such as the "Ale of Health and Strength" attributed in one recipe to Francis Bacon.[101] In 1640 John Parkinson, the apothecary and botanist, defined artificial wine as any wine made not from grapes but from "herbes, rootes, seedes or other parts," such as "Wormewood wine, Eyebright wine, Scammoniate wine." The making of both regular and artificial wine was not practiced in England, Parkinson said, but he suggested that they "might be paralleled almost, if the things were put into our new Ale, or Beere, to worke in them, as we use to do with our dyet Beere."[102]

The novel insight that was the foundation of Russell's project thus lay not just in the idea of artificial wine nor in the appropriation of indigenous medicines to treat Englishmen. Rather, it lay in Russell's lateral thinking that intertwined two observations. The first concerned the need to identify a drink for Virginian settlers, thirsty men laboring in a hot land, as beer and wine often soured or spoiled or were too expensive to import. A letter from George Thorpe in 1620 claimed that it was depression brought on by disappointed expectations that was killing Virginian settlers rather than actual disease. Not having alcohol to drink was apparently a particularly sore point; "more doe die here of the disease of their minde then of their body by having this country victuall ouer-praised unto them in England & by not knowing they shall drinke water." Thorpe claimed to have himself managed to make "soe good drinke of Indian corne as I protest I haue diuers times refused to drinke good stronge Englishe beare and chosen to drinke that."[103] Most settlers apparently did not agree with Thorpe's taste, however.

Russell's second observation concerned sassafras. Although its price in England had fallen drastically since the days when Sir Walter Raleigh sold sassafras for 1,000 to 2,000 pounds a ton,[104] sassafras was still an imported medicament and could never serve as a widespread replacement for beer for English laborers. However, it should be very cheap in Virginia, Russell assumed. Thus, Russell crafted his proposal: Virginian settlers could be sold an affordable drink made from local sassafras, a drink with added health benefits that would never be worthwhile to produce in England, but which in Virginia (he hoped), could more than make up for the lack of English beer. This project was based upon distance. It was distance that put English beer out of reach of Virginian colonizers but which also brought sassafras into their orbit. The potential for profit lay in the fact that thirsty consumers and powerful ingredients were already located in the same remote land.

At the core of period projects often lay some such ingenious concatenation of otherwise distant insights, observations, and phenomena. However, extrapolated over vast distances and populations, there were many ways such lateral and projective thinking might fail. While Russell could provide a trial of making artificial wine, and while figures in London like Sir John Brooke might have enjoyed drinking it, much about this proposed industry remained unknown. It was not known if it would be to the taste of Virginian laboring men. It was not known if, from the distance of England, Russell could indeed manage a large-scale industry based on collecting a native Virginian plant to serve a wildly fluctuating colonial population (soon to experience the 1622 Powhatan uprising). Although Mr. Russell offered to demonstrate the wine, such small-scale trials left much unknown.

Practices of Trial

The trial of a soap project in which Thomas Russell played a central role became a matter of major public contention.[105] This was the "New Sope" Hartlib had mentioned in his account of Russell, the "father of all projectors." The project began under James I with a 1623 privilege taken out by Andrew Palmer and Roger Jones and their assignees for "making of hard soap with the material called barilla, and without the use of any fire in the boiling and making thereof, and also of making of soft soap without the use of fire in the boiling thereof." A Spanish coinage, *barrilla* referred to salt-tolerant plants that were baked to produce soda ash (sodium carbonate) used in soap and as a flux in the production of glass. It was normally imported from Spain. The patentees promised to prepare their ash from "bean-straw and pea-straw, and of inland kelp and English barilla, and other vegetables fit and serviceable for the making of soap which otherwise would be of little or no benefit to the kingdom, but being thus used and employed will save the expense of many thousands yearly which are now expended on foreign commodities"[106] By making the soap in a new way using "motions" rather than fire, Palmer and Jones claimed to save on fuel. Finally, the patent mentions "an assay glasse for trying of their lye."[107] Thus, the project claimed to make soap from unused resources, through a process that consumed fewer resources, and in a way that could be better tested for quality control through a new instrument. According to the patent, this project concerned "regal prudence for the public weal" and the "interest" of the "public."[108]

The soap project combined several interventions drawn from diverse endeavors. The new soapmakers' experiments using the potash and salt contained in kelp represented a continuation of novel uses found for kelp. Russell had previously been one of the main proponents of the idea that kelp could replace urine in the alum production process.[109] The attention to assaying soap can be related to Andrew Palmer's profession as the assay-master in the Mint, where he may have first met Russell. In 1608, when Russell had attempted to purify the Scottish ore in the Tower, Palmer had privately assayed the ore for the benefit of Robert Cecil, 1st Earl of Salisbury.[110]

In 1624 investment in the project for "making of hard soap with berilia" was divided equally between (1) Sir John Brooke, (2) Sir John Bourchier, (3) Thomas Jones, (4) Roger Jones and Andrew Palmer, (5) John Coventry, and (6) Thomas Russell.[111] Sir John Bourchier, who had been one of the projectors from the failed alum works, had his debts to the Crown from the alum project forgiven and immediately reinvested in soap. An agreement between Thomas Jones and Thomas Russell, Esqs, who held by assignment two letters patent "for hard & soft soap" was also signed with Sir William Cockayne, who was to lay out 3,000 pounds to get the soapworks up and running (1,000 to purchase potash and 2,000 for instruments and wages). Thus, a third of this initial budget was earmarked for potash, rather than sourcing potash from kelp as had been proposed in the project.[112]

The Lord Mayor of London (then Sir John Gore) and the Court of Aldermen were notified by the Crown that "some gentlemen had propounded a new invention for the making of hard and soft soap of the materials of this kingdom only." Although his Majesty "thought their propositions reasonable," due to a complaint by the Soap Boilers of London, the king desired that the Court call "the said Gentlemen and the Soap Boilers, and in their presence put the new sort of soap to trial." The 1624 trial was not very favorable to the gentlemen's soap. Washing women "utterly disliked" the new soap. The report of the trial even raised doubts that the new soap was made "only of the materials of this kingdom."[113] Nevertheless, this negative finding did not affect the patent.

The patent was then renewed in 1631 by a group of twelve individuals, including Russell, the "Assistant and Assignee" of Jones and Palmer, on the condition that "upon Tryall it is found that the Soape made by them . . . is more sweete and usefull than other Soape used to be made."[114] This group of gentlemen soapmakers, including Russell, then incorporated under the name of the "Society of Sopers within the Citty of Westminster."[115] The application for incorporation referred to "certain gentlemen of ability" who were willing to "offer a trial of their soap."[116] They argued that their soap benefited the Crown by being made from entirely domestic materials and offered calculations of royal profit in their support.[117] This company promised to produce 5000 tons of soap a year, paying into the royal exchequer 4 lb. for every ton sold. It enjoyed the right of inspection of all other soap producers.

As controversy over the new soap heightened, the Crown continued to grant the company further and more invasive rights, such as forbidding the use of any oil but olive or rape oil in making soap, requiring that all soap and potash be first "tried by the Society of Soapmakers of Westminster," and forbidding the importation of potash. The Society could appoint four Overseers to inspect all soap in England and Wales and "to punish the gaynesayers and rebells." An "Office for keeping of the Patterns, and making of the Assay of Soapes" by the company was erected by letters patent, and "Francis Coningsby Esquire" was appointed "Assay-Master."[118]

Petitions to the Crown against the new soap referred to long-standing customs and established trades opposed to these innovations. The prevention of the use of whale oil soap would "wholly subvert the ancient trade of soapmaking, by which the kingdom has and shall be served better and cheaper than by the intended new inventions." The petitions called for "an indifferent trial" of the old and new soaps. They pointed out that as "citizens and freemen of London" they had always "enjoyed this right" of "the free vent or sale" of soap, whereas the new society was made up of "gentlemen not brought up or skilful in the petitioners' trade."[119]

According to one 1633 newsletter writer, "the lord mayor [Sir Ralph Freeman] was sent for to the court, where his majesty and the lords rebuked him for his partial proceeding in favor of the old soap."[120] They also signed a warrant "to bring a poor woman out of Southwark before them for speaking invectively against the new soap." Meanwhile, the "gentlemen soapboilers procured Mrs. Sanderson, the queen's laundress, to subscribe to the goodness of the new soap." In a surprise twist, the laundress turned against the gentlemen, claiming that neither she nor "most of those ladies that have subscribed" in fact would use the new soap.[121] One can only imagine the expressions of the assembled projectors and courtiers at the revelation by the courageous Mrs. Sanderson. While the King and the lords granted "that this new soap hath blistered the washer's hands, and done other mischiefs," they suspected that this was because the "soap was sophisticated with some other obnoxious matter to work that mischief." For instance, slipping "a small quantity of rhubarb or a glass of sack [wine]" in the barrel of new soap could have made the trial mistakenly show such harmful effects.[122]

Despite this testimony against the new soap, a royal proclamation came out in favor of it. It recounted in great detail the practices of trial of the soap in ways that might make historians of science in particular sit up and take notice, and thus I cite at length. Ralph Freeman, the Lord Mayor of London, and six others, described as "Gentlemen of quality of great Trust and Credit" submitted a "Certificate" of their trial. With an order dated December 6, 1633, they

went into Storehouses, where the new Soap is kept at Lambeth, St. Katherines
and the Strand, and from thence Wee took indifferently sixteen small Vessels out
of a multitude marked by the Assaymaster; We caused those sixteen small Vessels
to be brought to the Guildhall, and took one half Firkin out of those sixteen,
and sent for on other Firkin of the sequestered old Soap . . . and weighing out an
equal proportion or quantity of each sort of Soap, Wee caused two Laundresses
indifferently chosen to wash therewith, and Wee did plainly perceive that the new
white Soap did with a very small difference, lather much better than the old
Crown Soap did, and that the Clothes washed with the said new Soap (being
dried) where as white and sweeter. . . . and foreasmuch as a general complaint was
made, that the new white Soap did fret the Skin from their Hands that washed
therewith, and that it did spoil and burn the Clothes, Wee have according to your
Lordships further Order received Certificates from above four-score persons, of
which number there are four Countesses and five Viscountesses, and divers other
Ladies and Gentlewomen of great Credit and Quality, besides common Laundresses and others, who have a long time used the same white Soap, who generally certify that the new white Soap washeth whiter and sweeter then the old
Soap.[123]

However, they also turned in "divers Petitions complaining of the new Soap
with some two or three hundred Names set thereunto." These, however, were
found to be "clamorous" and written by people of "mean condition" or with "no
experience" of the new soap.[124]

The reason why such an account might well interest an audience of historians of science is that this trial has many of the hallmarks of a shift that we understand as crucial to the emergence of experimental science. Experience in a
general sense and experiment were once synonyms. Over the course of the seventeenth century, the meaning of experiment evolved to denote a particular
trial performed at a particular time and place in order to test something and establish facts.[125] Historians such as Barbara Shapiro have demonstrated how the
meaning of facts, as established in a court trial, originated in the law courts and
only then passed into the establishment of experimental findings as facts about
nature.[126] Trials of projects offered a central means through which this passage
from the legal to the experimental setting occurred.

According to sociologists and historians of science such as Steven Shapin and
Simon Schaffer, founding fellows of the Royal Society further developed such
practices of trial in ways that produced uncontentious matters of fact through
the use of multiple, trustworthy social superiors as witnesses. Employing a

distinctive "literary technology," writers of experimental essays such as Sir Robert Boyle developed a particular style for reporting upon experimentation through the intense description of every detail of the experimental setup, in ways that made readers feel as though they could "virtually witness" the event.[127] Shapin and Schaffer argue that such techniques were deployed by Restoration gentlemen philosophers in order to distance the findings of natural facts from contentious political, social, and religious disputes that had typified the Interregnum period. The experimental essay advertised an indifference to prejudicial dogmas or theories and a willingness to accept whatever the facts showed.

The account of the trial of the new soap bears many of the attributes associated with the reportage of the later experimental essay, such as the particular date and locations of the trial, details of sources and amounts of substances to be tried, an emphasis upon indifference (both the soap to be tried and the laundresses to do the testing were "indifferently" chosen), close observation, and the reliance upon the social credibility of witnesses. However, such practices of trial were elaborated in a setting that was not at all distanced from political and social disputes. Practices of trial developed in ever more precise ways because of, not in spite of, the highly contentious and socially divided nature of early Stuart projecting. As Koji Yamamoto has discussed, distrust encouraged knowledge practices that formed "the incipient practices of experimental philosophy."[128] Gentlemen soapmakers, not gentlemen philosophers, developed these practices of trial.

Professional soapmakers, laundresses, and members of the public sought redress through further trials and petitions. However, as had already been shown in the very first trial that found against the new soap but was nevertheless ignored, the gentlemen soapboilers by virtue of their social status could shift findings in their favor. The new soap company became one of the most contentious of the era, not only for its egregious abuses, but for the ways that gentlemen were interloping downward by invading a craft, which they were able to dominate not due to their superior knowledge of the craft itself, but due to their skills in negotiation and networking. The Company of Soapmakers of Bristol are a case in point. Bristol had served as a center for soapmaking in England for centuries, however, its soapmakers were restricted to producing only six hundred tons annually, divided among the forty-one-member company, due to the favor shown to the new Society of Soapmakers. Russell blamed the Bristol soapmakers' poor negotiating abilities for this measly allotment. The company had paid forty pounds in their defense "against the pattentees of Lambeth and St. Katherine's." However, in July 1634 it was reported to the company that "Mr. Thomas

Russell, one of the new Sossiety of Sopboylers at Wisminster did say . . . that the men who came upp [to negotiate] about the sop busnes" were either madmen or "fooles," or else they could have had "one thousand tons allowed them."[129] Russell himself (according to a July 1637 petition by one individual whom he had stopped paying) was in charge of making 1,815 tons of soft soap yearly at St. Katherine's.[130] This was a rather large percentage of the company's entire commitment to produce five thousand tons annually.

The story of the soap company became one of contention between laboring soap producers and the depredations of elite soap projectors, who used their court connections, capital, and political savvy to gain any result they wanted from a trial. The social divide between the old and the new soapmakers was patent. The title of one 1641 chronicle of these events, *A Short and True Relation Concerning the Soap-busines*, pitted "the soape-makers of London" against "the gentlemen that were the patentees for soape"; one group represented soapmakers by profession, the other gentlemen who merely took out a patent for soap, as well as for many other industries. The London soapmakers further accused the patentees of being "Popish recusants." They objected to many articles of the charter that these gentlemen patentees acquired, including their right to intrude themselves into "the trade, and take apprentices though they were never any."[131] As a result, "many Citizens of London were put out of an old Trade, in which they had beene bred all their time, and which was their onely lively-hood, by Knights, Esquires, and Gentlemen, never bred up to the Trade, upon pretence of Project and new invention."[132] These "Projectors," they claimed, "became oppressors of the Kings People."[133]

A Short and True Relation Concerning the Soap-busines accused the gentlemen of employing "much skill and endeavor" to "bring their white soape in use." Such manipulations, they argued, included a 1632 "tryall" by Sir Robert Ducie, then the Lord Mayor of London "by him made (which was made in his own private house, without the knowledge of the Soap-makers of London)," that produced a certificate affirming the superiority of the new soap.[134] Contention with London soapmakers led to a case in the Star Chamber that found once again in favor of the patentees and against sixteen London soapmakers, who were imprisoned.[135] The 1641 account accused the patentees of having "personally solicited this suit in Starre-chamber and disbursed monies therin, on the behalfe of themselves."[136] Thereafter, according to this account, the gentlemen soapmakers became increasingly more invasive, using the Star Chamber as their instrument. They were granted the right to "enter into houses" to search for soap.[137] Several Star Chamber orders called in "many Soap-makers of London . . .

without any Bill against them" to be questioned.[138] The heated situation engendered public protests. After "women and others in the street in open manner" protested that the new soap burned linen and hurt their hands, Sir Ralph Freeman, the new Lord Mayor undertook yet another trial that found in favor of the new soap.[139]

At last, the laboring soapmakers of London were granted their wish that the soap projectors be put down by the Long Parliament. The original venture of Palmer and Jones was declared "a deceatful Project" and its patent illegal. The new company of soapmakers was declared an illegal monopoly. The members of the company, including "Tho. Russell," were identified as delinquents that should be arrested. Those soapmakers of London who had had their businesses destroyed ought to receive amends.[140]

Because of the large cut taken by the Crown, the new soap company has been called "one of the most successful projects of the time from the point of view of royal finances."[141] It was very profitable to its projectors as well, and many other gentlemen and courtiers imitated its success by petitioning to form new craft societies or livery companies. The political costs of the new soap society were very high, however. Parliament's suppression of the gentlemen soapmakers was widely celebrated. The 1642 *The Proiectors Down-Fall, or Times Changeling* wondered how so recently, "all professions were counted infamous" in comparison to the wealth and power of a projector, but now "Projecting-pates" were weakened and their "brain-strings almost broken."[142]

Conclusion: Projects and Experimental Philosophy

In recent historiography, Russell has been positioned as a craftsperson or technician in contrast to social elites like Bacon who were political administrators and philosophers. In fact, Russell was a well-connected gentleman and a member of Parliament like his coprojectors Sir John Brooke, Sir John Bourchier, and Sir Thomas Chaloner. He was seen by the soapmakers of London, who were technical experts, as an interloping gentleman, who relied upon favor with political powers to intrude himself into the endeavors of social inferiors. Without ever having served as an apprentice themselves, "Knights, Esquires, and Gentlemen" like Russell began working in the trade and even assumed the right to take on apprentices, the soapmakers complained.

Russell was not a typical craftsman, but someone who combined manual—often disgusting—practices of many crafts with a wide range of other forms of knowledge. As the "father of all projectors," he not only fused practical and theoretical understandings of nature, or what Hartlib via Watts characterized as "mechanical" and "rational" forms of knowledge, he also deployed practices of argumentation, calculation, trial, negotiation, recruitment, financing, planning, polemics, and the ingenious concatenation of a host of varied phenomena. He scanned a variety of substances, from urine-soaked earth, to kelp, to sassafras, considering them as materials whose use was waiting to be discovered.

Russell reached across many fields in which he had no particular expertise. Many of his projects failed in spectacular fashion, to the great cost of both investors and the workmen laboring for "tickets." His relationship with some collaborators went down in flames. Nevertheless, the fact that he continued to attract investors and collaborators and even to gain royal employment speaks to the confidence that interlopers placed in his abilities to cross divides and engineer ingenious schemes.

As elite opportunists, projectors frequently crossed over into new undertakings, into the realms of others, and into enemy lands. They did so on a massive scale, not only geographically dispersed between Virginia and dung heaps of London, but also expanded numerically into calculations of immense profit. The magic of enormous numbers helped bolster their assurances. Huge sums were spent, lost and, occasionally, gained. Projectors often claimed that the resources or spaces they used were neglected by all and their interventions harmed nobody. They treated the riskscapes in which they projected their plans as empty spaces, from which, in case of failure, they could withdraw, no harm, no foul, and try again somewhere new. These spaces were not empty, however. They were filled not only with roving bands of rival projectors, but of customary practices, peoples and parts of nature that were not merely resources awaiting utilization. If we sought a metaphor to characterize the world of early Stuart interloping, it would not be a trading zone offering a space for mutually beneficial exchanges between philosophers and craftsmen. Rather, it would be a choppy sea upon which vessels, armed to the teeth, sailed in all directions and sought advantage wherever they could.

It is striking that Samuel Hartlib, who is currently so strongly associated with Interregnum politics, recorded such a positive view of Russell and his participation in the new soap project, given this project's royalist associations and the widespread opprobrium it generated. However, Hartlib was not an exclusively Interregnum figure. He moved to London early in Charles I's reign in 1628, and his daily notes or *Ephemerides* are full of reports of courtly activities and projects led by royalist figures.[143] Hartlib's account of Russell, the "father of all projectors" speaks to how a culture of interloping centered upon the early Stuart court continued to serve as a model for knowledge practices in later generations.

In particular, it speaks to the ways that "rational" and "mechanical" forms of thought were connected in emergent experimental philosophy.

"Experimental philosophy," although sometimes associated today with Francis Bacon's assumed joining of mind and hand, was not a term that Bacon ever used. It first began to appear in the 1630s among Hartlib's interlocutors to stress an approach toward knowledge based on what was useful, "real," and known through the senses. Fellows of the Royal Society later adopted this term, advertising their approach as that of "experimental philosophy," although in reality their approaches were diverse and differed widely from member to member.[144] The term "experimental philosophy" became popular without any clear examination, precise justification, or universal agreement about what it meant. The view that what could be sensed was more real, and that philosophy could be based in experience, was an oxymoron according to previously standard definitions of philosophy and of *scientia*, personified in the figure of Science we met in the introduction. No widely accepted, logical explanation of how knowledge could move from sense experience to philosophical explanation appeared; the term "experimental philosophy" was used loosely. There was thus a great gap between the standards of proof and argumentation that philosophy should have demanded and the types of loose reasoning, calculation, and trials that could in some quarters be considered experimental philosophy.

Hartlib used the term "promiscuously."[145] Sometimes he did so in the context of discussing projectors and their abilities as naturalists. For example, Hartlib and Watts discussed experimental philosophy immediately after their discussion of Russell. In 1634 Hartlib also included Watts on a list of people from whom he would like some written counsel concerning experimental philosophy.[146] In 1635 Hartlib characterized William Juxon, Bishop of London, as someone who took "great delight to all Experimental philosophy and Mathematical-Mechanical Inventions." Juxon criticized "Labourers and Mechanical Men" for not being "better Artificers with their heads, specially about Water-businesses and industries" and argued that the "Gentry" "properly should set upon Work's of that nature to advance some Public good." He also recommended a "Colledge for Inventions."[147] In addition to holding his ecclesiastical office, Juxon would become Lord Treasurer in 1636 and Lord Admiral in 1638, in which positions he had the opportunity to make use of his views on invention in the review and administration of many projects.[148] In 1648 Hartlib supported the efforts of Balthazar Gerbier to found an academy that advertised in print its teaching of experimental natural philosophy.[149] In 1649 Hartlib described another "excellent Inventive head" as "a meere Naturalist but a great lover and advancer of sense and Experimental Philosophy."[150] This was Edmund Wylde, a friend of John Aubrey and a founding fellow of the Royal Society. He "valued himself upon new Inventions of his own," such as "painted Curtains in Varnish upon silk."[151] Wylde was one of Aubrey's informants concerning Thomas Bushell.[152]

In all these cases, "experimental philosophy" was not a term through which Hartlib indicated a mutually beneficial relationship between a philosopher and a craftsman, but rather the interests of socially elite figures who themselves arrogated practices of invention and rather disdained the thought processes of traditional craftspeople in relation to industry. What is clear from the contexts in which this term developed is that the union of experiment and philosophy does not represent a harmonious exchange between a craftsman and a philosopher. Rather, experimental philosophy developed in relation to individuals who were neither of those two identities. They were interlopers who freely took from both.

CHAPTER 4

Statecraft

"Swimming between two Waters" in Global Policy

The Submarine, Agile Policy, and Failure

In 1620, as his friend Constantijn Huygens described, Cornelis Drebbel demonstrated a newly invented submarine on the Thames before a large audience that included King James. For three hours, Drebbel made the king, the court, and several thousand spectating Londoners hold their breath, as he submerged under the river. Just as the spectators began to think that he had sacrificed himself to his invention, he arose again a great distance away from where he had disappeared beneath the waves. He called for witnesses to try it out and testify that they experienced no discomfort under water; they went down, then up, sometimes just a little bit below the water's surface, then much deeper, as they chose. They had a source of light, and inside this "whale" they had no difficulty doing "what we humans are accustomed to perform in the open air."[1]

The submarine of Cornelis Drebbel outdid any other form of underwater travel of the time. Techniques for recovering underwater goods were often proposed. Ordinarily, they could submerge and arise but not navigate underwater. William Bourne had published a method "to make a Ship or a Boate that may goe vnder the water vnto the bottome, and so to come vp againe at your pleasure," but it could not navigate underwater. For that, one had to "rise and come vp aboue the water, and swimme as it did before."[2] As Abraham Beeckman wrote to his son Isaac from England in March 1620, Drebbel had a boat "with which he could voyage either above or below water, as he wished."[3] Diving below the waves spectacularly demonstrated Drebbel's ability to surpass natural borders. In a poem published in 1620, Balthazar Gerbier called Drebbel "nature's darling" (Naturas Troetel kint) and a "superhuman" (meer dan Mensch). Drebbel was completely unrestrained by the elements, since he "who lacks no Art, would dare to swim like a fish, to fly on the winds, to soar up to the Moon and into the depths of the sea, to sail without mast, tiller, oars, sail, or, yard-arm."[4]

The utility of such an invention, both to retrieve items from the bottom of the ocean and to act in covert ways in military situations was obvious in other similar proposals of the time. In 1605 a trio (Pieter Pieters, Jan Adriaensz Leeghwater, and Wilhelm Pieters) took out a patent from the Dutch States General for a "water art, by means of which one can go under water, stand, sit, lie down, eat, drink, read, write, sing, and speak, as well as repair or destroy bridges and sluices, fasten cables under sunken ships to pull them up, and search for pearls and other treasures from the bottom of the sea, or carry messages secretly underwater."[5] In 1628 the military engineer Robert Norton published an image of an underwater outfit designed by the Venetian engineer Niccolò Tartaglia (1499/1500–1557). Cannon were recovered from shipwrecks in this period, so some means of retrieving heavy weights from beneath the waves must have existed, possibly through the use of such underwater suits and diving bells.[6]

None of these, however, were able to navigate underwater and pop back up in expected locations, like Drebbel's boat, which greatly enhanced its ability to function covertly. As Huygens noted concerning the submarine, "It is not hard to conjecture what use this daring invention would have in warfare," especially if it could allow one to creep up on ships unseen and explode them.[7] Drebbel himself told Huygens several times that the ship could be used to attach petards, at the time used for breaching doors and exploding bridges, to the undersides of enemy ships.[8] Huygens deployed this as a further example of Drebbel's ability to reverse the workings of the elements, by mastering a fire that could function within water; "For that Daedalus [that is, Drebbel] so rules the powers of gunpowder that water cannot keep him in check, any more than air can," wrote Huygens.[9]

This somewhat sinister ability of Drebbel's boat to navigate unseen under the waves was described at the time as traveling "between two waters."[10] The expression in French of "nager entre deux eaux" (to swim between two waters) was a metaphor for political misdirection that gave the appearance of aiming in one direction while in fact switching course, catching everyone by surprise. Many years later, Gerbier would relate this metaphor to Drebbel's submarine and underwater bombs, turning these hidden means of destruction into a symbol of the inconstancy so often practiced in the period's "wicked" politics.[11] Indeed, as this chapter discusses, almost as soon as it was invented, the submarine was yoked into a devious plan with many hidden depths, a "great design" for Asia, that entangled commerce, conquest, piracy, diplomacy, and submarine exploration.

The Crown was quick to see the potential of Drebbel's submarine. There was even a story that Prince Charles himself had traveled in it and that a submarine had been sent as a diplomatic gift to Moscow "as a rare and incredible thing."[12] Given the dangers of underwater travel, it seems unlikely that a royal in fact took a submarine journey.[13] However, James and especially Charles were highly interested in the boat's design and potential. As Drebbel's associates, the Küffler brothers, reported in an interview with the French polymath Nicolas Claude Fabri de Peiresc in 1622, after Drebbel had demonstrated his underwater boat to James I in a form that could carry nine people, the king demanded that he make one hundred smaller ones, each able to fit a single person.[14] This would have been a veritable submarine army, but it never seems to have materialized. Nor was Drebbel's submarine deployed in Asia. This does not mean that submarine projects therefore dropped silently out of history, leaving no ripples behind.

Failed projects exercise agency in history. At their most basic level, the study of failed projects helps us contextualize endeavors that, in retrospect, did succeed. They allow us to resituate those endeavors as also once potential failures undertaken in a high-risk environment. Failed projects also shape history in other ways. Even if the stated goals of a project were never realized, the efforts to realize those goals often entailed materially destructive interventions in people's lives and natural and built environments. This was especially the case given the great scales of distance, time, and violence over which projects were attempted. Finally, failure was also a disruptive and productive force on a more intellectual level. A culture's approach to failure is part of its history of knowledge. For the interlopers, failure encouraged debates concerning the nature of risk and possibility. It furthered swift, iterative practices of recycling and remixing knowledge in new ventures. It generated a distinctively agile approach to policy. It also gave rise to conflict with more cautious figures, such as merchants, who aimed to limit risk and minimize failure.

This chapter suggests the stormy seas and transformed circumstances that even failure leaves in its wake. It explores several interrelated projects in which the Crown took an active role: the proposed deploying of the newly invented submarine to fish for shipwrecks and pearls in the Indian Ocean, building municipal waterworks for Agra and as a gift for the Mughal emperor Jahangir (1569–1627), and shifting the Persian silk trade from an overland route passing through the Ottoman Empire to a southern Iranian port. None of these were realized as they were proposed. Exploring them together, however, illustrates not only the interrelationships between many projects, realized and unrealized, domestic and global, but also the great number of projects that the Crown was considering acting upon at any one time.

The breakneck pace of projecting allowed for a large number of projects to be floated and rejected at any time. This was a type of "swimming between two waters." Projects were pursued and cast aside as opportunities arose and submerged. Even when abandoned, however, they often lurked in the shallows, always available to be dredged up again in some new form or at a later date when circumstances allowed. The continual possibility of such retrieval and revival not only sets any projects that "succeeded" or were realized within this broader, shifting seascape of possibility and opportunity. It also showcases the ripple effects of even the perception that some project or another was about to be launched.

The "Great Design"

The history of projecting, including failed proposals, can shed light on the history of empire as a whole as a part of the period's often contentious, internally riven, and bungled interloping. This history, particularly in relationship to plans for world domination, runs athwart to the most recent trends in the history of empire and colonialism. In order to avoid triumphantly teleological narratives of centrally propelled empires, current histories of Western Europe's overseas empires emphasize contingent, opportunistic, and multidirectional agency, often located far from the control of governments in Europe, with central roles accorded to non-European empires.[15] There was no way that a weak and disorganized island nation like England could compete with the centralized bureaucracy and well-organized military of period empires like the Mughals, the Safavids, or the Ottomans. Current histories of empire and colonialism thus downplay the history of centralized planning in Europe. Lyndal Roper, for instance, has stressed that English interests expanded overseas through the activities of individual "operators" to whom the Crown merely reacted.[16]

These points are well taken. In fact, they contribute to our understanding of early Stuart opportunistic interloping by small, risk-taking forces into dominant structures of power, often using underhanded means. Many such attempts failed, not without great costs or impact on later policy-making.[17] The "operators" studied by Roper were, in period terms, monopolists and projectors, and their interventions around the world were in their time inseparable from their numerous domestic ventures. The patent system encouraged individuals proposing particular interventions to frame them within much larger projects situated in national, international, and global settings. To gain privileges granted by royal prerogatives, projects had to demonstrate how proposals interacted with other industries to the benefit of the commonweal. This meant continually relating domestic industry to global concerns. Moreover, the same attitudes, personnel, and types of reasoning that made up contemporary practices of projecting also applied to colonial schemes, which were also granted through patents.

"Grand design," "Groote desseyn," or "Grand dessein" was the term in the early seventeenth century for a very particular type of project of even grander ambition than most: an interconnected, multi-plank plan for global intervention. The study of global grand designs might appear to make an antirevisionist argument about national state-building and the exertion of metropolitan European state power over the globe. Grand designs, however, were not only transnational in their aims, reshaping the domestic sphere as much as the foreign one. They were also transnational in their personnel, often utilizing the abilities of émigrés, like Drebbel or Gerbier, or individuals who fell between cultures, or who had acquired a degree of cultural embeddedness that allowed them to function as go-betweens, such as the young Bristol merchant and aspiring canal builder Richard Steele, discussed in this chapter, who joined the cast of the Mughal court artisans, much to the chagrin of the East India Company. Moreover, interlopers like Steele might also approach rival European nations, such as France, with the same project.[18]

Rather than representing the exertion of a confident central power, grand designs exemplify the logic of counterfactual ambition in projecting: the less likely to succeed, the more honorable a project was. They showcase how the most bizarre projects of the period were not the work of socially marginal craftsmen and inventors, but were championed at the height of power, here by King James, Prince Charles, and their most elevated courtiers. Drebbel may have invented a submarine, but it was his social superiors who rapidly wove it into a massive design for the total transformation of England's power in Asia, framed in terms of hope and probability.

Roper comments on the "remarkable resilience" of "seventeenth-century investors in the expansion of English interests outside Europe . . . that enabled them to weather" setbacks and move on to other opportunities.[19] This was an attribute not only of overseas investors, but of projectors in general, who frequently discussed the hazardous, adventurous, and merely probable nature of their endeavors and foresaw the likelihood of failure. Projects were not assured programs for success. They were mercurial, quick-changing forms of risk-taking, dependent upon the shifting opportunities of the moment and a wide range of other, competing projects. What seems like the resilience of colonial investors and projectors was developed in this riskscape of possible projects, frequent failure, and rapidly arising or re-emerging opportunities. Investors and projectors had to develop the skill of tacking quickly in new directions.

This chapter follows the threads of a number of projects in ways that, in their stops, starts, and entanglements, at times might seem somewhat bewildering to the reader. This jittery pace of projects, however, is germane to my argument here. If the number, scale, and interrelationships of these selected projects stupefies us, how much more must it have confused decision makers at the time, who had an even greater purview of possible projects? As the royal chaplain, William Loe, wondered about projects in a 1623 sermon, "Is any Skill branched into more Species, or is of an higher straine, that must attend so many, so mighty, so manifold occasions and occurrences?"[20]

Many, mighty, and manifold indeed were the aims intertwined in the grand project presented to King James in 1622 via his young Scottish favorite James Hay (1590–1636), newly created the Earl of Carlisle. Due to the vociferous objections of the East India Company to this proposal, it generated many pages of discussion in the records of the East India Company over many months, which allows us not only to explore the project in detail but also to bear witness to the tenacity of Prince Charles in particular, who not only championed this project at great length, but would continue to return to various parts of it throughout his reign.[21]

The Great Design: Royal Honor in Global Trade

Charles was engaged with questions of global trade from a very young age. In 1613 he already ordered an agent to travel to Iran to negotiate with the Safavids for trade in rugs, silk, and other goods.[22] Seeking to replace imports with new domestic crafts, King James, Prince Charles, and the favorite, Villiers, particularly supported the Mortlake tapestry works founded in 1619 by Sir Francis Crane. These tapestries were produced via a hierarchy of personnel stretching from Crane through to recruited international artists and skilled craftsmen. Crane aimed to sell these English tapestries to Persia.[23] When Crane suffered financial difficulties, he wrote confidently to James that he was sure the king wished for "the continuance of the business of the Tapestries, which in the eye of the worlde appears as a worke of your Majestie's greatness, and brings with it both honor to your Majestie and profitt to the Kingdome."[24] When Crane had previously written to Prince Charles and Villiers about his need for funds, the prince, in his "own phrase," as Crane reported to James, promised "to keep the fire goinge."[25]

The Mortlake tapestry works represented an expansion of national honor because it transplanted arts from abroad onto English soil. Such transplantation was conceptualized in the period as a form of conquest augmenting the internal riches and power of the country by seizing knowledge and sources of profit from another land. It would be a wonderful coup for royal honor if silken tapestries from Mortlake could then be vended back to the country from which the raw silk had been purchased. In another letter around the end of the reign of King James, Crane wrote to Villiers that he was about to meet the Persian ambassador (as Robert Sherley, discussed further below, claimed to be) in order to try "to establish some trade of our Comodity into those parts. By this you see that I was destined for adventures."[26] Crane's language of adventure and honor shows how much more was at stake in the tapestry works than simply profit. Many years later, Crane was still requesting that the East India Company send "a piece of tapestry or two" to Persia "to make experiment whether it will vend there."[27]

The 1622 "great design" reflected such views of transplanted industry, adventure, and honor, and Prince Charles in particular became enamored with it. King James, who wished to grant Jahangir a present the Mughal emperor could not disdain, was also on board. In April 1622, Sir William Heydon and Mr. Endymion Porter "both of them the princes servants," appeared before the East India Company to present this plan. Mr. Porter had a message to deliver from King James and Sir William had one from Prince Charles. As Heydon reminded the company, James had received "divers Messages and Letters from the greate Mogoll requesting such rareties as this kingdome affordes, for the which he will returne him with presents of that cuntry."[28] James was "determined to give him [Jahangir] the best satisfacion he can." He planned to send not just "some Jewells of valew," but also "some Inventions and particularly with that of convey-ing water into their houses in such a manner as will be a greate cooling and re-freshing in these extreame hotte Contries, and a benefit much desired by the Mogoll." He had already decided to send these gifts and had already chosen two of the prince's servants for this journey, of whom Sir William Heydon was one. He was merely acquainting the company with this plan in case it wished to make use of his servants, and not asking for its permission.[29]

Mughal emperor Jahangir boasted a far more extensive, technically advanced, and bureaucratically organized kingdom than that of James I. Jahangir was eager to illustrate this power imbalance in the theater of diplomatic gift ex-change, where he continually manifested his disdain for English abilities. As a representative of the East India Company reported concerning previous gifts to Jahangir, a gift of dogs was "only well liked" while other gifts were "disgraced and made ridiculous on purpose." Jahangir had his artisans reupholster an En-glish coach in Indian textiles. A gift of virginals and musicians to play them did not impress him. A painting of Venus and a Satyr even offended him, as it seemed to him to represent "a scorn of Asiatics." He asked for more dogs and "a horse of the greatest size."[30] Jahangir's request for English pets rather than En-glish arts was demeaning, given that Jahangir lost no opportunity to humiliate James through artistic and technical competition.[31]

A knowledgeable collector of European art, Jahangir had his artists contin-ually improve upon and thereby disgrace Western rivals.[32] One of his artists, Bichitr, transformed a portrait of King James, presented to Jahangir in 1615, into a sign of Jahangir's disregard. In Bichitr's miniature, Jahangir, seated on an in-ventive hourglass throne, favors a humble Sufi sheikh while ignoring both the Ottoman Sultan and King James, rendered in minuscule and fine detail. Ja-hangir's fanciful throne is perhaps inspired by the imaginative grotesques on the textile beneath him, which appears to be an example of the Indian carpet de-signs informed by European grotesques (figure 9) and developed in the court of Jahangir's predecessor, Akbar.[33] In other versions of this theme, Jahangir dis-dained an ancient Roman emperor.[34]

The art of cooling that James planned to send could have been Cornelis Drebbel's. In one of his displays of his ability to manipulate the elements, at the request of King James, Drebbel so cooled "the great Hall at Westminster" that the king and all the Lords had to hurry from the room to fetch their coats in the middle of the summer.[35] Samuel Hartlib would later refer to Drebbel's "re-frigeratory instruments for the summer and especially for hot places such as India."[36] As Drebbel is not mentioned by name in the description of James's gift,

Figure 9. Bichitr, *Jahangir Preferring a Sufi Shaikh to Kings* from the *St. Petersburg Album*, ca. 1615–18. James is the second figure from the bottom at the lower left, just above the artist's self-portrait and below the Ottoman emperor. Freer Gallery of Art, Smithsonian Institution, Washington, DC. Purchase—Charles Lang Freer Endowment, F1942.15a

it is also possible that James intended the waterworks project of Richard Steele, who had wished to build a municipal waterworks in Agra for some time, as is discussed further below.

While gifting the art of cooling was James's part of the proposal, the "princes parte in the ymployment was by itself and that he amed only at the weying upp and recovering of ships that with rich lading had been wrackt in those partes." "Cornelius Dryvet" (as the company butchered Drebbel's name in their minutes) had devised an "engine" whereof he could

> at anie time give the Companie satisfacion by waie of a demonstracon that the engine shall fetch upp anie waight and for the better sertch to find the places where theise ritch wracks are, as also to fasten hold with the best advantag for weying them up, there is a boate devised to go under water, where men maie live and if need be a man maie go forth and walke under water 20 or 30 yardes and use his armes to any kind of labour.

Here, once again, the promise of a demonstration of the engine was reiterated. Another assurance that Heydon made here was of the limited nature of the prince's aims, as he declared that the prince had no plans to engage in trade. Protesting somewhat too much, he also claimed that there was no need to suspect "anie desperate attempt uppon anie the shipps or places of that Cuntry," that might "drawe the Companie into danger, for they are only to follow such Instruccons as his Majestie should give them without medling with the trade."[37]

The East India Company's governor (the wealthy merchant William Holliday) answered that he had already heard about the "business" from several sources, "first by the Lord Marques Buckingham [Villiers] as from the king and afterwardes by the Prince," indicating the involvement of Villiers, who otherwise stayed behind the scenes of this project. The company was unwilling to entertain the scale of the design. If "it weare but a matter of presentes or the transporte of Engineers the Companie should be able to accommodate their passage in their next shipps." Heydon and Porter, however, countered that they needed their own ship. They needed a ship "of some good burthen" in order to raise sunken ships from the bottom of the ocean, and they would not be able to fit their efforts around the company's trading schedule. The company suggested that sending such ships would "prove dangerous to the Companie for they might perhaps attempt something to make upp their voyage (in case their first hopes should faile)." Thus, the possibility of failure was foremost in the mind of the East India Company, with the assumption that if the adventurers were unable to recoup their expenses in their unlikely attempt at the submarine mining of pearls and shipwreck, then they were likely to improvise with more established means of raising profits, such as piracy. Porter and Heydon "answered that such as were to be sent were so well knowen to the king, and both he and the prince would become answerable for them."

This assurance of personal royal responsibility proved cold comfort to the company. In return, one member of the company threatened that such a royal initiative would spur the company to abscond entirely from trading in the region, an outcome that King James certainly did not desire. If the king or the prince "would sett out anie shipping," the company would have to "yeild to it," but it were to be wished that they might hold off "for a few yeares," so that the company could fetch home its stock, "and then they would willinglie leave the trade to his Majesties good pleasure." Despite the seeming obsequiousness of the company's willingness to hand over trade to royal control, the implied threat was apparent and provoked a high-handed response. Porter and Heydon retorted that James would not wait: "his Majestie was resolved to send fourthwith neither was their coming to enquire anie thing of the Companie & touching the conveniencie of sending, but onely to acquaint them that it is the meere Act of his Majestie and the Prince." As the company continued to object and ask for the king and prince to delay, Heydon and Porter brusquely replied "that if they expected anie further sattisfacon, they must have it from the king, for their partes they could give no time of deliberacon."[38]

The governor of the company noted that their hesitation was based on their experience with the second Earl of Warwick.[39] Warwick, supposedly commissioned to apprehend pirates, in fact acted as one himself. He very nearly attacked the ship of the Mughal emperor's mother, only to be stopped in the nick of time by company ships.[40] Thus, the company realized that sending ships out of their control, particularly ones led by noble projectors, might entail violent interventions in waters they were trying to calm.

Heydon and Porter departed, and the company appointed a group to draw up a petition to the prince. It was but the first of an exchange of papers that flew back and forth between the company and the prince's servants. At a later discussion in the company, Holliday reported on a discussion with the king and the prince about the project. He noted that Porter had delivered a paper that contained "certeyne reasons of the proiectors that of necessitie they must send a shipp and pinnase of their owne for the more convenient effecting of their designes." The company prepared a response to the prince saying that "if this proiect procede it wilbe exceeding preiudiciall to the Company and that the Prince can have no assurance of the sucesse" of the endeavor.[41]

Two months later the governor reported back on a meeting "att the Counsell table before the Princes Highness and divers of the lordes whome the King had appointed to take consideracon of the prince his proiect for sending a shipp and a flatt bottom boate into the Indies with Inventions for the Mogull to fish pearles and to weigh such wrecke as have bene sunk in the Indian Seas." A written account of those deliberations was read, from which it appeared that "notwithstanding all obiections and opposicon the Company could make to the Contrarie, yet the prince insisted upon his resolucon." The prince offered financial sureties against his servants' engaging in trade. He also promised "they shall not attempt any hostile or piraticall Act." If his servants disobeyed these orders, the company would be allowed "to surprise the said shippes and take them into their owne power";[42] thus, the preventative to piracy would be piracy.

This paper detailing the prince's project for consideration might be the document now preserved in the National Archives, identified as "the designe which hath byn presented unto his Majestie and referred unto the right honorable Lords the Committees thereof." This text also refers to the plan as a "greate designe" and a "great attempt." It begins by addressing the two requests of the East India Company for a delay and against the use of ships independent from Company control. It denies both these requests, claiming "a necessitie for the sudden & present sending out of one great shippe of connsonance & burden, as well to serve for honoring of the kings name, & to be of force, portage, and sufficiency for safety of returning such profit as God willing shall be in the first attempt gathered together." Also required were two smaller pinnaces as "handmaydens unto this great ship." One of the pinnaces could speed "to returne his Majestie with such overture of possessions, & first profitt as shalbe made, the other to trade in the cuntrie from port to port" and to encourage a "friendly prospecte until the great designe be put in executon." A list of suggested commodities that could be sent out for a "quick starte," the proposal predicts, would "return three for one profitt, and that in the time of two yeres or thereaboute only." For princes, this list included jewels, gold and silver plate, and "fire lock peeces, pistolls, knives, embroyderies, saddles & furnitures for horses & men, caparisons, picktures Tappestries, cases off bottles of strong waters, sack & white wine" as well as "two or three peeces of ordinance to present either the king of Bantam, Marhatta, or Empire of Mattaram, all our good friends, all which will & maye stand us in stead."

The weaponry gifted to allies would prove useful because the ultimate aims of this project were quite violent, despite the prince's assertions elsewhere to the contrary. While this pinnace plied about from court to court, making immense profits and supplying allies with arms, the great ship would serve as a sort of floating embassy to the court of Jahangir. The ship would be used to "Court please & observe the kinge in those Indian parts" with "rarities & novellties" as well as "with severall sourte of musickes, of motions, & other slight toyes & delightes, wch will fasten them unto us, and bring them on our shippes board, and maye serve to welcome us on shoare wherby the safetie and lifes of such persons as shale be herein ymployed maye be the better secured & freed from danger."[43] After gaining ground in India through novelties, automatons, and instruments, the

> designes which are pointed at, and hathe byn perticularly delivered unto the right honorable the Lord of Carlisle are in brief theis which the Lord maye be pleased to inlarge, as he hath receaved them:
>
> 1. Ffirst takeing possession of Summatra
> 2. The attempt that maye be made on the towne & treasure of Acheine
> 3. Concerning the towne & wealth of Sciam
> 4. Prisalls of the China Jonnakes [junks] with whom we have noe comerce as also on the Spaniards & Portingalls & their adherents our enemies together with the laudable trade of Japann & the venting of our English cloth there, and hereby to recover the (almost) lost honour of our nation, in all parts of India
> 5. The fishing of Pearles in many parts of India, a thing of great hope and import, provided his Majestys engines made by Cornelius Dribble prove true, & may be had, soe that of all these designs here mencioned, there are great hopes and probabilities, not only to returne a present profitt, but an annual & everlasting treasure to his Majesty & his Successors for ever.

It was promised that the details of these inventions would be made very clear; "the parties at present employed" on such engines would leave "a large and ample demonstracon, as well to remain here recorded, as to inable those that shall follow in the great attempt which (God willing) is intended to be put in execucon."[44]

If only this document laying out all the details of the newly invented submarine survived, it would offer much-to-be-desired information concerning this famous invention. As it is, however, the way that the royal "engines" made by Drebbel were immediately integrated into this massive global project or "great design" already has much to tell us about how such awe-inspiring new inventions were considered from the perspective of the court. Notably, this project entangles together invasion, piracy, diplomacy, trade, and engineering. The design promised to restore English trading in Aceh, a strategically important mercantile center from which the English had been recently cut off, strengthen ties with the Mughals, weaken Catholic enemies, and discover an untapped source of revenue in submarine shipwreck exploration and pearl-diving.[45] It is interested not only in revenue, but in honor, and indeed in intertwining the two, as the decline in sale of English woolens abroad is here interpreted as a blow to national honor that this project could redeem. Finally, it is framed in terms of "hopes and probabilities," identifying the greatest such probabilities with what might seem to us the least likely of all these proposals, that is, deploying a brand-new and dangerous art of submarine navigation halfway around the world in a large-scale industry.

This plan was informed by insight into Jahangir's views of technology and James's concerns with achieving the double ends of honor and profit. It persuasively placed at the heart of its design notions of craft that had become fundamentally linked with statecraft. The first was the valorization of improving nature through art, and thus the need for countries to found industries and export manufactured goods. The fall in the sale of woolen cloth, so tightly identified with notions of English industry, was a disgrace to the land. Not only venting that, but also pleasing imperial Asian courts with advanced technologies like musical instruments, automatons, paintings, and tapestries were a way of displaying English power in interstate, emulative craft rivalry. Likewise, when Steele proposed a project for Mughal trade to the French, he stressed how "honorable and profitable" it would be to sell "manufactures, niceties, and the rarities that are made in Paris" to such a great prince as the Mughal emperor.[46]

At last, after five months of heated discussion between the royal servants and the company, and despite the company's strenuous objections, James signed a commission for William Heydon and Charles Glenham to make the voyage. Heydon and Glenham were described as "singulerlie furnished with . . . laudable industrious and hopefull meanes for acquiring of Riches and Treasure both at Sea and Land . . . without giving just Offence to any, by recoverie of wrecked Treasure Pearle and other Riches in the Seas, and by divers other ingenious Arts Inventions Workes and Manufactures." The commission also refers to "sundrie Letters Messages and Requests from the *Great Mougoll*, to gratefie him with some choice Arts and Rareties which our Dominions are famed to afford no less usefull then unknown in those Parts."[47] James's commission stressed Prince Charles's personal investment in the inventions described in the project, which were "perticulerly recommended unto us by the Favor and Affection of our said deerlie beloved Son the Prince."[48] In its claim of providing useful and unknown arts, as well as in its denial that it would give "just Offense to any," the language of this commission reflected that of period domestic projects.

Heydon and Glenham were to trade a specific list of goods determined by the Privy Council and to import "Gould Silver Pearles Bullion Jewells Novelties or other Commodities as they shall think fit and convenient." Unlike in the manuscript of the "great design" that specifically aimed at piracy upon Spanish shipping, in this public commission Prince Charles, Heydon, and Glenham were prohibited from preying upon "all Christian Princes and States." At the same time, however, they were offered grounds for plausible deniability; piracy upon Christian states was allowed in the case of "just Defence of their own Persons Shippes Vessels or Goodes." Restrictions against violence in such commissions were not taken seriously. Sir Thomas Dale (ca. 1570–1619), former deputy governor of Virginia, had also been enjoined against offering "injury or discurtesy" to any "Nation or People" in his 1618 royal commission to sail to the Indian Ocean.[49] Dale not only took prizes from the Portuguese and Chinese, but immediately upon his arrival in the region resolved to "mak warr" upon the Dutch; he allied with the King of Jakarta in an assault upon the Dutch fortress there.[50]

Heydon and Glenham were granted extraordinary powers. They were allowed to exercise martial law "in as large and ample manner as our Lieutenant Generall by Sea or Land . . . have had or ought to have." This phrase in the commission echoed the 1616 commission that had been granted to Sir Walter Raleigh, briefly released from prison for his last voyage to the Americas.[51] However, Heydon and Glenham were permitted far more latitude than Raleigh was for further, unplanned incursions. They were authorized to realize the project not only within India, but also among any "other Princes or States, People, Citties, Islands, Continents and Places whatsoever already discovered or not discovered" within the vast swath of the globe stretching from the tip of Africa eastward to the tip of America.[52] The project's design anticipated that it might sail into the unknown, reserving the power to take action as opportunities arose. This project exemplified how entangled the pursuit of honor—including through games of technological one-upmanship—became with global trade. As the commission James granted his servants for the Southeast Asian project specified, its aim was to "further extend the Honor of Our Name and Kingdoms."[53] This view ran counter to that of the company, which shied away from grand projects that were not assured of success.

The great design, after much deliberation and discussion, came to the very brink of being realized, until the sudden adventure of Prince Charles's journey into Spain to investigate his possible marriage to the Infanta suddenly consumed his attention (as well as that of nearly all the courtiers involved in this project) and rendered undesirable the piracy on Spanish shipping that had been part of the secret "great design". Other projects afoot at this time were also suddenly abandoned for Charles's journey to Spain. In 1624 the Küffler brothers described to Peiresc another ambitious plan of Drebbel's they claimed he had proposed to Charles. This one would centrally heat the entire city of London through an artificial sun, just as the New River Company had succeeded in bringing water to the city.

> The most recent and most excellent invention Drebbel discovered was an artificial sun, that is to say, a perpetual fire which would burn and illuminate forever. When the Prince of Wales [Charles] went to Spain [in 1623], Drebbel proposed to him that just as one has filled London with fountains by means of a small river, conducted by little pipes to all the houses, he [Drebbel] would make a fire on a little mountain near London, whence all the Londoners could obtain fire and conduct it to their houses, and with this fire they could boil and roast their meats without need for wood. . . . I believe that he only asked for 20,000 pounds sterling.[54]

This sounds like an unlikely project. However, the neat parallel to the 1622 design, with its plan for cooling Mughal palaces through channels of water, forces us to reconsider. A project to centrally heat London with a perpetually burning artificial sun should be no less improbable, at the time and to the prince, than the very nearly realized plan to explore the Indian Ocean with the newly invented submarine. Drebbel compared it to other ambitious, but realized projects, such as the New River Company which opened its waterworks in 1612.[55] His high price, the equivalent of about six million dollars today, was on par with similar undertakings.[56] In 1621, for instance, Salomon de Caus proposed providing fresh water throughout Paris, for which he was supposed to be paid 60,000 livres annually.[57]

The Küfflers reported that Drebbel planned to achieve central heating through an instrument that would concentrate the rays of the sun on a particular spot, where it would light a material that would burn without being consumed. This may be the "solar instrument" depicted in a now lost manuscript showing thirteen of Drebbel's inventions, including also his "refrigeratory instrument" for India.[58] Drebbel's investigation of the elements, and his abilities to reverse and control their qualities, such as heat and cold, promised abilities to transform climate, heating cold countries and cooling warm ones. It was this ability to override constraints of particular locales that made his technologies appear to Prince Charles and his courtiers to promise so many "hopes & probabilities" when expanded onto a global stage.

In his 1622 "great design," Charles wished to act as an interloper in many ways, intervening in the area of the East India Company's influence and trade, invading kingdoms, and even establishing an industry beneath the waves by drawing on Drebbel's boundary-abrogating technologies. Yet, interloping ran both ways. The magnificent stage upon which Charles wished to enact Drebbel's inventions, and the associations that were drawn between these abilities and royal and national honor, served to support Drebbel's own claim to authority and the importance of the role that his abilities played in the state. Such views spurred the Küfflers to air a significant criticism of Charles's Spanish voyage and the ways that it had led him to ignore Drebbel's most recent, greatest invention. They remarked that "the voyage that this Prince undertook prevented him from furnishing what would be necessary to have this miracle made," and this "voyage did great harm to the public."[59] The integration of an individual's limitless abilities into global great designs as well as the practice of thinking about individual projects in terms of public benefit could translate into claims to be able to speak politically.

Steele's Waterworks

While the discussions of the 1622 great design repeatedly mention Drebbel in connection with the submarine and its associated engines for lifting weights, no name is mentioned in connection with King James's desired gift of cooling the air in Jahangir's palace. Thus, it is not entirely certain that the invention that James had in mind was by Drebbel. Another candidate is Richard Steele, who following a journey to Persia in 1614–16, had presented a collection of five proposals to the East India Company around the end of 1617: (1) purchasing raw silk from Persia, a proposal supported by "the kinge of Persia"; (2) taking over from the Portuguese the trade in transporting the goods of "Persians, Turkes, Indians and Armenians"; (3) selling spices to the Mughals and the Persians; (4) placating merchants from Gujarat and Lahore who complained that the trade with the English was not worth a thousandth part of the trade with the Portuguese; and (5) "a water worke, which might give the Chief Cittye of the Mogores content, vent Lead, and bring both our land proffitt . . . for water in those hott parts is verry comfortable, and I doubt not but either the king will give a good gratification or the people of that Cittye [Agra] pay quarterly or yearly for that, as of that of Broken Wharff and this contenting the Mogor, willbe meanes to further any other buisines in his Countrye."[60] By developing a revenue stream for Jahangir from fees for the municipal water supply of Agra (as was the case for James in London) this proposal would curry favor with the emperor. Such imperial favor would then oil the wheels of business for all other trade there. In addition to presenting this proposal to the company, Steele also had an audience around this time with King James.[61]

As the Küfflers did in their description of Drebbel's centralized heating for London, Steele compared his proposal for Agra to London's recent hydraulic engineering projects. Steele referred to the waterworks constructed in London by Sir Bevis Bulmer in 1593 at Broken Wharf near St. Paul's, which conveyed water in pipes to houses in the area. The very fact that both Drebbel and Steele were working on cooling projects for India, and that both of their heating/cooling projects were compared to the new hydraulic engineering in London, illustrates how projectors interloped or "leapt between." They competed, developing rival and similar projects. They extrapolated from one project to the next, including from a project for London to one for Agra, or from a project for water to one for fire. They also attempted to emulate and appropriate knowledge from other cultures. The Mughal emperors already had access to chemical techniques for cooling the air as well as to sophisticated cooling waterworks for their palaces.[62] Northern Europeans like Drebbel and Steele, in proposing to supply a hot country with cooling techniques in which the Mughals were more experienced, participated in the game of imitation and one-upmanship of the Mughal court, which imitated and improved upon European arts.

There are certain similarities between Steele's 1617 proposals and the 1622 great design that make Steele a candidate for the involvement in the latter. They both had a five-point structure, although Steele's 1617 proposal was not woven together into a great design. However, his idea of using one endeavor to curry favor with Jahangir in order to enable further initiatives recalls strategies of the great design for using an automatons-filled boat in order to win over the princes of the region. Steele had great experience with Jahangir's taste in art and had sold pearls in the Mughal court.[63] He was also particularly optimistic about the ability to vend English woolens in chilly Persian northern regions; this optimism was the foundation of later hopes of exchanging raw Persian silk for English woolen cloth, or even for the silk-and-wool blend tapestries woven at the new factory in Mortlake.[64]

If Steele was involved in the 1622 design, it would help explain the company's immediate negative response to the proposal, as he had become a thorn in the company's side. Steele had been allowed to journey to India in 1618, with a great deal of lead for pipes and a staff of engineers, to attempt his municipal waterworks in Agra. Sir Thomas Roe (ca. 1581–1644) was sure it would fail. Roe had been sent to India as a royal envoy at the company's request and expense.[65] He worked to further the company's interests there. For appearance's sake, Roe claimed, he had to let Steele try. If he forbade a trial, he himself might stand accused of peremptory, rash judgment, foreclosing opportunities for profit: "if some of his projects are yet doubtful, some in my judgment infeasible, yet we must not disgrace them without trial, lest we incur the same censure of rashness which by it we would cast upon him."[66] Roe later explained his reasons for opposing this scheme, not least that "the King [Jahangir] and nobility have as excellent and artificial waterworks of their own as can be desired."[67]

Using the recruitment of craft talent, Steele was able to win favor with Jahangir independently of the company. He not only brought to India engineers for the waterworks, but also engaged a painter and a clocksmith. Agents on site continually wrote to warn the company that Steele was not acting to further the company's profits: "Richard Steele is arrived here with his Engineers, where he imployeth himself and them to his owne purposes. . . . I am confident he abused your worships at home nor intends you less abroad." Steele's project of engaging Persia through commerce in the port of "Jasques" (Bandar-e-Jask) on the Gulf of Oman was "a worke of impossibility," and his other project of waterworks was "meerely idle." While the lead for the waterworks had been purchased by the company "at your prime cost in England," the engineers Steele had brought were being employed by Jahangir, who wished to use them in making "some fountayne" rather than in order to profit the company.[68]

Sir Thomas Roe reported around the same time that the presence of Steele was inciting dissatisfaction among the company's other employees in the area, who complained that "a light braynd man that goes home and fills your ears with fables shall returne in better estate then they for paynfull service; you must pardon mee for my directness, he neither can nor intended to performe any of his great braggs." Steele's direct interaction with Jahangir was particularly unprofitable for the company, as it led to the devaluation of English knowledge. After employing Steele's artificers temporarily at low wages, Jahangir would appropriate their knowledge and dismiss them, thus doing enduring harm to the company's trade as they would lose control over yet another form of artistry and industry that they might have been able to sell to the Mughal court. As Roe reported, Jahangir "hath taken the woorkmen at dayes wages . . . The king is desirous of all new arts; will entertayne the Artificers and soone learne their skill and cast them off." This strategy made it difficult for the company to identify any new curiosities that might consistently interest Jahangir or his court as a present, or indeed any commodity that could maintain its commercial value. For a short time, "a New raretie or curiosities never seene here, and of smalle price in Europe," such as a glass showing many colors (perhaps a kaleidoscope), could achieve a high rate of return, but not for long: "they Imitate every thing wee bring. . . . these People will covett any thing; when they see it, disgrace it, and not come to halfe the price." Only jewels retained their value.[69]

In another letter, Roe reported that Steele's true end was achieving some personal reward from Jahangir by putting his personnel of craftsmen at the emperor's disposal to achieve some astounding work of art: "he hopes to doe a great worke in the kyngs castle, and it is not unlikely; to live upon the pay, and to gett by 5 artificers many a crowne with his owne purse, and at last . . . great reward, this was his plott."[70] Jahangir invited Hatfield, the English painter whom Steele had engaged, to his castle to paint his portrait. He allowed Steele, who could speak Persian, to accompany Hatfield as an interpreter.[71] Roe wished to send Hatfield home, but feared Steele would complain to Jahangir "if I cross his pleasure in paynting."[72] Steele even accompanied Jahangir as the emperor went "upon his progresse," going "with the plum[b]er and other artificers sent with the King in his travels to perform some of his projects."[73]

Steele's plan relied on his ability to recruit and manage a staff of craftsmen and thus it is not surprising that he entangled with the Mughal official whose job it already was to oversee the court artisans. During Jahangir's travels, Steele fell "into some difference or dislike with Mirmiran, a gentleman of the Kinge to whom the care of workmen was commended, was [by] the said gentleman dispossessed of his camels and other carriadge and left to his own provicion."[74] After being abandoned, Steele returned to England. He never completed his waterworks project, and he was "much condemned for his unworthy carriage abroad" by the company, who complained of his interloping in "a great private trade," as well as his "arrogating a higher title and place to himself than ever was intended."[75] It was perhaps then that James learned of Jahangir's interest in Steele's artificers and attempted to gift the Mughal emperor cooling waterworks for his palace.

After the 1622 great design also did not transpire, Steele turned again to the East India Company in 1623, presenting two of the same proposals he had proffered in 1617, the Persian silk trade and the waterworks for Agra. In their emphasis on the glory that these projects could bring England, these proposals likewise recall the 1622 great design's discussion of restoring English honor in Asia. Steele read aloud to the company "certayne observations of his owne in writing concerning the Persian Trade and the waterworks proiected by himself." He asserted that the waterworks would not only yield "10,000 pounds per annum," but also bring "glory to the English Nacon, and be infinitely pleaseing both to the greate Mogore and to his subjiectes." He claimed that the Mayor of Agra doted upon the project, and "imbraced it with such alacrity that he gave greate guifts both to Mr. Steele in particular and to all those that went over with him for the effecting of that design." The company remembered, however, Sir Thomas Roe's advice that it was "Dangerous" to employ Steele. It refused his proposals.[76]

Mercantile Caution and Noble Adventure in the Persian Silk Trade

The plan to deploy arts abroad in order to win both honor and profit ran counter to the ways that the East India Company operated, as illustrated in the contrasting receptions the company and the Crown gave to the other projects of Richard Steele. The East India Company entertained some projects that had been well proven through trial and were not too expansive. It sent several packing without even so much as a trial.[77] The company seemed to pride itself on its more circumspect view of projects in contrast to royal views. Examining one projector's claims, it "found nothing but airy conceits and impossibilities." It advised him to impart "his projects and mathematical inventions . . . to the State, who, no doubt . . . will recompense him."[78]

As Robert Brenner has discussed, a generational conflict in international trade divided "the great City merchants" who ran the East India Company on the one hand and younger royal patentees and projectors on the other. The older generation attempted to gain profits not by overriding constraints, but by setting them and thereby minimizing risk.[79] In his 1625 essay "Of Youth and Age," Francis Bacon seems to ruminate upon a perceived generational divide: "Young Men are Fitter to Invent, then to Judge; Fitter for Execution, then for Counsell; and Fitter for New Projects, then for Settled Business." Each age had its advantages and disadvantages, according to Bacon. The youth "Embrace more then they can Hold, Stirre more then they can Quiet, Fly to the End, without Consideration of the Meanes, and Degrees, Pursue some few Principles, which they have chanced upon absurdly," and do not hesitate to "Innovate, which draws unknowne Inconveniences."[80] The aged, however, "Object too much, Consult too long, Adventure too little, Repent too soone, and seldome drive Business home to the full Period; But content themselves with a Mediocrity of Successe."[81] If the motto of the youth was "move fast and break things," that of the older generation was "quit while you are ahead."

Yet, this was not just an age difference, but a cultural and social one. Such differing views helps explain the poor relations between King James I and Charles I with the East India Company.[82] Logics of extending honor promoted by the court often ran counter to limiting risk and making profit. For this reason, Nerlich argued, "around 1603 the East India Company declared that it would rather give up its planned trips than employ 'any gentleman in any place of charge or commandment' of a ship."[83] Sir Thomas Roe, who, despite his title as royal envoy reflected the company's views on projects more than the Crown's, was averse to employing gentlemen in the trade. When in 1617 Thomas Herbert (1597–ca. 1642), the stepson of Sir John Danvers (discussed in the next chapter) came to accompany Roe in the Mughal court, Roe wrote "Mr. Herbert shall be welcome as a voluntary, not a servant; gentlemen expect more, and do less."[84] Indeed, on his way to the Mughal court, Herbert horsewhipped and shot at one of the servants of the Mughal prince (who would later reign under the title of Shah Jahan) because the servant had refused to hold Herbert's horse for him.[85] Roe ordered company employees to "live frugally, soberly, like merchants."[86]

While Steele was a merchant, he assumed the position and rhetoric of somebody of a higher status. The Crown smiled upon bold ventures such as those of Steele that promised both immense power and profit as well as the winning of glory. However, his projects were not the products of his own adventurous spirit. In fact, Steele's project of engaging in trade directly with the king of Persia for raw silk originated in the plans of the Safavid Shah Abbas (1571–1629). In concert with a recent expansion of his territory and financial and military warfare with the Ottomans, Abbas centralized and expanded silk production and its sale. As English factors noted, through forced migration he moved nearly 30,000 Georgian and Armenian families eastward to found sericulture in new areas. He controlled the selling of the silk, via Armenian, Georgian, and Jewish merchants, in new ports. He attempted to engage Europeans directly in the purchase of his silk, sending various Armenian emissaries with silk samples to Venice between 1600 and 1632.[87] He wished to avoid the overland route through the Ottoman Empire by selling to European merchants in Persian ports, such as Bander-e-Jask on the Gulf of Oman. Abbas's massive restructuring of the silk industry had many ripple effects in other lands, including Ottoman and Western European efforts to introduce and expand sericulture.[88]

Abbas proposed the idea of selling silk to the East India Company's factor, although Roe had written to the company advising against it in 1617.[89]

Counseling against this shift, Roe advised, "The trade will not be turned from his ancient course . . . Great waters will keep their own channel."[90] In the name of Shah Abbas, Sir Robert Sherley (ca. 1581–1628), the self-proclaimed Persian ambassador, offered an "overture" to King James I for the English trade in Persian silks that would, he claimed, "bee advantagious for your Majesties particuler profit . . . and noe whit preiuditiall to your Subiects." This proposal included the promise that Abbas would supply twenty-five thousand armed men at his own charge if James wished to undertake "any action or enterprise" in the region. Sherley suggested that the Persian trade could be engaged in experimentally with no consequences: "as upon tryall you shall like of it, your Majestie may proceede, or leave of it at your pleasure."[91] In another set of propositions communicated to the Privy Council, Villiers, and King James, Sherley calculated that the Persian silk trade was worth 5 million pounds, of which his Majesty's custom would be 500,000 pounds "besides all our poor set on work, and great part of our native commodities vented."[92]

As a matter of state, Abbas's suggestion was presented to the Privy Council, which thought "so well of it . . . that it is concluded that a trial thereof shall be made."[93] In 1617 Roe took on the making of "a judicious experience [that is, experiment] of the profits and possibility of that trade."[94] Despite Roe's doubts, this trade at first succeeded through the combined efforts of Roger Steele, Robert Sherley, Edward Connock (the company's factor in Persia), and Abbas, who granted the traders many privileges. It soon stalled, in part due to an abundance of caution from the East India Company. Seeking security, the company attempted to stipulate conditions to Abbas in a contract, which he refused.[95]

Sir Thomas Roe's practices of carrying out a commercial trial were heavily weighted against incurring too much risk, either financially or in the loss of credit. In 1618, when Steele again pushed for the trade to Bandor o Jack, Roe wrote to the company, "I will do my endeavour to settle you in this trade, if I maye doe it upon such grounds as I may have Creddit by and you profitt." Roe would only attempt to move a "little supplie" and, "if we see the danger and chardge unavoydable," and no means to win a profit that could "beare your expense and hazard, then wee will tymely recall your servants."[96]

The risk-limiting practices of the company ran into direct conflict with the way that Sherley, as well as top courtiers, wanted to do business. Sherley's "buccaneering quasi-commercial proclivities" aligned with the interloping tendencies of the court, who found his proposals "exciting" while the company considered them "exasperating" and the work of someone "intent on damaging their interests through his special pull at Court."[97] Sherley's manner of operations fit the gambles that courtly interlopers interpreted as signs of greatness of spirit.

The project for contracting a trade for raw silk with Shah Abbas was revived multiple times by Steele as well as by Sherley. Despite many accusations by the East India Company of Sherley's villainy, James, Charles, and Villiers continually supported him. At Theobalds in July 1624, King James and his Privy Counsellors wished to know what the company thought of the "rich trade into Persia" discovered by Steele and "seconded by Sir Robert Sherley," through which English cloths could be traded for Persian silk. "To this was answeared that the vent of Clothes in Persia is better knowne to the Company then . . . either to Sir Robert Sherley or Steele, the Company factors having endeavoured to vent of Clothes all the Country over."[98] The company repeatedly tried to persuade the king that Sherley was not really an authorized Persian ambassador but merely a private adventurer, whereas they acted based on superior knowledge, painfully acquired through many repeated and careful trials.

The next month, the company was asked by the Lords at Whitehall to consider "the project propounded by Sir Robert Sherley." It tried to deploy the self-interest of merchants against Sherley's arguments. Merchants, they argued, would already have sought to maximize their profit by any possible means. Thus, "if there were any possibility of doing good by other ways than have been already found, the merchant would for his gain find it out." The company also claimed it improbable that "Sir Robt. Sherley or Steele whose hand is in this project" could possibly compete with the company concerning the trade, since the company employs "the ablest men they can get."[99] The company's arguments against Sherley and Steele assumed that avoiding improbability and drawing upon tried and trusted experience was a good thing and would be persuasive. It was not.

Despite the opposition of the East India Company, James, Charles, and Villiers were "determined to push" the Persian trade. They sought investors, and an inquiry was undertaken concerning "what profits can be held out, in order to draw in adventurers to join with the Prince, Duke, and other nobles therein."[100] At a consultation between Sherley, Steele, Sir Thomas Smythe, the shipwright William Burrell, and two other knights, it was agreed that making England the "staple of the Persian raw silks" was the "only means, to draw the greatest part of Europe's money hither."[101] Commissioners of the Navy investigated the costs and potential profit of this voyage, warning "There are to be no longer delays, or this hopeful trade will be in the hands of the Dutch."[102] The Crown and chief noblemen sought to establish their own trading company. The Venetian ambassador in England observed:

> The king, the prince and Buckingham, with other noblemen, are disposed to venture six of their own ships, that being the number asked for by the ambassador to send to lade the silk, which he promises to have transferred to this kingdom at the risk of the King of Persia, to be sent hence to all parts, and deprive the Turk of the advantage of this trade. But it would damage all the trade of the Mediterranean and of those who have marts in Syria. It is a long voyage from the Persian Sea to English waters; perhaps they will not find it so easy as they say.[103]

Within two months, the Court of the East India Company "was informed that somewhat is sett upon the Burse to invite a subscription for a new Persian Company."[104]

The governor of the East India Company was summoned to the Privy Council, where he found Prince Charles attending in person, along with Sherley, to discuss Sherley's "proiect." The governor was "askt if this Company will joyne in the proiect." Villiers questioned him closely. "My Lord Duke of Buckingham said the King will in his own shippes fetch home the whole silk of Persia, and demanded what hurt this would be to the Company?" The governor said that the company was "utterly unwilling" to adventure in this project, pointing out how Sherley's views of the cost and profit of the silk trade were quite incorrect.[105]

The East India Company felt much vindicated when, after their factors had decided to withdraw entirely from Persia, they received a letter from Shah Abbas that did not mention Sherley. This seemed to prove their frequently reiterated point that Sherley was not a Persian ambassador at all. They showed this letter to Villiers, who called them in to speak to King James. The company gleefully "took knowledge that some persons that carryed a busy hand in a designe for Persia, do now hang down the head, and do believe rather what the Company out of their experience had formerly reported."[106]

The company was to be disappointed, as it turned out that Charles did not care about Sherley's official status nor about the company's superior mercantile experience; he still approved of the project. Two months after James's death, Charles convened a committee, headed by Villiers. Since Charles aimed to "enlarge" the trade of the kingdom, he had taken "notice of certain overtures for settling a commerce with the King of Persia sundry times debated in the late King's time." He noted that "there have been of long time differences thereabouts amongst the East India merchants," and required the committee to investigate why the Persian silk trade had not yet been advanced.[107] The governor and some other members of the company were again summoned to "attend the Lords," and listen to a "long narracon" about the benefits of Sherley's plan. They were reminded that "the late King was so well assented thereunto that once hee had given consent to send out 4 of his owne pinnaces . . . for the furnishing out whereof money had been prepared." This trade was considered "exceeding advantagious to this kingdome." The governor replied "that he must make bold to represent the Company's former answer and resolucon to follow the trade as merchants upon such grounds as their experience hath discovered, neither were they willing either to bee driven out of their owne way." To which was retorted, that, whether or not Sherley was an ambassador, "the proposions tending to the publick good ought not to bee neglected."[108] Charles cast the pursuit of national honor and the public "good" through the adventurous enlargement of trade as trumping the concerns for safety, security, and limited risk held by merchants. This was a view that judged mercantile perspectives to be small-minded, blinkered and self-interested and that linked the public good to noble ways of doing business.

Charles never abandoned the idea of a trading company to rival the East India Company and other ways to interlope into their trade.[109] In 1633 he licensed Endymion Porter to practice piracy upon the ship of any state not in league with England and to "range the seas all the world over."[110] Two ships, the *Samaritan* and the *Roebuck*, sailed in 1635 on behalf of Porter and his association with two merchants. The *Roebuck* captured two Mughal vessels in the Red Sea. Despite an attack launched on the *Roebuck* by an English East India Company ship, the *Swan*, the *Roebuck* returned home victorious.[111] The president of the English East India Company factory at Ahmadabad, Captain Methwold, apologized to the Viceroy of Goa that the commander of the *Swan* "did not seize the *Roebuck* and bring her with him to India; but he is young, and the sight of the royal commission daunted him."[112] The company protested Porter's piracy, and Charles made a pretense of listening to their concerns, arresting and imprisoning one of Porter's associated merchants, whom, however, was immediately released on bail. Charles assured the captain of the voyage that he enjoyed "constant and continued gracious favour."[113]

Charles continued to foster Porter's interloping around the globe. In 1635 Charles granted Porter another patent for a trading association along with William Courten, a.k.a. Courteen (1572–1636), scion of a family of Dutch global merchants. Charles himself was credited with a share of 10,000 pounds, and the effort was characterized in the East India Company papers as "The King's Undertaking to Join in the Adventure to the Indies." One of the goals of this voyage was to seek out a northeast passage from Japan.[114] Courten passed away within the year, but his son, the physician William Courten (1609–55) realized the expedition.[115] Four ships and two pinnaces set out and plundered two vessels. East India Company servants in Surat were imprisoned in retribution.

In planning the Courten expedition, King Charles continued to tangle with the East India Company. Urged on by Endymion Porter, Charles wished to send along with the Courten and Porter expedition his nephew, Prince Rupert of the Rhine, and make him his viceroy of Madagascar. According to a draft letter from Charles to Rupert concerning Madagascar, he wished Rupert "to plant his Majesty's subjects there and make trial of so hopeful a project."[116] Rupert thought Madagascar could be made into "the ballance of all the trade betwist the East Indies and theis parts of the world."[117] Sir Thomas Roe wrote to Rupert's mother to persuade him to drop the "absurd" project.[118] With the strong opposition of the company to Rupert's participation that part of the plan was indeed scratched.[119] The Courten-Porter association, however, did set out for Madagascar with 140 settlers aboard three ships in 1644. Things did not go well; embarrassingly, the settlers had to be rescued by the East India Company.[120]

Conclusion: Ripple Effects

Royally supported projects for intervening in Asia involved a veritable parade of projects. These might seem to belie any sense of order or organization, and indeed, these projects illustrate undisciplining in the ways that they leapt between and loosely intertwined various objectives. However, a coherent set of interloping attitudes and knowledge practices does emerge from looking at projects in global trade from the court's perspective. These are thrown into relief when contrasted with an opposing, risk-limiting view—that of the East India Company. Interloping sheds light on how the Crown viewed global intervention.

From the viewpoint of projects, the global flow of power appears as an interplay of constantly morphing schemes, aims, and resources. These were not colonial programs centrally conceived or locked down by the English state. Rather, aspects of these schemes might have originated in the interests of foreign princes, such as Shah Abbas, might have been developed through cultural embeddedness in the Mughal court, or might be offered at any time to rival European powers, such as the Dutch or the French. Interlopers, who here included figures such as Steele, Sherley, Endymion Porter, and Prince Charles himself, sought ways to penetrate this flow and shift it to their advantage by cobbling together as many fragmented opportunities and pieces of intelligence as they could. Fueled in part by a confidence in the power of new inventions and in the ingenuity of political strategy, they understood the dynamism of these power relations to be much greater than that suggested by Roe's advice that "Great waters will keep their own channel." Quick-changing seizure of advantage meant leaping upon intelligence and projects such as those offered by Steele and Sherley, perhaps blending those with other courtly initiatives, such as the Mortlake tapestry works or Drebbel's newly invented submarine. New shards slipped into place as the kaleidoscope of political relations and opportunities constantly shifted.

This shifting ecology of projects and interests required rapidly tacking courses of action and sometimes the sudden abandonment of a project. Historians have interpreted such swift changes as Charles's personal flaw of vacillation, but quick-changing action was a part of the policy of the period. Charles had little choice but to "swim between two waters," given the other major powers involved: Jahangir and Abbas, both masters of statecraft who valorized emulative rivalry in industry every bit as much as James and Charles did, and who were endowed with far more efficient and disciplined tools of statecraft to make large-scale designs a reality. Using strategy and ingenious, underhanded technology, Englishmen could only hope to act as interlopers, navigating around the plans, ambitions, and animosities of other states, and finding ways to turn the global network of intersecting objectives to their own advantage.

When it came to inventions, Charles did not vacillate, but from a young age engaged in thinking about how to pursue them on an ambitious scale, ready for the opportune moment of their deployment. He may have dropped his 1622 "great design" involving the submarine when the voyage to Spain instead occupied his attention. However, he not only continued to involve himself in Asian trade in ways that ran into conflict with the East India Company. As discussed further in chapter 7, he also engaged Drebbel to work further on submarine boats and bombs, which were then deployed on an international stage in Villiers's campaign off the coast of France in 1627.

Although they were never realized, both the 1622 "great design" and the continual projects of Sherley and Steele demonstrate the personal involvement of Charles in global projects in ways that can shed light on later ventures. Roper, for instance, sees the Courten Association as formative for English imperial thinking, as its massive scale united ventures in Asia and Africa for the first time. This case, Roper avers, saw a "relatively large scale of government involvement," yet the Crown's role was still "reactive."[121] An analysis of the Crown's role in the

Courten Association that looks back to Charles's strenuous advocacy for an intervention in Asian trade, first on behalf of the 1622 project and then for a Persian Company in 1624 and again in 1625, would perhaps reconsider this view of reactivity. Charles's favoring of the Porter-Courten Association can be seen as in part a later realization of long-term aspirations.

Projects past, proposed, failed, or just beginning to be imagined created a horizon of possibility. Across this horizon, we recognize many of the concerns and practices we previously encountered in the case of domestic projects. Global projectors framed projects in ways that responded to arguments drawn from the reason of state. They justified their ambitious and invasive plans as a means for advancing the state in ways that redounded to the public good. Aiming for greatness, they saw their experimental, adventurous mode of enlarging commerce as a way to augment the power and glory of the state. They turned to practices of trial in global trade as much as in the domestic manufacture of soap.

Global projects entangled imperial rivalry into large-scale projects that were already concatenated from many elements for all the reasons that we have seen previously. Spurning the discipline of established crafts, projects combined far-flung forms of knowledge. They sought new discoveries and potential profit by crossing boundaries, emulating knowledge, and seeking out new environments, from the submarine world to Persian trade. Projectors seeking to recoup intellectual and financial investments continually designed new projects integrating aspects of previously failed or unrealized projects. Projects were also complicated by design as a means of avoiding claims that they harmed current interests. The more heterogeneous a project, the less likely it was to compete with an already existing endeavor. Global projectors had further reasons to develop grand, multi-plank projects. They sought to show how their multiple aims interacted with an international theater of political interests and fluctuating phenomena. In the extent and ambition of their projects, projectors claimed to expand the glory of the state, particularly in a "great design." This continual stream of quick-changing yet interrelated proposals created a riskscape, an aesthetic, an emotional palette, and a knowledge practice that together formed the culture of interloping.

CHAPTER 5

Transplanters of Empire
Forcing Nature and Labor

The intellectual practices of interloping spurred knowledge to make leaps across scales, between forms of knowledge, and past wide cultural differences, in ways that could encourage the circulation of ignorance as much as of knowledge. Such leaps allowed for poorly supported arguments, inferences, and assumptions. Interloping permitted extrapolation from small-scale trials to widespread application in far different settings. It allowed for biased selection, in ways that overlooked counterarguments, evidence, and harm to others. It also legitimated hasty action and rapid about-faces, with sudden abandonment of whatever was not proving advantageous. Confidence in the power of experimentation and risk-taking supported these loose and hasty intellectual practices.

Attention to interloping, and the habits of extrapolation and knowledge suppression that it encouraged, can help us revise our view of the relationship between colonialism and the circulation and growth of knowledge. The history of colonialism is one of few historiographical arenas where projecting is currently heavily emphasized, thanks to works such as Karen Ordahl Kupperman's *The Jamestown Project*. Kupperman repositions the origin story of ultimately "successful" settlements in the context of the history of many unrealized projects. She draws attention to projectors' intersecting plans for colonies around the world, from Ireland to the West African coast to South America to the Arctic, placing them within a welter of projects, including domestic ones, such as a failed attempt to grow orange trees in England.[1] Yet, her story has been one of the growth of knowledge, of how often ill-conceived projects, which as we can now see in retrospect were "bogus, ill-informed, or misguided," confronted harsh realities, leading to experimentation on the ground, the acquisition of experience and expertise, and thus survival.[2] However, even the seeming success stories not only entailed knowledge growth and survival, but also knowledge loss and destruction. Furthermore, Kupperman's commonsensical view of a trial-and-error approach cannot explain several bizarre features of colonial projects that can be better understood through the larger epistemic and moral commitments of interloping. This chapter resituates colonial projecting within a history of ignorance, risk, adventure, and avowedly improbable plans to shift natural parameters around the globe.

Kupperman argues that London-based investors in the Virginia Company were hardheaded figures who "looked to the next quarterly report as much as those in the twenty-first."[3] One would assume that these hardheaded investors looking for short-term gains would have valued the experience relayed to them from overseas and would have used that information in making investment decisions, approving modest projects, and rejecting ambitious ones that required a longer-term, less secure outlay. This was the case, for example, in the approach of the East India Company, whose body of investors even extensively overlapped with the Virginia Company. This was not, the case, however, in Virginia. The Virginia Company exhibited practices of adventurous interloping that the previous chapter contrasted with the East India Company's careful business sense. While the crown and chief courtiers sought to prod the East India Company into taking greater risks, the ambitions and political strategizing assumed by the Virginia Company led King James to view them with suspicion and ultimately to revoke their charter.

One of the reasons for this difference in behavior between the East India and the Virginia Companies may lie in the differing statuses that Europeans accorded to Asia and to America in their comparative global surveys of industry and advantage. According to Giovanni Botero, Asia far outstripped Europe, and even Renaissance Italy itself, in its cultivation of art, urbanism, and trade. As he wrote:

> There is not in all the world a Kingdome . . . that is either greater, or more populous, or more riche, or more abounding in all good things . . . than that famous and renowned Kingdome of China. . . . Artificers are infinit, and . . . Artes

are brought unto most excellent and high perfection. . . . To say the truth, wee Italians do flatter our selves too much . . . when we will preferre Italy, and her Cities, beyond all the rest in the world. . . . Our trade and traffique is but poore, in respect of the Marts and fayres of Cantan, Malacca, Calicut, Ormuze.[4]

By contrast, Botero discussed no cities in North America, although he greatly admired Cuzco.[5]

Challenged by the perceived superiority of Asian industry, the Crown-backed projects we explored in the last chapter aimed to rival these more developed kingdoms through mimetic one-upmanship in a defense of English honor. In order to advance England's position in the region, they engaged with the industries and policies of the monarchies there, by navigating submarine-like around dominant power structures, policies, and industries and attempting to undermine them. The merchants of the East India Company, as we have seen, who also recognized the might of the region's kingdoms, sought to avoid such contests of honor and piratical incursions into powerful domains.

Despite acknowledgment of Native artistry, ingenuity, and might, Englishmen often cast Virginia as akin to undeveloped and (in their eyes) underutilized common lands in England that might be "improved." As they did in England, they thus felt a much greater latitude to enter the situation and to transform it. Moreover, their ambition was not merely to transform Virginia into a new England. Rather, they hoped to leap past England's state of development to make Virginia, as well as England, into a new China and Persia. The prestige project through which they hoped to accomplish this was the cultivation of silk in Virginia, a project doomed to failure for the high degree of sophisticated work and investment in materials and resources, such as specialized silkhouses, that it required.

Company officials, in particular those in the faction of Sir Edwin Sandys (1561–1629), pushed sericulture on the colony despite constant communication from the colony to London noting the many obstacles that stood in the way of success for this project in Virginia. Supposedly hardheaded investors ignored first-hand experience and repeatedly demanded that colonial settlers attempt this ambitious project despite settlers' arguments about its impossibility. As Kupperman describes the situation, "This transatlantic correspondence has the quality of people shouting past each other; it is clear that leaders in Virginia, like leaders in London, could not understand why their opposite numbers refused to comprehend simple reality."[6] The "most determined promoters, such as Sir Edwin Sandys, were the worst because they persisted so energetically with unworkable projects."[7] As Kupperman concedes, "The remarkable thing about Jamestown is that the investors and the colonists did not simply walk away from the project . . . Despite a record of unremitting problems, and even in the absence of any concrete hope for future success, the backers determined not to give up."[8] The mystery of why the investors would continue to push for silk in the face of mounting negative evidence is even deeper than it appears in Kupperman's account, for as Ben Marsh has detailed, the quest to produce silk in North America continued, despite frequent failures, all the way through to the nineteenth century.[9]

Promoters of colonial projects in Virginia can be better understood, I argue, through the larger history of ignorance in the period and through how colonial projecting intersected with epistemic shifts and emergent ways of knowing. Early Stuart projects did not specialize in the pursuit of single aims. Due to interlopers' attention to how phenomena related, to practices of global survey, and to thinking across categories, the sericulture project was not just about silk. It entangled together debates over common property and the commonweal, risk, enforced labor and transplantation, genocide, acculturation, acclimatization, honor, rivalry, the global economy, and profit with epistemological issues concerning experimentalism and probabilism.

Company members were not refusing to comprehend simple reality; they were refusing the notion that reality could be comprehended simply. The disconnect between ambitious schemes emanating from London and the harsh realities on the ground was not merely a case of the metropolitan investors falsely asserting their superior knowledge over settlers.[10] Rather, they asserted their superior sense of the lack of knowledge, of an ignorance that might mean that the boundaries between possibility and impossibility could be drawn quite differently than was currently assumed. They were operating in a context of much greater unknowns than just the outlines of coastal America or the nature of its climate. Rather, the context of colonial ignorance in which the Jamestown project played out intersected with a larger scale of ignorance in the history of knowledge as a whole. The shape of the universe and the nature of matter itself was up for grabs. By exploring the language of experimentation and probabilism in colonial writings, we can see how larger epistemic shifts entangled with and were coproduced by attempts to intervene in the world through the advancement of empire.

This chapter explores the experimental epistemology that colonial projectors in England promoted. The epistemology explored here was not developed on location in overseas colonies, but in the London region in global collections and gardens. Experimentalism supported sometimes improbable extrapolations from limited trials to large-scale applications in far different settings. In this movement of scale from a small, often domestic trial to widespread application, promoters of colonies staked a claim to their participation in larger national interests around the world while advancing their own private interests.

Promoters of overseas colonies drew upon the literature of reason of state in developing political languages and a body of economic ideas that connected individual enterprise to national standing in a global economy.[11] They utilized comparative analyses of global resources and national characters to analyze areas of strength and weakness and to recommend opportunities for intervention and transformation. They often suggested transformation through technological or economic means, such as overriding natural constraints and transplanting art, nature, and labor. Promoters also saw their task as one of cultural transformation, as they believed that Englishmen needed to be prodded toward greater individual risk-taking and enterprise. Tracts and sermons prodded the public to adventure, to experiment, and to invest in probable projects, without waiting for certain evidence of success. The public global collection and garden of the Tradescants in Lambeth concentrated parts of nature and forms of art from around the globe, in ways that made them accessible to interloping intellectual practices that connected various phenomena in projects. The strangeness, wonder, and heterogeneity of the collection supported the cultural work of period tracts and sermons exhorting the public to push past assumed boundaries on knowledge and power.

In this chapter, we will explore works promoting colonization that encouraged conjecture, risk-taking, extrapolation, swift action, and probabilistic experimentalism. We will then visit two private estates outside London whose owners followed through on these imperial visions by developing their properties into experimental sites for knowledge capture, transplantation, and empire: Sir John Danvers's in Chelsea and Sir Nicholas Crispe's in Hammersmith. Although they ultimately chose to support opposite sides in the Civil War, Danvers and Crispe shared many similarities. Both of them had traveled on the Continent, transplanting what they learned there, such as garden design and brickmaking, first to their own estates in England, and then further afield in global colonial ventures. They were both knighted. They were both elected to Parliament (although Crispe was almost immediately rejected as a monopolist) and both pursued similar ventures, such as alum production, brickmaking, colonialism, and enforced and enslaved labor. By bringing together disparate endeavors, they maximized advantage both for themselves and for the national interest, they argued. They situated themselves as hopeful adventurers, willing to engage in risky investments and hazardous projects through their own investments and actions, and thus to prod their countrymen into greater ambitions, activity, and industry on behalf of the nation.

Opportunity, Risk, and Reason of State

Kupperman notes that colonial failure was often blamed upon the poor morals of the colonists, generally their sloth and desire for easy living.[12] Yet, one prominent criticism launched at both colonists and domestic doubters of Virginia was not just laziness, but more specifically, "diffidence," or a lack of courage. Colonial propagandists held out a model of courageous risk-taking, berating Englishmen for their lack of enterprise and projecting attitudes. The merchant Robert Johnson, in one of the many works he penned promoting Virginian colonization, accused the industry of the English for falling far short of the Dutch. Johnson, who was "deputy governor of the Bermuda Company, a director of the Levant and East India Companies, and treasurer of the Virginia and East India Companies" translated Botero's *Universal Relations* in ways that promoted several overseas projects.[13] In his own 1609 tract promoting Virginia, Johnson used the political language of the state, whose wealth and therefore strength was dependent upon each individual subject taking entrepreneurial initiatives in order to better his own personal estate and thus the state as a whole. Johnson shamed his fellow countrymen, calling on "each well minded man" to "lend his helpe to heale and cure such staines and scarres in the face of our state" that "may very well make us blush."[14] Alacrity in the pursuit of industry meant holding out hope for future discoveries, despite all odds and many experiences of failure.

For example, wrote Johnson, in "searching the land, there is undoubted hope of finding Cochinell, the plant of rich Indico, Graineberries, Beauer Hydes, Pearles, rich Treasure, and the South sea, leading to China, with many other benefits which our day-light will discouer."[15] This list traces a continuum from the most likely discoveries, based on prior experience, to those that were the most ambitious, conjectural, and still unknown. The dyes and hides with which Johnson began were already known to exist elsewhere in the Americas, and thus further search might well uncover more. With "rich Treasure" we are beginning to enter the realm of less likely probability, further magnified by Johnson's suggestion that the long-sought northwest passage to China might be found, or further, not even specifiable "benefits." Yet, despite the wide continuum of knowledge that this list traverses, from relatively well-known cochineal to still unknown advantages, Johnson painted them all as equally full of "undoubted hope." By this he does not necessarily mean that success in realizing each of these aims was undoubted, but that a general practice of "searching the land" for advantage would lead to something, even if these particular aims might fail. They were thus all full of hope.

In addition to his contrast between the Englishmen and the Dutch, Johnson also unfavorably contrasted his countrymen with the bravery and magnanimity of Spaniards. When Christopher Columbus first approached various courts, including England's, due to "his poore apparell and simple lookes, and for the noueltie of his proposition," he "was of most men accounted a vaine foole, and vtterly reiected: saue that the Spanish better conceiuing then some others, beganne to entertaine and make vse of his skill." This "highly commendeth the wisedome of Spaine: whose quicke apprehension and spéedy addresse, preuented all other Princes." Just as courtiers would urge (as discussed in the previous chapter), that the English had to pursue Sherley and Steele's projects of a Persian silk trade immediately in order to prevent the Dutch from seizing the advantage, here Johnson argued that the speed with which the Spanish crown acted in response to Columbus's proposal allowed it to preempt all other powers. This speedy, risk-taking action at an opportune moment led to Spain's lasting domination in large parts of the world. The result of this wise investment could be seen more than a century later. The Spaniards' "greatnes of minde arising together with their money and meanes" had made them the lords of Christendom.[16] By contrast, Johnson criticized "the blind diffidence of our English natures, which laugh to scorne the name of Virginia, and all other new projects, bee they never so probable, and will not beleeue till wee see the effects."[17] Taking action at an opportune moment, without waiting for evidence of success, was worth the risk.

That same year identical language appeared in a sermon dedicated to the Virginia Company and attributed to Robert Gray, rector of St. Bennet Sherehog. Gray's sermon joined a slew of sermons in 1609–1610 promoting colonization in Virginia. They are often discussed in the context of religious justifications for colonization.[18] Here I point out how Gray and other preachers also invoked the language of projecting and the reason of state. Gray further elaborated upon England's historic refusal of Columbus's proposal, offering a political, statist analysis for why the structural condition of a state might engender differing attitudes toward risk. The English especially scorned Columbus as an "idle Nouellist." Although some "thinke it was because of his poore apparell, and simple lookes," the real reason lay in "the improuidency & imprudence of our

Nation, which hath alwayes bred such diffidences in vs, that we conceit no new report, bee it never so likely, nor beleeue any thing be it never so probable, before we see the effects."[19] At the time, according to Gray, the country was not as "ouercharged with swarmes of people" as it would later be, and thus "neither [was] it necessarie for anie man to beléeve reports, though probable, nor to follow strange proiects be they neuer so likely, so long as he hath home inbred hopes to relie vpon, and assured certainties to satisfie his future expectation."[20] The lack of demographic pressure at the time was not pushing Englishmen toward risk-taking, and they thus indulged in proven courses of action.

The ability that Gray showcases to identify the disadvantages in seemingly benign phenomena, and conversely, the benefits to be found in seemingly dire circumstances, was a classic example of the nonintuitive political cunning of the time. For example, one Jacobean project for exporting idle children to work in colonies in New England blamed "the happiness of our universall peace" for increasing the population, creating a surfeit of children who "runne idly upp and downe, in breaking of hedges, gathering of woolle, pelting of shepe, and in harvest tyme gleaning and filtching of corne."[21] Thus, in Gray's case, the seemingly happy demographic situation of England in the fifteenth century led to it not pursuing unlikely projects, to its lasting regret over a century later. Relying upon "assured certainties" was hardly desirable. Since they had not been pushed by necessity, Englishmen failed to reap the advantage of being the first European invaders of the Americas, according to Gray. Their complacent sense of well-being, however, was a mirage; relinquishing such advantage only deferred the moment when England would face other existential crises posed by the enemy. Continual seizure of advantage, even in seemingly peaceful and happy times, was always a necessity.

Opportunities appeared at critical moments. Other preachers promoting Virginia in 1609 framed colonization as a matter of political "opportunitie" similar to the moment when Columbus approached England with his offer.[22] They exhorted their audiences not to be like the Englishmen of old who had declined Columbus's proposal. William Crashaw framed Virginia as a means to "inrich our nation, strengthen our navie, fortifie our kingdome, and be less beholding to other nations for their commodities" as well as to "wippe off the staine that stickes upon our nation since (either for idleness or some other base feares, or foolish conceits) we refused the offer of the west Indies, made unto us by that famous *Christopher Columbus.*"[23] Daniel Price suggested that "the dull and unworthy Sceptickes" who had dissuaded Henry the Seventh from

supporting Columbus had been reborn as the "scandalous and slanderous De-
tractors" of Virginian colonization."[24]

The period's reason-of-state thinking found in projects undermined notions
of stable well-being inherent to the commonweal. There would never be a time
when all would be right in the world. Certainty could not in fact be relied upon.
Opportunities needed to be pursued quickly, at the right moment, even before
conclusive evidence was available. Constant innovation, risk-taking, and augmentation of power was necessary simply to keep up in the context of international rivalry. The Virginia Company warned that "other politique Princes and
States . . . for their proper vtility, deuise all courses to grinde our merchants . . .
and to draw from vs all marrow of gaine by their inquisitiue inuentions."[25] If
England would not colonize Virginia, other nations stood at the ready to do so,
just as England once lost out to Spain on the opportunity of Columbus's offer.
"Let no man therefore be ouer wise," the company warned, reminding readers
that "Henry the seuenth by too much ouer-warines, lost the riches of the golden
Indies." When an opportunity presented itself, it had to be pursued quickly;
"Occasion is pretious, but when it is occasion."[26]

Trial Gardens: Transplanting Nature and People

Publications sent to colonists by the company encouraged a risk-taking attitude,
urging them to doubt accounts that claimed certainty and that offered general
rules, and to undertake their own experiments, adapting knowledge to new and
particular circumstances. For instance, in 1620 a guide to sericulture by the royal
keeper of silkworms, Jean Bonoeil, was translated and printed by the Company
and distributed in hundreds of copies to colonists in Virginia. In this work,
Bonoeil related a highly detailed process, "commended by some Authors," for
generating silkworms from a rotten calf that had been fed on mulberry leaves.[27]
Kupperman cites the process as an example of how "the expertise on which investors relied could be problematic."[28] According to her account, it was the
ignorance of the bogus nature of this process that misled London-based investors to promote absurd projects like colonial sericulture. However, the epistemic
context of Boneoil's process is more complex than a dichotomy between mythical processes and real experience would suggest.

Boneoil's process relates rather to the changing nature of epistemic genres.
An ancient recipe format, enshrined in a voice of authoritative command and
not designed to highlight probability, was evolving into new textual forms that
specified ignorance.[29] The process of strangling the calf that Bonoeil related was

an example of a merely conjectural recipe, and it would have been recognized as such at the time. That it was doubtful was obvious to Bonoeil, as he went on to offer a reason to give it some credence, as well as a risk-benefit analysis that made its trial worthwhile. Since "wee see daily many creatures come of putri-faction: this is no improbable thing, and therefore is worth the triall, to save the labor and danger of sending Silk-worme seed by Sea."[30] In other words, Bonoeil did not claim this process as reliable expertise, but rather as worth a trial. Many improbable ventures were tested in trials for projects, as we have previously seen. Here, however, rather than a public trial for a project undertaken by an assigned committee or a professional assay-master, Bonoeil suggests that individuals en-gage in a domestic culture of seemingly unlikely trials as a means of advancing the Virginian enterprise as a whole.

The domestic pleasure garden was one site where we find cultures of trial that pushed the natural bounds of possibility in service to empire, in part through the manipulation of climate. Numerous works, from Cook's *Matter of Exchange* to Drayton's *Nature's Government* have connected gardens and the global botanical searches to imperial "improvement," control, and the circulation of knowledge.[31] Peter Harrison has described how the pleasure garden "represented on a small symbolic scale" what "widespread agriculture and colonisation in-stantiated at a more grand and literal level."[32] Here I point out that the plea-sure garden was not intended to be merely symbolic, but rather to serve as an experimental site where knowledge could be developed for extrapolation to far distant locations on far greater scales. How did such confidence arise in this presumed scalability from the domestic garden to the colony?

These experimental sites were precisely the sorts of ambitious, globally ori-ented gardens that the Neostoic philosopher Justus Lipsius criticized in his *On Constancy*, also set in a garden. Lipsius's walled retreat was a garden space de-signed for the cultivation of apathy rather than emotion. There, he influentially criticized "that sect . . . who . . . hunt after strange herbs and flowers, which hav-ing got, they preserve and cherish more carefully than any mother does her child." Gardens should be "for solace and not for sloth." They were a place of "quietness, withdrawing from the world, meditation, reading, writing."[33] Sir Thomas Tempest (1594–after 1652) drew verbatim from Lipsius when he accused "Flower-mongers," of making gardens into "sepulchres of Sloath and Idlenesse" when they "soe greedily and ambitiously hunt after strange and exotic Plants and Flowers, and no lesse anxiously nourish and cherish them, then a most tender Mother would her dearest Child."[34] The gardens explored here represent just those gardens of strange and curious plants that Lipsius and Tempest criticized.

Greed and ambition, emotions anathema to the Neostoic gardener, fueled garden experiments and their deployment in global projects.

The early modern garden has recently begun to receive the attention as an experimental space accorded to the laboratory.[35] The period plant collector paid huge sums to source curiosities from around the globe, to grow them in new environments through an intensive investment in resources, and to further attempt to transform their size, shape, color, and smell through breeding and grafting. Contemporary horticultural techniques acclimatizing global plants and transforming them showcased an ability to transgress seemingly any climate or particular natural state. Curious grafting techniques, linking powers and virtues across different plant species and even between plants, animals, and inanimate objects, were a prominent feature of contemporary books of secrets and horticultural treatises.[36]

Just as small-scale courtly curiosities and limited trials were quickly spun out to large-scale utilitarian projects, curious forms of horticulture and experimentation with exotic plants were seen as a way of testing the boundaries of possibility that might later underwrite utilitarian colonial projects on a grand scale. Interlopers were eager to deploy the seeming malleability of nature that gardening demonstrated. Thinking across scales was crucial here in several ways, both in theorizing how global nature might be transformed by transplanting it across vast distances to metropolitan London gardens, and how the knowledge gained from such transformations might be again further deployed.

Political thinking supported this extrapolation from the minute flowerbed to global scales. In his contribution to Samuel Hartlib's *Legacie of Husbandry* in the Interregnum, the colonial entrepreneur Robert Child drew upon the "Discourse on England" found in the late sixteenth-century *Political Treasury* discussed in chapter 1 (a work Child referred to as the "French author" since he read it in that translation) for a retrospective overview of how much England's economy had improved since Elizabethan times. Child believed that England's standing in world trade had begun improving "about 50 years ago," that is, around 1600, "about which time Ingenuities first began to flourish in England; the Art of Gardening began to creepe into England."[37] He held up as models the gardener John Tradescant the Younger, "who dayly raiseth new and curious things," and the apothecary and gardener John Parkinson, who taught the most sophisticated garden arts. Tradescant and Parkinson showed how everyone, through the cultivation of ingenuities and curiosities, might influence the nation's balance of trade.[38]

This seems an odd, almost irrational argument. How could the cultivation of garden curiosities influence England's standing in the global economy? Child praised the intensive attention that gardeners bestowed upon wondrous horticultural curiosities, lavishing care upon individual pots of horticultural stars in elite settings using highly specialized tools, costly artificial environments, and large amounts of labor. The specific arts such gardeners cultivated overrode natural constraints, acclimating plants to new environments and transforming their size, color, shape, smell, and taste. The natural magic of the garden suggested a still poorly understood frontier of human ability to extend the boundaries of possibility. Botanical writers such as John Parkinson often described the ways plants could be improved through transplantation away from their natural habitats.[39] The idea that nature could be improved through transplantation became a veritable maxim of what was known as "garden philosophy" in the mid-seventeenth century, as ways to civilize so-called savage plants, in a context rife with projects aiming to "civilize" human populations through transplantation.[40]

If gardening experiments in overcoming climate could be further extended, then the curious arts of pleasure gardening might enable radical shifts in colonial possibility, increasing the ability to transplant nature, art, and people around the globe. The precise relationship between a small trial or study in an elite pleasure garden and its later application on a grand scale in a vastly different colonial setting was often hazy, more suggested than defined or followed through. Thus, this is not a story about more precise forms of knowledge accumulated through colonial experiences and the transatlantic circulation of knowledge. It is a story about how colonial itineraries supported the suppression of knowledge and the maintenance of knowledge in states of ignorance, imprecision, and mere (if powerful) suggestion.[41]

According to popular versions of the reason of state, value for the state could be extracted from the act of transplantation, of both people and parts of nature, as a way to shift natural parameters through industry. As Giovanni Botero had written, just "as plants cannot grow and multiply as well in the nursery . . . as in the open ground where they are transplanted," so too, "human beings do not propagate as successfully when enclosed within the walls of the city where they are born, as they would in different places to which they are sent."[42] Contemporary policy took this relationship between human and vegetal transplantation very seriously as a justification for the sometimes enforced movement of populations.

Colonial labor projects offered a perfect opportunity to pair complementary phenomena for maximum advantage by transplanting England's overpopulation to what colonial promoters considered underpopulated lands. Robert Johnson described how the swarms of people in England endangered the vitality of the entire plant, "for it fares with populous common weales, as with plants and trees that bee too frolicke, which not able to sustaine and feede their multitude of branches, do admit an engrafting of their buds and Siences into some other soile, acounting it a benefite for preseruation of their kind, and a disburdening their stocke of those superfluous twigs that sucke away their nourishment." Transplanting excess population was thus "no new thing, but most profitable for our State."[43]

Forms of violence that were enacted upon plants, such as uprooting, grafting, pruning, and forcing blooms and fruitfulness out of season, legitimated the enforced interventions into human populations, such as removal and resettlement, as beneficial to the survival and productivity of the entire plant, or commonweal. Sir John Davies, Ireland's attorney general, offering justifications for the English resettling of the recalcitrant Irish, argued in 1610 that in enacting such a policy, King James would imitate "the skilful husbandman, who doth remove his fruit trees not with a purpose to extirpate and destroy them, but that they may bring better and sweeter fruit after the transplantation."[44] As a pamphlet promoting the Earl of Berkshire's colony on the Amazon claimed, wild nutmegs had been found there "which no doubt may in short time bee brought to a more fuller perfection, by either cropping the old trees and dreaning the waters from them, or by transplanting the young trees, as by experience is commonly seene, how that nature is much helped by art and industry."[45] It was not only trees that could improve by transplantation, according to the pamphlet. People could too. The author of the pamphlet wrote that "want of true knowledge of a plantation . . . causeth such to live heere like plants, which many times prove lesse fruitfull in their naturall soile, then when they are removed to places better liking them." He was thus providing such languishing human shoots with "the knowledge of a most hopefull plantation newly undertaken by the right honorable the Earle of *Barkeshire*."[46] This metaphor worked both ways, with plants described as various types of colonizers and colonizers as different sorts of plants.[47] Nor is it obvious that this relationship between the two was merely metaphorical, as porous early modern notions of the relationship of the human body to its environment suggested ways that transforming the diet, climate, and even aesthetics of a locale might affect its human inhabitants.[48]

In early Stuart England, opinion appeared to shift from Botero's suggestion that people improved when transplanted from the nursery to the field, to the idea that open spaces allowed for unhealthy, luxuriant growth while intense cultivation might better civilize humans through the concentrated investment of

industry. Such belief in the civilizing effects of intense garden cultivation perhaps explains why, when Sir Thomas Dale, deputy governor of Virginia, granted
every settler in Jamestown three-acre plots of land to cultivate in 1614, these
grants were called gardens rather than farms, as was the "Common Garden"
Dale set up to yield revenues for the Virginia Company. The treasurer of the
company, Sir Edwin Sandys, praised Dale's efforts in reclaiming "almost mirac-
ulously those idle and disordered people" and reducing them "to labour and an
honest fashion of life," in part through work on the "Garden."[49] By the end of
Dale's deputy governorship, this "Garden" was yielding an annual profit of 300
pounds, with fifty-four "Servants" working in it, although two years later, it had
evaporated, with only "six Goates" left.[50]

Showcases of Possibility: Curiosity Collecting, Horticultural Ingenuity, and Conquest

Global gardening was deeply intermeshed with networks of exploration and conquest that transplanted resources to England, studied and transformed them,
and transplanted them outwards again. Showcase gardens of horticultural
oddities, as well as the curiosity collections that were often associated with them,
thus entangled colonial trade with the ways that curiosity collecting in general
questioned received categories and systems of nature. Collections of monsters
and other strange phenomena that appeared to transgress natural divides spurred
the doubting of an inherited body of natural knowledge and encouraged new
experimentation.[51] In England the very beginnings of curiosity collecting in the
holdings of Sir Walter Cope, John Tradescant Sr., George Villiers, and Job Best
connected tightly to global trade and colonial settlement.

Interlopers made a habit of surveying the world, or, as Johnson put it, "searching the land." They sought everywhere forms of knowledge, parts of nature,
and human populations that could be taken from elsewhere and worked up into
a project. Curiosity collections and global gardens offered a condensed store-
house of the world's art and nature for interlopers to survey.

John Tradescant Sr. began his career as a plant collector in 1610 for Robert
Cecil, Earl of Salisbury, and for Sir Walter Cope.[52] He also collected plants while
traveling on other missions, joining, for example, Sir Dudley Digges's 1618 diplomatic mission to Russia and volunteering to serve on the anti-pirate Algiers expedition of 1620–21, where he collected two sorts of "Barbary" apricots.[53] His son
went even further afield, traveling several times to Virginia and Bermuda. The
two of them set their ground-breaking Ark, the first public collection in England,
within their global garden in Lambeth. Their collection was especially rich in the

plants, animals, weapons, habits, and works of art of Virginia, the placename that occurs by far the most often in the catalog of the Tradescants' collection.[54]

Tradescant Sr.'s turn from plants toward the collection of all sorts of curiosities was intertwined with his service to Villiers. In a 1625 letter written from Villiers's property, New Hall, Tradescant wrote to the Secretary of the Navy, asking him to negotiate on behalf of the duke with

> All Marchants from All Places but Espetially the Virgine & Bermewde & new-found Land Men that when they [go] into those Parts that they will tak Care to furnishe His Grace Withe all maner of Beasts & fowells & Birds Alyve or If Not Withe Heads Horns Beaks Clawes Skins Gethers Slipes or Seeds Plants Trees or Shrubs Also from Gine or Binne or Senego [Upper and Lower Guinea, Benin, and Senegal] Turkey Espetially to Sir Thomas Rowe Who is Leger At Constantinoble Also to Captain Northe to the New Plantation towards the Amasonians With All thes fore Resyted Rarityes & Also from the East Indes Withe Shells Stones Bones Egge-Shells With what Cannot Com Alive My Lord having heard of the Dewke of Sheveres & Partlie seene of His Strang Fowlls Also from hollond of Storks A payre or two of yong ons Withe Divers kinds of Ruffes."[55]

Captain Roger North was a creature of Villiers's who had been granted permission by James to found a new colony on the Amazon in 1619; when the Spanish objected and North wound up in the Tower, it was Villiers who secured his release.[56]

Tradescant wrote another undated letter addressed to the "Merchants of the Ginne Company & the Gouldcost," specifically "Mr Humfrie Slainy Captain [Nicholas] Crispe & Mr Clobery & Mr John Wood cape marchant" with a list of thirteen desired categories of objects peculiar to Africa. This list of desired objects from Guinea was apparently intended for the collection of Charles I. The Guinea Company monopolizing trade to Guinea, Benin, and Senegal, had been patented in 1618 but by the second half of the 1620s was being controlled by a group of associates, Slany, Cloberry, Crispe, and Wood, who were also trading privately (and illegally) in the region. In 1631 Charles I would grant this group of merchants, along with Sir Kenelm Digby, a new exclusive patent.[57] One of the main criteria for selection on Tradescant's list was strangeness.

> All ther strang sorts of fowelles & Birds Skines and Beakes Leggs & phetheres that be Rare or Not known to us / of All sorts of str[a]ng[e] fishes skinnes . . . Great flying fishes & sucking fishes withe what Else strang . . . Of All sorts of Shining Stones or of Any Strange Shapes / Any thing that Is strang.[58]

Tradescant also included on this "habits weapons & instruments" and especially "ther Ivory Long fluts." His own collection would include three types of
"Knives from Ginny," "Bracelets from Guiny," "a Ginny Lanthorn," "Ginny
drinking cups made of birch," "Plates made of Rushes from Ginny" as well as
evidence of both the slave trade and of privateering in the form of an "Iron
Manacle taken in the Spanish-Fleet."[59] Tradescant listed Sir Kenelm Digby
among his donors, along with "Captain Wood," John Slany, secretary of the
Newfoundland Company and brother of Humphrey Slany, and "Mr. Cleborne,"
possibly William Claiborne, a business associate of William Cloberry.[60]

After Villiers's assassination in 1628, Tradescant Sr. settled in Lambeth on a
property belonging to the manor of Vauxhall and began developing the global
garden that would come to number about 1,800 varieties and would later also
house his collection of curiosities, the Ark.[61] Less than a mile away from Tradescant's garden, the riverside property of Vauxhall was acquired by Charles I in
1629 as a center of invention (as discussed in chapter 7). Vauxhall would also
come to hold a significant (if private) research collection of models, while Tradescant's Ark would hold objects donated by inventors working at Vauxhall.
Charles I also hired Tradescant, appointing him Keeper of the Gardens, Vines
and Silkworms at Oatlands Palace in 1630.[62] There he was responsible not only
for the gardens but for the continuing attempts to raise silkworms in the "silkworm house" built by Queen Anne in 1616.[63] Charles continued to employ Tradescant as an agent in global collecting. A 1633 letter from the secretary of state
to the court of the East India Company requested the company to deliver a list
of "varieties" "to John Tradescant to be reserved by him for His Majesties
Service."[64]

Charles I and Villiers appeared high on the list of donors to Tradescant's collection. According to a 1638 visitor to the collections, Villiers had given Tradescant a "beautiful present" of "gold and diamonds affixed to a feather by which
the four elements were signified."[65] Tradescant also continued the strategy developed under Charles and Villiers of collecting from merchants and ship captains. Although his catalog was dedicated to the College of Physicians, it listed
just five physicians among its donors, in comparison with eleven ship captains.
Tradescant's donors included "William Curteene [Courten] Esq.," a major financier of global expeditions and of the Crown, whose royal-backed trading
venture to Asia and Africa was discussed in chapter 4. It also included Courten's
young associates, "Mr. James Boovy [Boevey (1622–1696)]" and "Mr. [Jacob]
Pergins [Pergens (?–1681)]," as well as Courten's father-in-law "Mr. [Moses]
Trion"[66]; Captains John Weddell and Richard Swanley, respectively the com-

mander and vice admiral of the Courten trading fleet; and Captain Adam Denton, factor of the English East India Company in Pattani (in modern Thailand).[67]

Gardening, curiosity collecting, and global trade went hand-in-hand. The merchant Peter Mundy, newly arrived home from India, visited Tradescant's collection in 1634. There he spent the whole day perusing

> beasts, fowle, fishes, serpents, wormes (reall, although dead and dryed), pretious stones and other Armes, Coines, shells, fether, etts. of sundrey Nations, Countries, forme, Coullours; also diverse Curiosities in Carvinge, painteinge, etts., as 80 faces carved on a Cherry stone, Pictures to bee seene by a Celinder which otherwise appeare like confused blotts, Medalls of Sondrey sorts, etts. Moreover, a little garden with divers outlandish herbes and flowers, whereof some that I had not seen elsewhere but in India, being supplyd by Noblemen, Gentlemen, Sea Commaunders, etts. with such Toyes as they could bringe or procure from other parts. Soe that I am almost perswaded a Man might in one daye behold collected into one place more Curiosities then he should see if hee spent all his life in Travell.[68]

Munday also stopped by the collection of "Mr Job Best," which was "well stored with the like."[69] In building this collection, Mr. Best could draw on help from many relatives abroad; his father, Josias Best (1596–?), was an East Indian sea captain, his brother Nathaniel was (an at first captured and later willing) resident in Aden, and his son Thomas lived in Persia. Best himself had served from the early 1630s as clerk of Trinity House, which was responsible for overseeing many maritime affairs. He resigned in 1641, however, to fully commit to his new career as florist and experimentalist, with many of his experiments documented by Samuel Hartlib. In a letter to Hartlib, Best describes himself as someone who "long strove after . . . the mistry of flowers."[70]

Like Job Best, the merchant Alexander Marshall, a friend to John Tradescant Jr. and resident in his home in 1641, abandoned his trade to become a gardener, florist, artist, experimentalist, and collector, specializing in living insects. Early modern collections included not only the arts and costumes of "sundrey Nations," as Peter Mundy observed at Tradescant's collection, but also humans and human remains. A visitor to Marshall's collection in 1653 observed "a thousand sorts of Insects and a head of an Indian-boy which still shewed as if it were alive."[71]

Such practices further intertwined the transplantation of curiosities and people. In 1597 "Charles, a boy, by estimacon x or xii yeares old, brought by Sir

Walter Rawlie from Guiana" was christened in Chelsea Church; he was likely resident on the Chelsea estate of Sir Arthur Gorges, Raleigh's cousin and the commander of the ship on which Raleigh sailed that year, and the collaborator with the collector Sir Walter Cope on a project to found a general office of commerce at the New Exchange discussed in chapter 1.[72] In 1613 two unnamed Virginians were buried who had been resident in the house of Sir Thomas Smythe, treasurer of the Virginia Company and governor of the Somers Isles Company.[73] In 1611 Henry Peacham complained about the "rude vulgar" gathering at the royal menagerie in St. James Park to goggle at exotic animals like African guinea hens and the Javan cassowary presented to King James in 1610 by Robert Cecil.[74] In 1615–1616 the Powhatan youth Eiakintomino was added to this display.[75] In 1619 George Thorpe (the brother-in-law of Sir John Danvers, a councilor of the Virginia Company whose garden is analyzed in this chapter) employed a "virginian boy" as his amanuensis. He was "almost surely the 'Georgius Thorpe' whose baptism at St. Martin in the Fields occurred on 10 September 1619 and whose burial, seventeen days later, lists him as 'Georgius Thorp, homo Virginiae'—another American life cut short abroad."[76] The portrait after life of a twenty-three-year-old "American from Virginia" that Václav Hollar etched in Antwerp was likely sketched while the artist was working in London (figure 10).[77]

Growing Ambition: Opening Up the Vista

The gardens and collections of early Stuart London concentrated the world. Deborah Harkness studied similar sites of urban collecting and gardening during the Elizabethan period. She argued that such sites enabled amical collaborative exchange and careful empiricism.[78] By contrast, the global collections of early Stuart London justified the grasping, often violent conquest of nature, knowledge, and people. The assembly of diverse natural objects, crafts, and specimens of human ability encouraged the spinning out of global projects through associative thinking. Wonders that countered assumed categories or that seemed to accomplish the impossible emboldened viewers to spurn boundaries and limits. Gardens and associated curiosity collections were assembled to evoke and magnify emotions. The emotions with which such "strange" things were considered did not stop at wonder or an aesthetic and philosophical appreciation for the variation and unpredictability of nature. Rather, these collections were also approached with greed and ambition, as Lipsius had claimed.

Gerhard Oestreich interpreted the seventeenth-century formal garden, with its "artistically clipped trees and hedges" as yet another expression of discipline

Figure 10. Wenceslaus Hollar, *Portrait of a Man from Tsenacommacah (Virginia), America, at 23 Years of Age*, etching. Antwerp, 1645. Rijksmuseum. RP-P-OB-11.592. Public domain

and dispassionate control.[79] Here, I offer a different interpretation of the formal garden aesthetic: the opening of the garden from the enclosed, walled gardens of the Elizabethans to sweeping vistas shifted the role of the garden from a safe retreat from the turbulence of the world to projection upon a passionate theater of possibility. Rather than seeking to bend nature mechanically against itself to their will, gardeners investigated underlying forces of nature and sought to draw them out to manifest in material form. This was a view of nature infused with passions and appetites, sympathies and antipathies, and other hidden forms of natural knowledge best known to the gardener.

The abilities of Tradescant and other global gardeners of the period to nurse up and transform exotic species in hothouse environments long served as evidence for the human ability to change climate and nature. As one contributor to Samuel Hartlib's *Legacy of Husbandry* later opined, since the Deluge, all plants in use were improved from their status as weeds through human "contrivances: so John Tradeskin by Lambeth, by the advantage of putting his Trees, and other Plants into a warm house in winter or a stow, nurses up those things faire and fragrant, which would without that help either dye or be dwarft."[80] Tradescant does seem to have been interested in researching how varieties developed from wild originals: "his 300 varieties of the 50-odd main species of European cultivated flowers often included an original wild strain besides the varieties that had been developed from it."[81] Yet, horticultural knowledge, for all its excitement and achievements in the period, was hardly a formally developed discipline nor one grounded in well-established causal explanations for observed phenomena. This was a form of knowledge that drew attention to distant horizons and how much more knowledge remained to be explored.

The western village of Chelsea, full of grand estates, river views, and fresh air was at the forefront of England's "gardening revolution."[82] There, Sir John Danvers introduced the vista and Italianate garden design to England in a celebrated garden that opened in 1622. Danvers, a servant of Prince Henry's privy chamber for seven years, a member of Parliament, and a client of George Villiers, was a typical early Stuart gentleman projector, investing in forestry, urban redevelopment, alum works, brick- and tile-making, lime, and above all, colonial plantations. He was a major figure in both the Virginia and the Somers Isles (Bermuda) Companies.[83]

Danvers's neighbors were colonizers and experimentalists whose properties also boasted impressive gardens. Lionel Cranfield, Earl of Middlesex, a prominent member of the Virginia Company and patron of projectors, lived next door. On behalf of Villiers, Danvers's allies in the Virginia Company pushed in the

Commons for Cranfield's impeachment. After Cranfield's fall, Villiers acquired his Chelsea property. His widow long remained there after Villiers's assassination in 1628. In front of Villiers's estate stood the property of the sea captain Sir Arthur Gorges, who would be succeeded by his son, also Sir Arthur Gorges, one of the four chief aristocratic interlopers in New World colonization of the 1640s.[84] Between Gorges House and the river stood Lindsey House, where the royal physician Sir Theodore Mayerne resided from at least 1639.[85] Mayerne is most famous today for his deep interest in the fine arts, whose techniques he recorded in a large number of manuscripts. In fact, his interests spanned all forms of craft; Hugh Trevor-Roper has wondered why Mayerne experimented with dyes, as he "was not himself an artist or a craftsman."[86] In the early Stuart culture of interlopers, Mayerne's wide technical interests were shared by many of his similar social status, including his neighbor, Danvers, with whose family Mayerne was close.[87]

Danvers's relative, John Aubrey, FRS, described his property at length. Danvers's "very elegant and ingenious house" included a room engineered to "meliorate" the sound of an "excellent organ."[88] Guests could arrive by boat and climb a staircase to the elevated central hall.[89] Out one window in the great hall, ships sailed by on the Thames, while a garden vista stretched out the other, culminating in a grotto and a banqueting house with stained glass windows and summerhouses on either end.[90] The house and garden were innovatively designed as a unit, and throughout, views were engineered to elicit emotional responses from visitors. One entered the garden by a double flight of steps, as Aubrey tells us, walled "to hinder the imediate pleasure and totall view" at first glance. The sudden opening up of the vista would function to surprise and awe visitors used to the Elizabethan aesthetic of the *hortus conclusus*, or enclosed garden retreat. Later, a descent "into the darksome, deep Vault" of the grotto affected the visitor "with a kind of Religious horrour."[91]

In contrast to the view of the garden as a retreat from the world's turbulence and a safeguard from the passions, Danvers's garden elicited rapid-fire emotional responses through pleasurable distant vistas and horrific descents into darkness. Upon this well-designed stage, the star performers were the wonderful plants, from enormously expensive and collectible gilliflowers (carnations) and tulips in a vast array of shapes, patterns, and sizes to strange new horticultural curiosities just introduced from the Americas. Danvers was known as "one of the greatest Gilliflower Men in England having the largest and fairest Gilliflowers."[92]

Danvers's ability to source global horticultural curiosities, such as the sensitive plant, made his garden a widespread topic of discussion. Danvers's friend,

Sir Francis Kynaston, singled out his sensitive plants as one of the most "admirable" things he had ever seen in his life. Kynaston's description of this plant exemplifies the extreme attention that such horticultural curiosities garnered in the period. The seeds of these plants, wrote Kynaston, "being brought from the West Indies were set in a pot of composed earth by his excellent gardiner, John Gilbanke."[93] These seeds were likely collected by John Tradescant Jr., who voyaged to Barbados and Virginia in 1637 "to gather up all raritye of flowers, plants, shels, &c."[94] According to Kynaston, the plant could not overwinter in England; thus Gilbank's expertise and care in developing the "composed earth" for this precious plant was especially valuable. This pot of Gilbank's became an object of widespread commentary and depiction. Of the 650 plants that Alexander Marshal depicted in his *florilegium*, only the sensitive plant was in a pot.[95] John Parkinson also observed several sensitive plants, "in a pot at Chelsey in Sir John Davers [*sic*] Garden," depicting the plant along with a curious hand touching it.[96] He returned to Danvers's garden two years in a row to test the plant's sensitivity and discussed it with his friend John Morris and with Jan de Laet (1581–1649), a director of the Dutch West India Company and a friend of Sir John Danvers.[97] When the Dutch diplomat Lodewijk Huygens (1631–99) visited the garden in the 1650s, he observed the "pot where in summer the *herba mimosa* [the sensitive plant] comes up" in a mirrored summerhouse for rare plants, where there was also a gilliflower being forced into bloom in the winter.[98] In the much later guide for gardeners composed by John Evelyn, FRS, starting gilliflowers early in a hotbed was still a "precious seacret."[99] Even though the sensitive plant itself was dormant and all that Huygens could see was the pot, this was exciting enough. The technologies of hotbeds, summerhouses, special composts, and forcing flowers were the tools that gardeners deployed to showcase a mastery of seasons, climates, and distant geographies.

Francis Bacon, as Aubrey recalled, "much delighted" in Danvers's "curious pretty garden at Chelsey."[100] As even Charles Webster (who tended to focus more on useful knowledge and thus on agriculture rather than horticulture) pointed out, perhaps "more than any other subject, horticulture illustrated for Bacon the potentialities of the search for *magnalia* [that is, great deeds often presumed to be impossible] in nature."[101] The transplantation and transformation of plants was among the most desired abilities of humankind, located at the furthest edge of natural magical ability.

The transformative setting of Danvers's garden proved a meeting place for developing both domestic and global schemes. John Aubrey described how Danvers kept an eight-page printed pamphlet about mines and improvements of their entrances and ventilation by Thomas Bushell, Bacon's follower and Danvers's "acquaintance," "nayled . . . to his parlor wall at Chelseay, with some Scheme: and I beleeve is there yet," Aubrey reported much later.[102] Danvers was sought out for advice concerning patents and inventions.[103] He had an opportunity to exercise his skills in this arena also through Virginia Company deliberations. For example, a defensive engine or fortification proposed by "Mr. Englebert," likely the well-known engineer William Englebert (?–1634), was one of the many projects the company considered. As was their usual practice, this was delegated to a subcommittee of individuals with presumably sufficient experience to review this subject, including in this case, Henry Briggs and Sir John Danvers. Danvers delivered his opinion that he thought Englebert "to be a fitt man and suffcient for that hee vndertaketh."[104] Although meeting locations of the Virginia Company in the 1620s seem to alternate between the more centrally located houses of Sir Thomas Smythe, Sir Edwin Sandys, and John Ferrar, subcommittees on occasion met for discussion at Danvers's house.[105] In later years, the poet George Wither (1588–1667) also recalled how when he came to visit Danvers's "house in Chelsey, I found with you many Gentlemen of the Sommer Islands [Bermuda] Company."[106] After the dissolution of the Virginia Company, Danvers repeatedly asked his old associate John Ferrar (1588–1657) to stay with him at his Chelsea estate where he was attempting to revive some of their old Virginian projects, so that "our conference may begett the best result & direction for our further labors or Endevors."[107] John Donne, who failed to win a secretaryship for the Virginia Company, but did preach an important sermon to it, was also resident in Danvers House in 1625.

Garden Improvement, Land Survey, and Sovereignty

This context can help us understand why the most ambitious of colonial projects embraced gardening as a powerful means to advance plantations.[108] Colonial promoters described the lands in Virginia as empty and waste, akin to the wild forests and common lands in England that should be enclosed and developed for maximum profit. Furthermore, rather than simply developing these lands agriculturally, such promoters advised bringing horticultural skills to bear in ways that might transform nature and increase profits exponentially through innovative projects. This was particularly the view of the faction of Sir Edwin Sandys in the Virginia Company, who allied with Sir John Danvers and John Ferrar. Ferrar spent several decades authoring works promoting Virginia and the ways that the colonies could be transformed through the skills of gardeners.[109]

For example, in a 1622 pamphlet sent to Virginia and Bermuda by the Virginia and Somers Isles Companies, Ferrar addressed colonists in the voice of "Nature herself," urging them to apply the skills of the garden to their colony.[110] "Nature" argued that she purposefully did not supply her commodities in an already perfected form to the settlers, since that would but "breed mine owne contempt, and nurse your sloth." Instead, she requested the assistance of two handmaids of "Art and Industry," and in particular, not "Husbandry at larg" but the "skill of Gardening." Through the "Arte of skilfull planting, grafting, transplanting, and remoouing, the bad wilde plants are wonderfully bettered."[111] Merely "transplanting and remoouing wild plants" bore immense power to "domesticate and inable them," so that, for example, cultivated sarsaparilla developed by Spaniards was worth twice as much as the wild plant. Likewise, the mulberries and vines native to Virginia could be improved through transplantation. "Nature" repeatedly asserted that "the Garden Art of planting" was the "best Art" to "aduance the Plantation and Planters."[112]

Ferrar deploys the category of nature here in a very strange way. Setting this command to cultivate and domesticate the garden in the mouth of Nature ran counter to more common contrasts between art and nature. This odd equation between the natural and the cultivated allowed Nature to voice illogical views about Native peoples. By identifying industry with nature (rather than as its opposite), and by claiming that Native peoples lacked industry, Nature claimed that they were therefore unnatural, and thus also lacked any sovereign rights bestowed upon them by nature. They were, she declared, "most unnaturall, and so none of mine." Since they "know no industry, no Arts, no culture, nor no good use of this blessed Country heere" they were "naturally borne slaves," and therefore, there is "a naturall kind of right in you . . . to direct them aright, to governe and to command them."[113] Here, Ferrar extended the justifications for the colonial appropriation of land based on the argument that a "natural right" existed to land that was either uninhabited, or where no claims to property and possession had been made (as discussed in chapter 1). Ferrar extended these arguments about natural law from the land to the people. By associating Nature with industry, in a very strange way according to contemporary categories, Ferrar claimed a natural right for settlers not only to claim and cultivate the land, but to enslave the people.

Other allies of the Sandys faction likewise extended arguments justifying infringement upon the liberty of indigenous peoples and others based on improvement through industry. In 1622, at the request of Sir John Danvers, John

Donne (who was a friend of Magdalen Danvers, Sir John's wife), preached a sermon to the company at a moment when they were facing the king's deep displeasure at the poor progress of their projects, and at a moment of existential crisis for the colony following the Powhatan uprising. Donne urged colonial promoters not to give up in the face of unrealized projects: "Bee not you discouraged, if the Promises which you have made to your selves, or to others, be not so soon discharg'd . . . Great Creatures ly long in the wombe."[114] The company presented Donne with a gilt-bound copy of their 1620 *A Declaration of the state of the colony and affairs in Virginia*, which claimed a flourishing state for the colony and advertised hopes for sericulture in particular.[115] In turn, the company purchased six copies of Donne's sermon, likely to send on to the colony.[116]

Some modern interpreters of Donne's sermon have differentiated his message, which stressed efforts to educate and convert the Powhatans, from the more nakedly genocidal rhetoric of many of his peers. Some have even read in it a message of love for common humanity because Donne defended the company's colonialism as a service to the commonweal, which he extrapolated to international and global terms, and thus in terms that would include the Powhatan population. Donne drew on arguments that were at the time deployed to support projects of enclosure of domestic common lands. Promoters of enclosure considered these areas wastelands because they were not, they claimed, fulfilling their potential in serving the commonweal. Inhabitants of these lands in fact drew many resources from them, as Eric Ash has discussed in the case of the Fens, but in ways that were ignored in projects promoting surveying, enclosure, and improvement.[117]

Donne extrapolated from the debates on enclosing domestic common lands to debates over indigenous sovereignty and colonial possession. Donne argued that it was a "Law of Nature and Nations" that any land left "derelicted . . . becomes theirs that will possess it." But beyond this, if inhabitants "do not in some measure fill the land, so as the land may bring forth her increase for the use of men," then the land would also be considered available for the taking. The landscape needed to be not just populated, but filled and developed in order for the people to retain their sovereignty. It would become available for the taking for anyone who could maximize the land for use. He extrapolated from the interest of particular states in encouraging individual improvement of private property for the advancement of the state to the interest of all the world to improve all parts of the globe for the advancement of humankind. Just as "in particular States, *Interest reipublicae ut quis re sua bene utatur,* The State must take order, that every man improoue that which he hath, for the best advantage of that State," so too, Donne reasoned, "*Interest mundo,* The whole world, all Mankinde must take care that all places be emproved, as farre as may be, to the best advantage of Mankind in generall."[118]

This has been seen as a remarkably benevolent response in the wake of the Powhatan uprising, by extending the perspective of a national interest to embrace the interest of all humanity, and thus including the Powhatans within a framework of international law. This interpretation does not recognize the ways that Donne here in fact arrogated far more powers for colonial intervention and further embedded power differentials within concepts of improvement, progress, and civilization itself, than was usual. As Christopher Tomlinson pointed out, the "slippage" from previous arguments that it was legal to occupy uninhabited land to the idea that it was legal to occupy inhabited land that wasn't being fully used "was highly expedient."[119]

The first legal maxim that Donne cited ("*Interest reipublicae ut quis re sua bene utatur*") not only could be interpreted in a far more limited fashion, but it was also often critiqued in its domestic English context as a damnable, unchristian practice of improvement through which landlords squeezed every possible profit from their land and its tenants. Just as claims to serving the public interest could be and were read skeptically in the period, so too were legal justifications for improvement. Improvement often meant the increase of profit for landlords through the enclosure of common lands and the racking of rents. Preachers fulminated against these practices from the pulpit.[120] Critiques of improvement often invoked the obverse legal maxim from the one Donne had quoted: *Interest reipublicae, nequis re sua male utatur*, it is in the interest of the state that nobody do harm with his property. Godfrey Goodman cited this principle when criticizing the way landowners "haue so much improued their estates, ra'ckt their poore tenants" through enclosures that they impoverished the kingdom.[121] It was such practices that, rather than developing the land, rendered it waste and its inhabitants shiftless.

Rather than augmenting the powers of the state through the advancement of each individual's property, improvement in fact depopulated and weakened the kingdom, argued Robert Powell, a lawyer from Wells. Enclosure had the power, Powell alleged, to transform "the quality of the people; from good Husbands, it makes them houseless and thriftless, puts them in a course of idelnesse . . . So as they become aliens and strangers to their nationall government, and the kingdome by that meanes in a manner dispeopled and desolated."[122] This debate over improvement was cast as dueling legal principles. With a high hand, "The Devill Covetuousness," pressed two questions, "Is it not lawfull for mee to doe what I list with my owne, to pull or let down my owne houses, or alter the property of my owne soyle, *&c?*" and "Shall I not buy and purchase as much as I am able to pay for; And shall I not make the most of my owne?" The answer to both was "*Interest reipublicae, nequis re sua male utatur*; The interest of the Common weale hath such a power over the actions and estates of men, that no man must abuse or mis-imploy the talent of his meanes."[123] A fortiori, nobody could, through the use of his property, harm "All, God and man, of men all sorts, from the highest to the lowest." Those who sought "to improve his private, and impaire the publike good" contravened this dictate and lacked Christian charity.[124]

A royal chaplain, William Loe, in a sermon defending royal projecting, railed against improvers on private estates. "How dare those Measurers, and improvers of their land grind the faces of the poore toyling, sweating, laborious husbandman with rackings, and raisings of rents?" Pairing the notion of the rack as a torture device with the idea of "racking up rent," Loe declared that "A Jesuiticall spirit first deuized these improvements." Loe compared efforts to squeeze the utmost profit out of the land to devilishly clever Jesuit attempts to stretch the pious upon the rack and to turn the screw as much as possible. He described the way that "improvers" studied their land by making miniature scale models of it to survey for all possible ways to extract more income from it: "Make a modell (say they) of all your land to a Mole-hill, that so you may lie in your bed, & see in a view every field, closse, grove, meadow, acre, and head land in your Farme, whereby you may set it to the utmost advantage; For it is lawfull for you to make the most of your owne."[125] This, declared Loe, was in fact an example of an illegal project as it did not agree with the principles of Christian charity.[126]

Loe refers to practices such as those discussed by the surveyor John Norden, an advocate of the enclosure of lands and social engineering of its inhabitants. Norden's *Surveyor's Dialogue* began by addressing the criticism of his profession as one through which men often "lose their land: and sometimes they are abridged of such liberties as they haue long vsed in Mannors; and customes are altred, broken, and sometimes peruerted or taken away by your meanes: And aboue all, you looke into the values of mens lands, whereby the Lords of Mannors do rack their tenants to a higher rent and rate then euer before."[127] A farmer, the critic of the surveyor, characterized the minute study of the land as serving violent interventions into the lives of many.[128] Surveyors were the "the cords whereby poore men are drawne into seruitude and slauery."[129] The surveyor retorted that by letting his lands decline in value through a lack of a survey, the landlord "offendeth God, deceyueth the King, and defraudeth the Common-wealth" and also proved himself "unworthie to oversee matters of State and Common-wealth."[130]

Thus, in the debate over improvement, at stake was not just whether or not improvers had realized their promises to civilize so-called uncivilized lands.[131] Rather, at stake was what improvers really meant when they spoke of use, advantage, or civilization. As Elly Robson has argued, today we should view skeptically the claims that improvers were making to the social benefits of their proposals. Supposedly empty lands were in fact inhabited by individuals who considered the landscape according to epistemologies starkly opposed to those of the improvers.[132] This was true of inhabitants of customarily common lands in England, and all the more true for Algonquins in Virginia. In both cases, the land was described as unused and the population therefore as removable according to the same period arguments for improvement.

In his sermon before the Virginia Company, Donne drew upon heated period debates related to surveying, enclosure, and improvement in England that entailed conflicting understandings of freedom, bondage, sovereignty, and the nature of the good life. By extrapolating this debate to a global scale, he further raised the stakes. According to Donne, it was in the interest not just of the kingdom, but of the entire world that all lands be improved to their maximum extent. Donne thus further amplified the moral high ground that the Virginia Company attempted to secure, and the company enthusiastically applauded his message, even though it entailed, like Ferrar's argument concerning the natural slavery of the Native inhabitants of the land, a major leap in logic that was riven with internal contradictions. Donne had simultaneously attempted to argue that only those who could develop a land to its maximum use enjoyed the right to it. At the same time, he forgave the company for not fulfilling its promises and for failing to reach its goals. By his own logic, the poor progress the company made in fulfilling its aims to develop profitable commodities and to maximize the use of the land should mean that they had lost their claim to the region. Indeed, this was the view toward which King James was tending.

The Sericulture Project

Virginian sericulture became a test of national industry and its ability to live up to the economic advice concerning the transplantation of nature and art advocated by Giovanni Botero. It was strenuously promoted by King James both for England and for Virginia.[133] It was a test that the Virginia Company failed.

The silkworm, *Bombyx mori*, is domesticated and cannot survive in the wild without tender human care. It requires specific housing, which, in England's climate, must be heated. It was believed to be sensitive to smells, loud noises, and many other disturbances. The Crown lavished resources on silk production. Queen Anne's elegant silk house at Oatlands was designed by court architect Inigo Jones, skilled individuals were recruited to serve as Keeper of the Gardens, Vines and Silkworms there, and King James repeatedly sent live silkworms thence to Virginia, although they just as constantly died en route.[134]

Colonial sericulture never proved successful, but its constant resuscitation "left a rich legacy of projecting colonial possibilities."[135] Many obstacles stood in the way of Virginian sericulture, not least the use of the wrong species of mulberry (the native red mulberry is not the preferred species of *B. mori*), the inability to keep the domesticated silkworm alive, and continual hopes that silk could be spun from the huge, gummy cocoons of the native cecropia moth (*Hyalophora cecropia*).[136] In their 1620 account of the colony, the company described Virginia's "innumerable store of Mulberrie Trees" and "some Silkwormes naturally found uppon them produceing excellent Silke some where of is to be seene."[137] Promoters of colonial sericulture argued that the existence of a native mulberry and a native silkworm in Virginia was providential; even if these were the wrong kinds, they could be transformed through industry and experimentation akin to the garden arts of the time. In this, they encouraged personal initiative and trial that differed from the manner in which King James wished colonial sericulture to be carried out.

King James I ordered the Virginia Company to see that the directions he was sending "bee carefully put in practice throwout our Plantations there, that so the worke may goe on cheerfully and receive no more interruptions nor delayes." This royal letter was printed in the 1622 sericultural work the Company sent in many copies to Virginia.[138] Also included was a letter from the "Treasurer Councell and Company of Virginia," reminding the "Governour and Councell of State in Virginia" that anyone who failed to proceed "as by the Booke is prescribed" should "by severe censures and punishment, be compelled thereunto." Yet Ferrar, in his conclusion voiced by Nature discussed above (which was appended to the same sericultural compilation), stressed individual experimentation rather than rule-following, as part of a more general shift toward probabilistic, experimental forms of knowledge.

"Nature herself" told settlers that "you must not thinke it to be absolutely necessary, to be so superstitious in curiously following all the booke rules and written precepts."[139] No rules could apply to every situation, and particularly not to the new and shifting environment of the colony: "all generall Rules euer admit some exceptions, and varie according to some particular circumstances."[140]

Colonists therefore should experiment with a wide range of variables Nature suggested, and accordingly "make your exceptions" to the prescribed rules.[141] "Nature herself" especially urged settlers to pay attention to the variability of climate, and just as gardeners tenderly nourished up their plants, make minute adjustments. Cultivators of silkworms needed to observe "the nature of the Clymate . . . how and in what one Clymate differs from another, as also the season of one yeere, altering from another in cold, heat, drought, or moysture." According to such variation, the cultivator had to adjust

> with discretion . . . the manner of the lodgings, the qualities of the windes, to be let in, and kept out . . . if it be a cold season, to vse more artificiall heat, for the cherishing the Wormes; if it be a verie hot season, to let in the coole ayre and the windes, as much as may be to refresh them. . . . If it be a moist time . . . to vse drying heats and perfumes . . . to qualifie the moist and the ill season.[142]

Ferrar recommended that, as in Sicily, the silkworm's lodgings be heated with stoves, "which by this Art may correct the ill site and temper, and qualifie the Ayre well" which was necessary in the "the cold, moyst and shady woods" as Ferrar imagined the Virginian wilds to be.[143] Here, Ferrar drew on the account of Sicilian sericulture related in the 1610 *Relation of a Journey* by George Sandys (1578–1644) and incorporated by Robert Johnson in his second edition of Giovanni Botero's *Universal Relations*; Sandys then went to Virginia, where sericulture was his main concern.[144]

While Lipsius had criticized those who nursed up strange plants "more carefully than any mother does her child" of "sloth," Ferar claimed the opposite. Those who did not undertake such minute and labor-intensive transformations of nature could be considered slothful, and therefore unnatural and fit for ordering and enslavement. In fact, Ferrar compared all those who resisted "this worthy enterprise" of sericulture to the Native population and its "ignorance, sloth, and brutishnesse." This move recalls many of the preemptive accusations that projectors launched at assumed opponents, such as Sir Arthur Gorges's contention (discussed in chapter 1) that anybody who opposed his general office of commerce could be assumed to be malignant and noxious to the commonweal. By laying claim to an existential role for an industry or project in the state, and even more so, in Ferrar's case, in civilization itself, projectors upped the ante of their proposal, and thus magnified their incrimination of likely skeptics. In contrast to such doubters of projects, according to Nature, those who pursued new proposals displayed "the industry of some noble and heroicke spirits (borne to immortalize their names and nation)."[145]

Labor Projects

Making colonial silk meant experimenting not just with materials and techniques, but with labor. Raising silkworms required skilled and intensive labor, yet that labor could not command much in the way of wages, or else the silk produced would not be competitive on a global market. This was one reason why success in raising silkworms by figures such as the royal gardeners in palatial silk houses could not translate to the colonies, at least not until sericulturists turned to large-scale enslaved labor.[146] The Virginia Company advertised the silkworms that King James had sent to the colony in an account of ships, people, and commodities sent to Virginia in 1619. Settlers were counted along with the ships for a total of 1,261 people. A separate category calculated labor that had been sent to the colony, for a total of 650 "Persons for publicke use." These included ninety "Young maides to make Wives," a hundred "Boyes to make Apprentices" and fifty "Servants for the publicke."[147] Like silkworms themselves, humans were resources for use by projectors.

Enterprises from soap to saltpeter to sericulture often entailed labor projects as evidence of their benefit to the commonweal. Projects for domestic industries frequently asserted an ability to employ the unemployed even when such projects in fact disrupted industries and put people out of work. For instance, the 1634 royal proclamation in favor of the gentlemen's soap company discussed in chapter 3 avouched that the project would employ many people of the realm "who are now grown to bee very numerous and populous and many of them poore for want of meanes of imployment."[148] A major argument in favor of colonization was the way it could empty England of excess population.

King James had suggested "sending diverse dissolute persons to Virginia" in October 1619, and the company agreed but wished for a slight delay until January.[149] By November, the Lord Mayor had already given an order to keep "dissolute persons" in Bridewell, awaiting their transportation to work as servants in Virginia.[150] His Majesty was not satisfied with any delay, but wished "50 of the 100 shipt away with all speed," although the company pointed out that they would have to be transported in several ships for fear that if they were carried in one hold, they would mutiny.[151] The Knight Marshall promised Sir John Danvers that if these men could be sent immediately, he could "furnish them with such persons of what quality and condition they desire." A group visited Bridewell to choose "the persons to be transported for his Majesty."[152] These were the fifty "Servants for the publicke" mentioned on the list above.

It was Sandys who had proposed sending "the Young maides" and "Boyes." The latter were abandoned children from the age of twelve upward.[153] According to the company's plan, the children would work as apprentices or servants until the age of twenty-one; when they were grown, they would be placed as tenants on the public land.[154] In 1620 Sandys presented a "Proiect of special importance" for the "better menaging of their affairs in Virginia, and good advancement of the Plantacion there," which was approved of.[155] This included "advancing" the plantation through "the getting of monnye beingge the Synews and moving Instrum[en]te in these greate Actions."[156] Plans for the strengthening of such sinews included sending settlers and "setting upp the best and richest Comodities."[157] Sandys wished to send another "100 young maydes to make wives," in addition to those that had recently been sent, "100 Boyes" to serve as apprentices, and "100 Servants."[158] Silk was the foremost commodity Sandys promoted, in addition to oil, wine, hemp, flax, soap ashes, potash, pitch, tar, fishing, and salt production. Sandys later suggested that those colonists who best succeeded in producing the most desirable commodities, such as silk, would have first choice from among the children that were to be sent to Virginia.[159] As an inducement to investment and emigration, the company advertised their plan to send additional maids, boys, and servants, as well as a "great store" of silkworms and "men skillfull" in their care.[160]

In fact, an additional fifty-seven women, ranging in ages from sixteen to twenty-eight years old, were sent in 1621. Women could be had for 150 lb. of tobacco a piece. According to the parchment subscription roll for investment in this project to export women, it was the lack of "the comforts without which God saw that Man could not live contentedlie noe not in Paradize" that was hindering the colonists' development of "Staple Commodities." Subscribers included Sandys, Nicholas and John Ferrar, and Danvers.[161] In 1621 Sandys also submitted a proposal to Parliament "to send away a great number of our poor people in to Virginia," a bill on which the Virginia Company asked Danvers to work.[162]

In 1620 Danvers described a project to Villiers, whose patronage he much valued, "for advancing the plantation of Virginia," which set out a highly detailed plan for how criminals would be detained and set to work in Virginia. Danvers's horrific project proposed "peopling" Virginia via the "seizure of Persons intending & affecting to live loosely & enforcing them to Apprentice" in the colony. Condemned criminals, who would have to shave their heads and wear identical clothing, could also be sent. By ridding England "of that sort of people," this project, "would be exceeding pleasing to his M[ajes]ty." Danvers also envisioned that many would be eager to invest in a "project," importing "a number of such apprentices to cleare ground & perform works of the hardest labor."[163]

Another related Virginia Company project was the founding of a college, ostensibly to Christianize Powhatan children, but also to develop a labor force. Established in 1619, Henrico College taught Powhatan children to work in agriculture and industry.[164] It was to be paid for by exported labor. The "Persons for publicke use" sent to Virginia in 1619 included a hundred tenants for the college lands and fifty men whose "labours" on college lands would pay for educating thirty Powhatan children.[165] The management of the college was assigned to a small committee of the Virginia Company, including Sir Dudley Digges, Sir John Danvers, John Ferrar, the alchemist Dr. Francis Anthony, and three others. They sent George Thorpe, a gentleman of King James's privy chamber, a silk enthusiast, and the brother-in-law of Danvers, to oversee the college.[166] This meant overseeing labor. Thorpe reported that he had overseen the planting of ten thousand grape vines on the college lands by the "tenants at halves," who contributed half of their labor to the college.[167] The production of silk there was also encouraged.[168] Even after Thorpe and seventeen of the "College people" died in the Powhatan uprising of 1622, a new committee for the college was immediately established that again included Danvers and Ferrar. The company published two pamphlets concerning the school.[169]

In response to the uprising, the Virginia Company considered several new projects. "The project and offer of Captaine John Smith" proposed continual "ranging the Countries, and tormenting the Salvages" with the help of a hundred soldiers, an idea that Sir John Brooke especially supported.[170] An alternative project proposed keeping the Powhatans (just barely) alive, in part due to their usefulness in preparing silk. Captain John Martin suggested a "manner howe to bringe in the Indians into subiection w[i]thout makinge an utter exterpation of them." It included the prevention of the Powhatans from developing any stores of food or other wares by having "some 200 Souldiers on foote, Contynuallie harrowinge and burneinge all their Townes in wynter, and spoileinge their weares."[171] One of the reasons Martin entertained for keeping the Powhatans alive was their greater capacity for skilled labor in silk, in comparison to English settlers. Many were impressed with the beautiful, silky objects the Powhatans made from the plant known as silk grass (*Apocynum cannabinum*).[172] The "natives," wrote Martin, were "apter for worke then yet o[u]r English are, knowinge howe to attayne great quantitie of silke, hempe, and flax, and most exquisite in the dressinge thereof ffor o[u]r uses."[173]

Garden Philosophy

Although London investors persevered with the colony and ambitious plans for transplanted nature and art, despite continual failure and existential threats to the colony's survival, their tenacity came to naught when James I revoked the Virginia Company's charter in 1624. However, during the Interregnum, the regicide Sir John Danvers became a powerful politician, sitting on the Council of State and appointed to a 1650 committee for establishing a council of trade, which positioned him to advance colonial sericulture once again. A flurry of publications reviving and expanding Virginia Company arguments for sericulture from the 1620s appeared on the presses, including *Virginia's Discovery of Silke-Wormes*, which drew largely on Bonoeil's tract, and *Virgo Triumphans*, a work dedicated by one Edward Williams to Parliament and the Council of State.[174] The main text of this work, according to the preface, was penned by Danvers's old associate, "Mr. John Farrer [*sic*]," and it drew extensively from Ferrar's 1622 text appended to Boneoil's treatise. Ferrar did not write using the voice of Nature in this text. Rather, he presented Virginia as the "Eldest Daughter of Nature" who was willing to offer her "Embraces" to the "Gentlemen" willing to colonize her.[175]

In *Virgo Triumphans* Ferrar further expanded his arguments concerning the need to experiment in garden arts for colonial success. Dilating on his comments from 1622, Ferrar reiterated the need for settlers to experiment themselves in ways that might differ from prescribed rules. He also had some new advice to dispense concerning how to arrange the labor of the project. It only took one master who had become knowledgeable in sericulture to direct "hundreds under him." Ferrar suggested certain tactics that might encourage sericultural competition in order to "quicken" the "observation and diligence" of those who were "ingeniously disposed."[176] Advertising the "honour and profit" of sericultural knowledge could "disperse seeds of emulation and diligence, since every one would imploy himselfe seriously to engrosse and appropriate to himselfe the reputation and advantage in the victory."[177] He also advised that Native children should be employed raising silkworms. His views here directly contradict his previous characterizations of the Powhatans. Here he suggested that Native children could "easily runne through this curious, but not heavy labour." In an indication of the ways that projectors made use of collections of Virginian objects, Ferrar noted that "the Indian is naturally curious and very ingenious, which they shew in all their works." Moreover, employing Native children would make them serve as hostages assuring "Indian fidelity."[178]

Ferrar much lamented how the previous injunctions from decades ago had not been followed by settlers, since the intervening time could have allowed for so much improvement upon nature. For example, if colonizers had planted thousands of the native silkgrass, as they had been asked to do, by this date "in all probability by transplantation it may thrive beyond comparison larger, and the skinne of it growne more tender and delicate, [and it] may arrive to some equality with the labour of the Silke-worme."[179] He thought it highly likely that transplantation of the native mulberry could also so improve it, that it might enable a double harvest of silk each year. One of the drawbacks of sericulture is that silkworms only spin cocoons once a year as their life cycle is timed to match the leafing out of mulberry trees. While the silkworm eggs lie dormant for nearly a full year, the expensive equipment and skilled labor required to raise them must also remain unutilized. Ferrar reasoned that such a double harvest must already exist in China and Persia, even if "we are at present ignorant" of it, since otherwise it seemed impossible to him for those states to produce such great quantities of silk to sell.

Although this was all highly conjectural, Ferrar asked "what should discourage us from delivering such conjectures to a tryall, since the examen of it is not without probability, nor the discovery without an extraordinary certainety of profit?"[180] As in the case of Bonoeil's conjectural process for generating silkworms from a strangled calf, Ferrar suggested here a cost-benefit analysis. There were reasons for considering this conjecture probable, and if it were proved to be so, it would certainly bear much profit, and thus it was worth a trial.

As Robert Child did in his contribution to Hartlib's *Legacy of Husbandry*, Ferrar held up the recent success of metropolitan London gardeners as a sign of what might be possible in the large-scale transplantation of resources: "What multitude of flowers have our late Gardens in England seen non native to this soyle or Climate?" he asked.[181] Oftentimes, fruits flourished more when transplanted than in their native climes, "who see often times the Colonies in a happier degree of prosperity then the Mother, for Fruit and Flowers."[182]

Such "designements," such as the slow improving of silkgrass and mulberry trees through transplantation, required "time, curiosity and industry." However, there was a certain investigation through which the way to such large-scale transplantations could be made "passable and easie." This was the "uncabinetting and deciphring of Nature, Garden Philosophy." This form of knowledge allowed gardeners to peer into Nature's most secretive and intimate cabinets, cracking her codes. Through such "Garden Philosophy," any "harsh disposition in the World" could be refined.[183] The "removing and transposition of Wild

Plants" alone could "engentile" them, although Ferrar was not sure why. Perhaps it was since the "nature of Plants, as of men is desirous of Novelty and peregrination." Perhaps the "gratefull novelty and allurement of a well cultivated soyle" made the plant set aside "the old savagenesse and indomestication of its first seat and nature" and adopt a new quality from its cultivated environment.[184]

Ferrar admitted that "Garden Philosophy" offered a form of knowledge that included much that was not known. He conceptualized a form of philosophy based on experimentation and observed phenomena that denied philosophy's traditional requirement of rational, causal understanding. This knowledge only offered hints at possible explanations and no certain conclusions. Nevertheless, Ferrar arrogated for it both the authority of philosophy and the promise of widespread utility.

Such garden experiments offered "allurements . . . for those whose delights only are interested and denoted to this retired activity." However, extrapolations drawn from the investigation of curious plants in the pleasure garden could then be applied on a much broader scale: "those who looke further will finde . . . this pleasure to be attended with an inestimable profit, and one of the most certain returnes in nature."[185] The silkworm itself offered an example of how concentrated forms of natural secrets—"cabinetted" nature—could serve as points of intensive investigation that might reveal knowledge to be applied more broadly. The silkworm was a tiny wonder, into which Nature "abbreviated all the Volume of her other Miracles into this, her little, but exact Epitome, like that Artist who contracted the whole body of Iliads and Odysses into a Nutshell."[186] Ferrar refers to a famous description by Cicero of a tiny edition of these works. Such an ability to surpass normal scales of operation made minuscule works frequently collected in the curiosity cabinets of Ferrar's day; Tradescant's Ark held many astonishingly minute works, such as "Halfe a Hasle-nut with 70 pieces of housholdstuffe in it."[187] Ferrar's description of the silkworm in these terms as a tiny cabinet of nature illustrates the for-profit and extractive perspective with which colonial promoters contemplated seemingly useless wonders in period collections. Through careful and minute observations, experiments, and adjustments in a highly sophisticated setting, the silkworm could be "uncabinetted," forcing nature to reveal her secrets and offering the cultivator a concentrated, single-point access to larger knowledge of nature. Ferrar offered an extractive view of the cabinet of nature. The "richest provinces in the World" were those that nature had created as "her Cabinets of excellency," such as China, home of the silkworm. These cabinets would be opened, their contents extracted, and made Virginia's own.[188]

Ferrar encouraged his audience to "looke further," beyond current assumptions of possibility and profitability. The "too fond a Reverence wee pay to antiquity," Ferrar complained, led many to reject all new proposals. Such obedience to the commands of the ancients was symbolized by the cliffs enclosing the narrow Straits of Gibraltar. Known to the ancients as the columns of Hercules, these stony eminences delimited the safe, known world of the Mediterranean from hubristic, risk-taking journeys beyond. According to ancient legend, the columns of Hercules sported a warning, "no further [ne plus ultra]." Ferrar equated reverence for antiquity with "the *ne plus ultra,* the *Hercules* Pillars of Wisedome, beyond which there were no passage."[189] Just as the new garden designs of Danvers's house and elsewhere extended the vista to distant prospects, Ferrar encouraged contemporaries to think about the boundaries of transforming knowledge in ways analogous to the title pages of Bacon's works (although Ferrar does not mention Bacon). Bacon's title pages repeatedly showed the ocean of unknown, future knowledge extending past the columns of Hercules and far beyond what the eye could see. Ferrar's quest to persevere in a risk-taking venture and to push colonial projects continually *plus ultra* came at a great and unequally distributed cost, and one that continued long into the future.

According to Virginian colonial promoters in both the early Stuart and Interregnum periods, the means through which colonial knowledge could be advanced beyond the assumed limits of possibility lay in the rivalry and ambitions of each settler to "appropriate" honor and advantage to himself, as John Ferrar put it in his scheme to advance sericultural knowledge through emulation. Some scholars have interpreted the political arguments that promoters of overseas ventures put forward as an expression of Justus Lipsius's views concerning Neostoic greatness. According to these views, it was the *studia humanitatis* that offered "a commonly shared language of participation" that stressed the responsibility of individual citizens to participate in governance, and thus in the foundation of new commonwealths in colonial settings.[190] However, colonizers' emphasis on wealth, commercial rivalry, and the advancement of private interest as the basis for the greatness of the state does not in fact reflect the ideals cultivated by humanistic study in general nor Lipsius's Neostoic concerns in particular. Rather, it draws on the economically oriented and novel reason of state developed by Giovanni Botero, which was further elaborated into a language of interests in the early seventeenth century, in part in the texts of projects themselves.

Interests offered an alternative mechanism through which individuals took initiative in colonialism. Rather than participating politically as citizens in a

shared commonweal or good, projectors positioned themselves as canny operators on behalf of the state. The reason-of-state literature painted appetites, commerce, industry, and innovation in the arts as engines of internal growth and
augmentation of the powers of the state. It was this link between the fulfillment
of individual desires and the augmentation of the powers of the state that formed
the commercial ideology of colonialism. This was why desires did not need to
be disciplined out of the colonizing population but rather strategically cultivated. As individuals tried conjectural experiments and searched hopefully
through the land, they collectively pushed past assumed boundaries of possibility and advanced both knowledge and empire.

Sir Nicholas Crispe: Experimenting for Slavery

Sir Nicholas Crispe (1598–1666), a future fellow of the Royal Society, claimed
that he was constructing the building blocks for England's national interest in
Africa through the groundwork for global enslavement that he laid on his Hammersmith property. He argued that his pursuit of personal profit served the
nation and thus deserved public support and reward, even when it intervened
upon the freedom and lives of others. However, he did not take kindly to it when
other interlopers attempted to do the same to him.

We have already encountered Crispe as a merchant in Africa who collected
curiosities for Tradescant. He was also very active in the alum industry we explored in previous chapters. He invested in both the East India Company and
English colonialism in Ireland.[191] Crispe tied his own estate more directly to
colonial settings than did Danvers, making it an experimental ground for the
literal building blocks and trade goods of colonialism. In so doing, he continually transplanted arts from abroad to his own soil and entangled a number
of different forms of knowledge together, from glass, brick, and chemical production, to schemes for enforced and enslaved labor. He set up copperas works
near London on a Deptford property that would later belong to John Evelyn,
also a future fellow of the Royal Society.[192] Crispe also experimented with a
new sort of ship dock of a "vast design" at Deptford, although his first trial
broke, flooding "the neighbor-fields with drowning of some cattle and spoil-
ing of some houses."[193] Overlooking such collateral damage, Crispe did not give
up and continued to work on dock design into the Restoration.[194]

Crispe was an established private trader in West Africa, journeying there
himself before 1625.[195] In 1631 Charles I granted a thirty-one-year trading license
in Guinea to a group of interlopers and other privateers; this group was led by
Crispe, and also included Sir Kenelm Digby, another future fellow of the Royal

Society. According to their patent they controlled the right to any commodity traded in West Africa.[196] The African trade deployed especially large and well-armed ships, such as the 400-ton *Crispiana* equipped with forty guns.[197] Crispe and his associates expanded their trade from redwood to ivory and gold, but did not loudly advertise their interest in the slave trade. Evidence, including iron shackles inventoried at the stronghold of Fort Cormantin in Ghana, shows that the company was engaged in the trade.[198] Crispe funded the building of Fort Cormantin between 1632 and 1634, which later became the center of the slave-trading of the Restoration Royal Adventurers into Africa (figure 11). Renamed Fort Amsterdam, remains of the structure still stand.[199]

According to current estimates, the English shipped 33,695 enslaved Africans between 1626 and 1650, surpassing the amount shipped by the Dutch.[200] Many interlopers into the African trade were engaged in the slave trade. The Guinea company fought off these competitors, but since Crispe and his associates also acted as interlopers, the boundary between official and unofficial trade "further blurred."[201] For example, Nicholas Crispe blocked the slaving voyage of one interloper, John Crispe, in 1638 in a vessel, the *Star*. However, the very next year, the *Star,* whose owners now included Nicholas Crispe and John Wood, sold

Figure 11. Fort Cormantin, 1668–70. Rijksmuseum. RP-P-1907-2822. Public domain

slaves in Barbados.[202] Nicholas Crispe had long worked as an interloper and privateer at the same time as he held stock in official trade monopolies. In the 1620s Crispe likely had interloped in the African trade monopoly of the Company of Adventurers of London trading to Gynney and Bynney, which he had also joined.[203] In 1645 Crispe obtained a commission to equip no fewer than fifteen ships of war with the power to take prizes making him, according to his enemies, "Admiral of the Sea-Pirats."[204] Sir Kenelm Digby, like the other members of the Guinea company, had also promoted privateering "during the French and Spanish wars of 1625–30."[205]

Crispe investigated many technologies that he intertwined in his projects. He established the brickmaking industry in Hammersmith, using the bricks to build both his own riverside house there and Fort Cormantin. He also investigated dyeing in ways that drew on both his chemical enterprises and his access to African goods. The royal physician Sir Theodore de Mayerne (1573–1655) spent many years experimenting on dyes and interviewing many informants, and Crispe was one of them.[206] From him, Mayerne copied a recipe using copperas, alum, and the "red wood of Guinea . . . commonly used in dying of wool."[207] Crispe's project for "Growing and Making" the red dye madder in England employed, he claimed, "above a thousand persons."[208]

Even such a wide-ranging individual as Alexander Marshall, the great insect collector, florist, gardener, artist, and merchant, accused Crispe of "projecting upon too many particulars at once."[209] It seems that others also thought that Crispe's style of projecting was overly ambitious. When in 1662 John Evelyn entertained the Duke of York and others at his Deptford property, they viewed some of the "grounds about a project for a receptacle for ships to be moored in, which was laid aside as a fancy of Sir Nicholas Crisp."[210] In a somewhat more laudatory fashion, Crispe was remembered for the "activity" of his "designing spirit."[211] Supposedly Charles I had said of him, "he was a man of a clear head, that by continual Agitation of thoughts went on smoothly in his business, sticking not at any difficulties."[212]

Around 1649, in a remonstrance to Parliament preserved in the Hartlib Papers, Crispe offered a retrospective analysis of his career. This document reveals how Crispe related ideas of private and national interest to profit, risk-taking, and honor in ways that legitimated violence. He stressed his willingness to adventure, to take risks, and thereby to appropriate global trade in ways that kept it out of competing hands. He claimed that of the group that had been given the patent in 1631 "to Discover a Trade for Gould in Africa," only he was so "zealous and industrious" as "to adventer upon soe great a hazard, and uncertenty as the first voyage was."[213] Crispe saw his building of Fort Cormantin as a major contribution he had made to England's honor and profit. Once again emphasizing qualities of adventure and gallantry, he criticized his fellow countrymen for possessing "neither Invention, Industry, nor ressolucion to adventure theire Estats in things soe hazardable." As a result, the "Industrie, and gallantry of this Nation, haht been soe embased and laid lowe" since the time of the explorers such as Sir Francis Drake. The Dutch were building forts all around the world, while "our Nation, skimeth vp and downe the world for the present yeares profitt neuer layeing foundation stones for future substantiall structurs."[214] Crispe described how John Wood, in order to build Fort Cormantin, negotiated a treaty with a local king that allowed them to subject "the adiacent people vnder our commaund obleigeinge them to pay vs a Tribute of all the Fish they tooke and other services, greatly to the honor & advantage of our Nation."[215]

Crispe integrated the African trade into a wide array of other industries that related to it, all developed on his own property using his estate's private staff. He described how he had planned and developed the building of Fort Cormantin from the ground up, making the "Tymber frame" of the fort on his own property in Hammersmith. He sent it already framed up to Africa along with "Bricks Tymber Lyme, lead, Iron and other Materialls, with a set of brick makers for making of Brick in the place" as well as "Lyme burners Bricklayers Carpenters, Smiths, and all other necessarie Artificers for that worke."[216] He proposed that his building of Fort Cormantin was deserving of "some marke of gratitute, fixed vpon him by his nation, that may remaine for his honor to posteritie" given that it was "the principall care, of all Nations to study all incouragments and rewards to those persons, whose Industrie doth Naturallise forraine Mines and Treasures, by Trade, as if they were within the Bowells of theire owne soyle."[217] Industry allowed for appropriation of the advantages of other lands into the powers of the state, argued Crispe, as though it had transplanted them to its own territory.

While Crispe considered African treasures as extractable only metaphorically from England's soil, he proudly described his efforts to compete with other nations through the transplantation of their knowledge to his own estate. This was in response to a series of trade stratagems the Dutch employed. The Dutch monopolized the supply of beads necessary for trading in Guinea, and even "precontracted with all the makers in Holland Venice and other parts." Crispe's only solution "was to draw over some of the cheife Bead makers and sett up the worke in England." He built a house for that purpose in Hammersmith, and trained "his Coachman, Waterman and other servants imployed in the same, to do the Mysteries of that worke as any straungers."[218] Meanwhile, for stringing the beads he employed "both the houses of Bridwell [houses of correction for vagrants and beggars], and a multitude of people in all parts of the Towne," including "Children of seven or eight yeares old."[219] Recent archaeological investigation of Crispe's property confirms that forty-three different types of beads were produced there, showing that "some experimentation went on at Hammersmith and some unique varieties were produced."[220] According to Crispe, this successful enterprise was then attacked by the glass monopolists "Sir Robert Mansfield [a.k.a. Mansell] and his lady [Elizabeth Roper]," who claimed that beads lay in their patent for glassmaking. Crispe was furious, alleging that the Mansells had no "understanding the nature and quallitie" of beads, nor of the changing winds of fashion in Ghana, that is, "the severall changes which every yeare the Neigers new fancies affected."[221] To avoid infringing on the Mansell patent, Crispe was forced to pay 200 pounds a year.

Crispe was an interloper in many ways. Crispe did not wish to respect the Mansell patent for glass production. Yet, he vigorously attempted to fight off other intruders into the African trade, arguing that "Interlopers" brought "mischeifs and confusions" to a trade that he claimed advantaged the public.[222] He naturalized the wealth of other countries, as if it was found in England's "owne soyle," not only by transplanting the Venetian art of beadmaking to Hammersmith, Hammersmith-produced bricks to Fort Cormantin, and African gold into his own pocket. He criticized the royal permitting of the Courten Association (discussed in chapter 4) "contrarie to privelleges" granted to the English East India Company. This was perhaps in part due to his membership in the East India Company, but it might also be because Courten had opposed a monopoly of trade with Morocco granted to Crispe and his associates by Charles I. Courten had "incorporated trade with both Morocco and Guinea with his overall East Indian interloping project" from its start.[223] While he himself was termed the "Admiral of the Sea-Pirats" by his enemies, he complained about how others were also allowed "by Authoritie to committ great Piraces vpon the Natiues," to the endangerment of the Company.[224]

Throughout his *Remonstrance,* Crispe set the issue of protecting his investments against others in terms of interest, both of his own and of the public. By erecting a castle, he settled "a Nationall Interest in Africa."[225] His own interest in the trade was "fastened with a triple Cord . . . Of Law and equitie," of the "grand merrit of the discovery and settlement," and because the "publique Interest and good is most conserved and advanced by his inioyinge of it."[226] According to Albert Hirschman's *Passions and the Interests,* such new seventeenth-century forms of reckoning by means of interest displaced older, more violent, chivalric modes pursuing glory, ambition, rivalry, and honor. However, Crispe combined arguments from interest with the pursuit of adventure and glory. His interest, he claimed, deserved support from the public precisely because he was willing to run hazards that other more risk-adverse merchants were not.

Crispe remained convinced throughout his life that he deserved a public reward for his contributions. In his 1666 will, he gave detailed instructions for his funeral. Attendees were to be given his portrait with the inscription "This is the Effigies of Sir Nicholas Crispe . . . who ffirst discovered and setled the Trade of Gold in Affrica and built there the Castle of Cormentine by which he lost out of purse above a Hundred Thousand pounds above all returnes from thence." He hoped this "Nation (when I am dead) will make Compensation or amends to my family forsoe great a service done to my Country at my soe great losse which will be better understood in the future."[227] The "Trade of Gold" was a euphemism, given that at that point in time Crispe had invested 2,000 pounds in the Royal Adventurers into Africa, which explicitly traded enslaved people and which centered upon the stronghold of Cormantin until its capture by the Dutch in 1665.[228]

Despite his many allusions to the protection of his own and the public interests, Crispe's arguments and trade practices did nothing to abate violence. This view of Crispe's appropriative and violent practices is sharply at odds with how he was remembered in later biographies and histories. Later generations, as he hoped, saw Crispe as a model of industry and improvement to the benefit of the public interest. The 1750 *Bibliographica Britannica* called Crispe "a man of enterprizing genius, ever active and sollicitous about new inventions and discoveries, and, which very rarely happens, wonderfully industrious and diligent about things he had brought to bear." It particularly praised his colonial perseverance: "When the trade to Guinea was under great difficulties and discouragements, he framed a project for retrieving it . . . and to give a good example, as well as to shew that he meant to adhere the work that he had once taken in hand, he caused the castle of Cormantyn upon the gold coast to be erected at his own expence."[229] A critical history of projects can thus offer revised accounts of how concepts of interest, invention, activity, and progress interrelated in the early modern period with violence, appropriation, and an ethos of daring adventure and conquest.

Conclusion: Appropriation over a Distance

Speaking about the appropriation of knowledge, culture, nature, power, and other resources in the early modern period might sound like a twenty-first-century anachronism. Appropriation, however, was very much an early modern idea, with a wide range of uses. In the seventeenth century, appropriation primarily referred to the requisitioning of men and materials for the military. Among other things, it also referred to the "appropriation of plants," that is, either the identification of certain virtues with certain plants, or the grouping of a plant together with another's identity, or the affiliation of certain plants with certain planets or parts of the human body.[230] It could also refer to a contested claim and theft from another, as in John Parkinson's assertion that the double daffodil, first found in England in the garden of John Tradescant, had been appropriated by another florist, "as if he were the first founder thereof" and named after himself.[231] Similarly, John Ferrar had argued that rivalry in the development of sericultural knowledge would make everyone wish to appropriate honor and advantage for himself.

None of these senses of appropriation referred to today's sense of appropriation across wide differences of culture or class. However, this notion of appropriation as a claim to resources and power across a social gulf was emergent in the early modern practice of scanning the world for resources and attempting to make them one's own through transplantation across vast distances. The concentration and juxtaposition of many forms of art and nature in period collections and gardens excited viewers with the potential fungibility and miscibility of regions of the world and parts of nature. Transplanting nature, industry, and people from afar offered a means to augment one's own land, naturalizing global resources through appropriation. In this way, argued Crispe, Ghanaian treasure was emboweled within English soil.

Writing about the possibility of a northwest passage from Virginia to China, Ferrar posited that access to foreign riches would inevitably lead to the appropriation of nature, art, and even people, as China transmigrated into Virginia. In Ferrar's account, discovery of resources inexorably led to appropriation.

> By this means what wealth can there be in those richest provinces of the World, in those Countries which Nature created for her Cabinets of excellency, which we shall not discover? What discover without a power of Appropriation? What opulency does China teeme with which shall not be made our own . . . ? And by this recourse all those curiosities of Art, in which those Eastern Nations transcend Europe, will bee conveyed to us with their persons.[232]

The promises of projects held out far distant objectives as though they were in reach, engaging the appeal of distant objects to activate subjective desires. Ted Porter has argued that rigidly quantitative forms of knowledge offer a "technology of distance" that separates knowledge from personal qualities such as inherited power or charisma, thus making it appear objective.[233] Here, projects offer a very different technology of distance. Distance allowed interlopers to overlook the negative effects of their experimental extrapolations. Just as modern economic formulas often selectively leave out factors such as environmental costs and unpaid, domestic labor from calculations of economic growth, the rhetoric of projectors calculated the possibility of future profit and its benefits for humankind in massively blinkered ways. The loose forms of reasoning of period projects, and the vast leaps taken both laterally among disconnected phenomena and forward to projections of future profit, allowed many concerns to be skipped over and silenced.

Acting across epistemic and physical distance not only proliferated new knowledge mergers and hybrid forms of practice, but also stimulated ignorance and the suppression of knowledge. Distance allowed for the disknowledge with which promoters of projects shrouded the harmful effects of their proposals or the memory of previous failures. The story of colonial projecting is not one of on-the-ground experiments leading inexorably to expertise and greater possibilities for success, in contrast to the lack of knowledge in the metropole. It is one of agonistic interests battling at cross-purposes, and often voicing ingenious but ultimately specious arguments for why one intervention, appropriation, or enforcement was a public benefit while another was not.

Seventeenth-century projectors diverged sharply from the local, intimate cycling of knowledge and material resources of Elizabethan England found in Ayesha Mukherjee's account of "dearth science," or attempts to "make shift" through experiments and contrivances in times of great need.[234] Mukherjee distinguished sharply between the hyperlocal and "communal" focus of Elizabethan dearth science, based on a "circular and reciprocal" ethical pattern, and "linear and 'improving'" seventeenth-century approaches.[235] Given the ways that knowledge and its effects remained within communities in Mukherjee's account, distance could never serve for off-loading of responsibility in the way that it did for projectors. In particular, according to Mukherjee, in the sixteenth century, "[b]lindness to local climates and ecologies was cast as being a form of waste which could disrupt the harmonious circulatory processes of nature and human knowledge."[236]

Early Stuart interlopers considered the ability to master climate and transplant resources over great distances a key tool for advancing England in the global scramble for advantage. In efforts to recruit investors and obtain privileges, projectors exhorted their audiences to disdain failure and take on ever greater risks. The massive expenditures they sought and not infrequently won seriously question the notion that this was a particularly thrifty time.[237] It was one, rather, that used the claim of service to the public interest in order to amplify its consumption of resources, often appropriated from others. By suppressing the consequences of their actions, projectors claimed to benefit the public, as well as themselves. They assumed a moral high ground, as noble, risk-taking adventurers leading society to a better place where those meaner of soul did not dare to tread. Elite projectors imagined themselves as new epistemic conquistadors whose willingness to filch all forms of knowledge, from Native medicaments to soapmaking, only proved their greatness of soul.

The intellectual practices of "leaping between" allowed for wild extrapolations of scale, the dismissal of contrary evidence, and frequent internal contradictions. Many interlopers eagerly overran the limits of others but were incensed when they experienced the same. They cast themselves in the position of supporting the commonweal or public interest, while accusing their rivals of self-interest. For example, Robert Johnson in his 1609 *Nova Britannia* defended the idea of pursuing improbable projects and offered several reasons why the apparent restraint of liberty, such as the enforced transplantation of people, did not in fact infringe upon liberty but rather benefited the whole. Yet, in the same text, he also argued against allowing a certain type of person into a colony because of the way they infringed upon the liberty of others. These were people "shaken out of their honest courses, and now shifting by their wittes." Having abandoned the limits of traditional trades, they disregarded the customary rights, privileges, and professions of others. They constantly beat their brains, seeking "under colour of good pretence to the Common-wealth to infringe our auncient liberties, and would (if they were not mette withall and curbed by authoritie) make a monopoly to themselues, of each thing after other, belonging to the freedome of euery mans profession."[238] In short, these were projectors. This avid promoter of colonial projects urged colonial authorities to steer clear of them.

CHAPTER 6

Active Knowledge

A Turn against the Liberal Arts

Rebelling against Constancy

This chapter explores several early Stuart educational experiments and proposals in the undisciplining of knowledge, from the efforts to imitate the military and mechanical exercises of London's civic militia by the young Prince Henry and Prince Charles to the noble academy founded by Sir Francis Kynaston to which Charles I himself subscribed. In its very origins, discipline was pedagogic, premised upon students learning the knowledge of a master and obeying the teacher's authority. How then, could an undisciplined form of education be developed? The forms of knowledge patronized in the early Stuart court denied the authority of the liberal arts and sciences, and mixed together corporeal and intellectual forms of training. In so doing, Henry and Charles followed the lead of their father in his rebellion against his humanist education and Lipsian constancy in particular.

In his published emblem book dedicated to Prince Henry, the schoolmaster Henry Peacham advocated the maintenance of constancy in the midst of the world's turbulence, with a citation to Lipsius's *On Constancy* (figure 12). In a stormy ocean, "a mightie Rock doth stand," signifying *"MANLIE CONSTANCIE."* In stark contrast to this impermeable stone, a ship burned down on the ocean. This was the ship of *"OPINION* tossed up and

AMID the waues, a mightie Rock doth ſtand,
Whoſe ruggie brow, had bidden many a ſhower,
And bitter ſtorme; which neither ſea, nor land,
Nor *IOVES* ſharpe-lightening ever could devoure:
    This ſame is *MANLIE CONSTANCIE* of mind,
    Not eaſly moou'd, with every blaſt of wind.

Neere which you ſee, a goodly ſhip to drowne,
Herewith bright flaming in a pitteous fire:
This is *OPINION*, toſſed vp and downe,
Whoſe Pilot's *PRIDE*, & Steereſman *VAINE DESIRE*,
    Thoſe flames *HOT PASSIONS*, & the *WORLD* the ſea,
    God bleſſe the man, that's carried thus away.

Vide Lipſium de
Conſtantia.

Z I. *Præcocia*

Figure 12. Minerva Britanna (London: Dight, 1612), 158. Folger Library. Call #: STC 19511 Copy 1. Used by permission of the Folger Shakespeare Library under a Creative Commons Attribution-ShareAlike 4.0 International License

down, / whose Pilot's *PRIDE*, & Steeresman *VAINE* Desire, / Those flames *HOT PASSIONS*, & the *WORLD* the sea."[1] Peacham's views were destined to be ignored, given the Crown's poor view of Lipsius. James particularly criticized "that Stoic insensible stupidity that proud inconstant Lipsius persuadeth in his *Constantia*."[2] Lipsian constancy did not discipline early Stuart forms of governance or knowledge. Rather, forms of knowledge encouraged by the early Stuart court resembled the burning boats of passionate voyages far more than the stony mount of apathetic Lipsian constancy.

James's criticisms of Lipsius were part of a general rebellion against his childhood education at the hands of the stern humanist George Buchanan. Surpassing even the Lipsian model of kingship, Buchanan set a "vertiginously high . . . bar on princely self-discipline."[3] According to a Senecan standard of stoic kingship, James was educated through immersion in Greek and Latin literature to be an emotionless, ambitionless, perpetual student of his teachers. He dramatically turned against this education. While prizing his own status as an author, he advised his sons to steer clear of bookish pedants, as he came "to criticise Buchanan's claim that a liberal education provided the best preparation for kingship."[4]

Instead, James sought an education for the princes that would stress imperial power, ambition, and martial prowess. He warned Prince Henry not to look like a "stupide pedant."[5] He also urged him to always write in English, "for there is nothing left to be said in Greeke and Latine alreadye."[6] He thought "exercises of the bodie," such as running, dancing, fencing, and horseback riding, the most suitable education for a prince.[7] He ridiculed a scholar such as Hugo Grotius as a "pedant, full of words."[8] Peter Heylyn admonished Prince Charles in 1621, "Action is the life of a Prince; speculation of a Scholler."[9]

Charles was celebrated for cultivating forms of knowledge found outside of books. He was "no great scholar; his learning consisted more in what he had seen than what he had studied: his judgment was good and better than most of his ministers."[10] "With any Artist or good Mechanick, Traveler, or Scholar he would discourse freely; and as he was commonly improved by them, so he often gave light to them in their own art or knowledge" for "few Gentlemen . . . knew more of useful or necessary learning." His "proportion of books was but small," since Charles had "learnt more by ear, than by study."[11]

At first glance, this elite interest in forms of practical knowledge might appear to be a perfect illustration of the idea of "trading zones" in knowledge (discussed in the Introduction). According to this model, clearly delineated skills and types of expertise belonging to discrete social roles are exchanged for mutual benefit, thus creating newly collaborative and hybrid forms of knowledge. What we find here, however, instead of collaboration, is imitation, rivalry, and appropriation. Forms of knowledge pertaining to different social roles muddled, as elite interlopers reached downward into craft domains and as practitioners interloped upward, claiming status and authority from below. This chapter explores various educational initiatives in which such social and epistemic boundaries were undermined.

The Artillery Garden

A case in point is the way both Prince Henry and Prince Charles imitated the training developed by a civic militia. According to a long tradition, whose exact origins are hazy, citizens of London might voluntarily practice martial arts in what was called the Artillery Garden or Yard (now Artillery Lane in Bishopsgate); this practice, however, had lapsed by the start of James I's reign. Efforts were made to revive it in the 1610s and 1620s, and these were hugely successful. "Citizens of London" were permitted "to exercise arms in the Artillery Garden" by James I in 1612.[12] The Artillery Garden became a center for the study of martial drills, inspired by the military discipline exercised by the troops of Maurice, Prince of Orange, informed by Justus Lipsius's studies of ancient Roman discipline. Numerous publications cast the Artillery Garden as a site for martial discipline.[13]

The Artillery Garden, and its imitation by the princes, might thus seem to suggest that the princes were heeding Peacham's advice after all in imitating Lipsian constancy and the forms of military discipline it inspired. Yet, discipline here does not mean dispassion nor the mechanical following of rules. Individual soldiers operating tricky early modern firearms needed to be "nimble" and adaptable, while commanders developed shifting strategies and covert operations.[14] Artillerymen required "Soundnesse of iudgement, Sharpeness of wit, Quicknesse of conceit, Stoutnesse and courage of minde, Vndauntednesse in danger, Discretion mixed with passion, Prudence, Patience, Ability and Agility of body."[15]

The imitation of the civic Artillery Garden by the princes Henry and Charles represented an eagerness to interlope downward into forms of ignoble knowledge. In this endeavor they went even further than their father had ever intended. The conflicting uses of the Artillery Garden made it a place of contention and social confusion. The volunteer militia practicing in the Artillery Garden articulated a "concept of civic chivalry and a calling to be members of a military elite based upon participation in the artillery societies," which "manifested a tendency to undermine the privileged status of aristocrats."[16] James I was suspicious of the civic martial spirit reviving in the Artillery Garden, and Prince Henry and Prince Charles were all too enamored of it.

This civic military training also represented a denial of the authority of a liberal education and the elevation of mechanical, corporeal, and practical forms of knowledge. Ben Jonson wished that the education offered in the Artillery Garden could be expanded into a more far-reaching institution to "instruct the noble English heires / In Politique, and Militar Affairs." Alas, "Tempestous Grandlings," according to Jonson, expected that such training would impugn their honor: "what's he dare tutor us? / Are we by Booke-wormes to be awde? must we / Live by their Scale, that dare doe nothing free?"[17] The liberal arts were supposed to offer freedom, but such noble sons, according to Jonson, thought learning servile, and something to be relegated to "Clownes," "Tradesmen," and "poore Nobilite."[18]

Both Prince Henry and Prince Charles admired the training taking place in the Artillery Garden and sought to create their own versions for themselves and the nobility. Henry founded his own training ground in Westminster, confusingly also called the "Artillery Garden," and often mistaken for the one in Bishopsgate.[19] The prince's academy possessed a library holding "all such books as related to arms, chivalry, military affairs, encamping, fortification &c." as well as a large armory, parlor, and dining room.[20] After his brother's death, Prince Charles wished to revive this establishment, if in a less textually oriented way. In 1612 Sir Edward Cecil had identified in Utrecht a "set of models of all engines and artillery fit for an army . . . fitting subjects of study for the Prince," as showing by demonstration rather than theory the "verie practice of everie thinge, either defensive or offensive."[21] In 1616 a "Dutchman" was brought over to set up this demonstration of engines and artillery for Prince Charles, who proved "pleased with the curious models." Charles detained "the Fleming" for a month to teach him about the models.[22] Charles also visited the Artillery Garden in Bishopsgate. At the end of December 1615, the elderly Welsh cannoneer James Kinnon, a lover of the "Arts Mathematick," keeled over while entertaining the prince there. As his humorous epitaph noted, the Fates had "halfe" fulfilled his wish to die in battle. A "Prince and Armie stood about him round," even if it was only old age that dealt his final blow.[23]

By October 1616 "the gentlemen of the Temple [one of the four Inns of Court]" were "erecting a new artillery garden," with the prince serving as "General" of their group.[24] It is unclear whether this was located at Prince Henry's Artillery Garden or at the Temple.[25] Prince Charles continued to be involved in his Artillery Garden for at least a decade. As the Venetian ambassador Girolamo Lando reported in 1622, although no learned scholar like his father, Charles "has a school of arms which belonged to his brother and frequently works there upon out of the way mathematics and methods of encamping, being very interested in inventions."[26]

Meanwhile, the original Artillery Garden remained contested ground. Despite its customary use by drilling civic militias, it belonged to the Crown and was used by the Ordnance Office for testing guns and teaching gunnery. In 1624 and again in 1626 officers of the Ordnance repeatedly attempted to defend the use of the Artillery Garden for the sole use of their own "Scholars," who were taught gunnery either by master gunner John Reynolds or by the lieutenant general of the office, Sir William Heydon himself.[27] When in 1626 Charles I commanded Sir William Heydon to oversee the new inventions of Cornelis Drebbel and Arnold Rotsipen, he noted that they might need to bring ordnance into the Artillery Garden for the sake of their trials.[28] The Artillery Garden was also contested within the Ordnance Office, as it remained ambiguous whether this ground belonged to Reynolds or to Heydon. All these contestations remained an issue during the tenure of Sir John Heydon as lieutenant general after the death of his brother in 1627.[29]

The roles of the various permutations of the Artillery Garden, and its relationship to the Crown, were never fully distinguished. In 1631 the courtier Endymion Porter, who was also a captain of one of the trained bands of London, offered certain propositions to Marmaduke Rawden (1582–1646), captain of the Artillery Garden militia, for the repurposing of the Artillery Garden group into a "Royal Regiment of selected Guards."[30] Porter served in this ceremonial capacity, greeting the ambassadors of the King of Poland and the Emperor of Morocco with an honor guard.[31] In 1633 Charles I granted a charter to the voluntary militia in the Artillery Garden under the name of the "Artillery Company of the city of London," and in 1641 his sons enlisted in the company, with Charles the Prince of Wales taking on the title of "Generall" of the "Captains and Souldiers exercising Arms in the Artillery Garden."[32]

The apparently ultimate expression of a disciplining society, the citizens' military training seemed to James I to arrogate authority from the hereditary military nobility in ways that he saw as dangerous to the monarchy. His sons' eagerness to formulate a training based on those military exercises represented an unintended consequence of James's own rebellion against the liberal arts as proper training for princes. The negative views of the liberal arts promulgated during the early Stuart period encouraged novel ways of reconsidering the boundaries between gentlemanly liberal arts and ungentlemanly mechanical ones.

In an anonymously published pamphlet setting out an argument for the ennobling influence of royalty upon the mechanical arts, Edmund Bolton recounted how it was in order to communicate to London the "most excellent titles to reall nobilitie in the Citie . . . of wits and letters" that institutions of learning like "the Artillery-yard, and Gressam [*sic*] Colledge were instituted."[33] Bolton rather misrepresented the aims and order of events in the founding of these two institutions. Gresham College (founded in 1597), although intended to provide the City of London with an education in the liberal arts, largely failed in that mission.[34] Instead, it served as a center for the investigation of forms of knowledge not traditionally part of the liberal arts, such as naval engineering or alchemy. Meanwhile, the Artillery Yard or Garden, instead of communicating nobility to the city, was an initiative founded by Londoners and mimicked by royalty. Both Prince Henry and Prince Charles interloped downward, copying the manual, military training of commoners to a greater degree than their father would have liked.

These views also encouraged interloping upward. After the death of Prince Henry, in addition to continuing Henry's investments in military training, Prince Charles also quickly took over the role of patron to several surveyors, beginning in 1612.[35] When Aaron Rathborne dedicated *The Surveyor* to him in 1616, he closely identified the prince, in his lofty, all-seeing position, with surveying: "you now plac'd where all is in your view / And, being the rule of what the people doo, / Are both the scale, and the surveyor too."[36] Rathborne thus interloped upward by associating the prince with his own profession and by proudly featuring on his title page the image of the artisan (Artifex) at work using a sophisticated azimuth theodolite (in contrast to his less technically equipped rival, shown in the bottom scene), while casually dominating the protesting figures of folly beneath his spurred boots (figure 13). Rarely does one find such an image of a working artisan in violent triumph. Is this supposed to suggest, one wonders, Charles I himself in the guise of a surveyor? Rather than offending Charles, such triumphant attitudes expressed by artisans were well calibrated to appeal to Prince Charles, who esteemed practitioners whose ambitions could serve his own. He was eager to get their instruments into his own hands. In fact, new educational proposals that hoped to win Charles's support would need to snub the liberal in favor of the mechanical arts, and many of them included some version of the training offered in the Artillery Garden.

Figure 13. Aaron Rathborne, *The Surveyor* (London: Stansby, 1616). Rijksmuseum, RP-P-1982-1225. Public domain

Bolton's Proposed Academy

In the late 1610s Bolton presented Villiers (a distant cousin) with a "Project for an Academy Royal in England" that would restore Prince Henry's academy for "learning of the Mathematiques, and Language; and for all kinds of noble exercises, as well of arms as other."[37] Villiers proposed the idea to the House of Lords in 1620/21, but nothing came of it.[38] Bolton's plan for an English academy, which he sought to revive in "at least ten separate efforts" between 1618 and 1628, was in fact an attempt to restore two different and competing precedents: Prince Henry's martial academy and the Elizabethan Society of Antiquaries, founded around 1586, but abolished in 1604 due to James I's suspicion concerning its possible political ends.[39]

Bolton repeatedly tried to make the case that these diverse aims were intermingled, and that both connected to a wide array of different forms of knowledge and to action in the world. He defended an Elizabethan notion of classical scholarship as supplying useful instruments for political action and global intervention.[40] Thus in 1626 he insinuated that Villiers was harming himself by not recognizing the utility of such scholarship. The duke's adversaries in Parliament, Bolton warned, were drawing upon a personnel of "fresh orators . . . smart poets . . . wise historians . . . searching antiquaries" and planning to use all their skills against him, while the duke had neglected "the generously and soberly well learned" and thus had endangered himself.[41] Bolton's arguments that antiquarian studies were required for an active, noble education proved unconvincing to Villiers and his contemporaries. Instead, art, curiosity collecting, new inventions, engineering, global exploration, and projects offered an explosive mix of excitement, profit, and glory that commended itself to Charles, Villiers, and other elite courtiers far more than did the restrained pursuit of "language and letters," that is, the *studia humanitatis*.

Two of the main subjects to be taught in Bolton's proposed academy were to be heraldry and ancient history. In his 1610 treatise of heraldry, the *Elements of Armories,* Bolton presented the study of heraldry as a form of "philosophie," full of noble "theorems on the interior," and swathed on the exterior in "those enwrapped circles of ingenuous sciences which the learned do entitle Cyclopaedie."[42] Bolton took the scholarly notion of an encyclopedia, understood according to a false etymology as the "circle of ring of learning," and tried to claim that all "ingenuous" or noble forms of knowledge were interconnected through it. As a demonstration of this interconnection, his treatment of heraldry embraced the most farflung topics, from penguins to atoms.

In a 1611 prospectus of projects for raising revenue that Bolton addressed to Henry Howard (1540–1614), Earl of Northampton, Lord Privy Seal, Bolton also made a bold claim to the usefulness of learning. He complained of the common opinion that learned men should not be consulted in political affairs, but rather used as background painting, "thrust out into the Compartiments or Lanskeps of action, for furniture, and ornament, rather then for use." This opinion might be true concerning "shallow Syntaxis-men, and school-clercks," but it was not true for those grown in the "genuine seedplots of that true, and substantial Wisdom," formed by "the knowledg of things divine, & humane."[43]

Countering those who thought ancient history useless, he sought to demonstrate its practical use. He dedicated a 1624 history of the infamous Roman emperor Nero to Villiers, showing how even Nero's corrupt example offered suggestions for contemporary enterprise. For instance, Bolton took a rather moderate view of Nero's attempt to cut through the Corinthian Isthmus with a canal. Many ancient historians including Suetonius and Pliny described this as a horrific, failed effort relying on political prisoners and slaves who died en masse.[44] The remnants, which include a relief of Hercules carved into the hillside still visible today, were pointed out to seventeenth-century travelers. The travel writer George Sandys, observing the ancient cuts into stone from the Corinthian canal, described it as an "audacious project" whose "contrivers" were "men of wit and impudency to attempt by Art what Nature had prohibited." The scars on the landscape of this unfinished effort to dig a canal served as a "lasting monument of overweening hopes, and franticke prodigality."[45]

In contrast to these common critical views of Nero's canal project, Bolton called it a "great worke . . . worthy of an emperour."[46] The example of Nero's canal project allowed Bolton to dilate upon the Isthmus of Panama, a project studied and rejected by Phillip II of Spain, and the possibility of cutting through that and thus saving a voyage of "many thousands of miles about"; others would also draw a comparison between the idea of a Panamian canal and the ancient attempt in Corinth.[47] Bolton suggested that the ambition of such canal projects was not in and of themselves a reason for their rejection. However, Nero's plan was not well laid out, and its failure attested to the absolute necessity of learned counselors in designing projects. Bolton opined, it is "a princely thing for princes to desire to excell all men in doing nobly, as they excell all men in sublimity of place: but to erre (as this prince did) in the object of endeavors, and in the meanes of atchievement is miserable, and unlearned."[48] Grand projects, surpassing the ambitions of ordinary men, were praiseworthy goals for princes. Grandiose aims required, however, learned counsel in order to reach their noble ends.

In 1627 Bolton reissued his history of Nero, this time addressed to King Charles. Reminding Charles of his plans for an academy, he listed the "points touching royall office, and affaires, which glitter in their golden threds throughout the historicall webb." One could pluck from that web knowledge of "Maisters in wisdome, best States-men . . . Studies of all sorts" and even "Forraigne Trades, namely, East-Indian."[49] In yet another effort to advance his academy project, *The Cabanet Royal*, addressed to Charles, he attempted to cater to Charles's interests in collecting by describing his proposed academy as though it were a royal collection. He also widened his conception of who would be included, from the historians, antiquaries, and heralds he had earlier proposed, to a wider cast of characters that might appeal better to Charles, "not for the service of leters only, but some for land and sea in arts of warr, & nauigation; some for theyr skill and iudgment in the excellencies of musick, paincting, and drawing, and other for other things."[50] He attempted to argue, as he had in his 1610 work on heraldry, that all these forms of learning were connected in the "golden gemell-rings" through which "ingenuous knowledges . . . interembrace."[51] Gemell-rings (from *gemellus*, L. twin) are friendship rings whose many intricate strands interlaced and often joined in clasped hands. However, his design obviously still laid by far the most emphasis upon ancient history. In fact, he argued that "ancient histories" were to be preferred to the study of the present because "they are more remote from present sence, and interesses." Ancient history would serve to "breed manhood and martial spirit" more than the study of the modern world, whose corruption gave rise to "foxes degenerous" rather than to "generous lions."[52] In other words, ancient history distanced learning from the courtly reason of state that was so immersed in shifting political events, rumors, schemes, and stratagems.

Bolton was attempting to have his cake and eat it too. At the same time that he sought to show that ancient letters were useful to the present, and intertwined with pragmatic forms of knowledge for sea and land, he also tried to argue that the advantage of studying ancient history of the past lay in the way it distanced the student from the current culture of cunning tricks and political interests. Neither of these arguments were well aimed. The idea that the study of antiquities, history, and letters could be politically useful might confirm rather than allay royal suspicions of the potentially subversive political purposes for which it might be used. Furthermore, the notion that the study of antiquity was further removed from political cunning and interest merely suggested that it could not in fact play much of a useful role in the framing of projects, which Bolton claimed that it could. Projects relied precisely upon a knowledge of interests as well as a deep acquaintance with quickly shifting political circumstances.

The Public Amphitheater

Bolton's proposals competed with other forms of martial and active knowledge, many of them promoted by Endymion Porter, Bolton's brother-in-law and his main avenue to Villiers. Porter was on the list of worthies Bolton wished to include as a founding member of his academy, however, they seem to have had a falling out. Recalling that Porter had once promised to help him, Bolton, with "the affection of a loving brother," sent Porter "a scroll, containing some suggestions for the writer's benefit, which Porter had torn in pieces."[53]

Porter supported a competing project for the mixing of diverse forms of knowledge: a massive public amphitheater devised as a form of public education and entertainment. The promoters of the amphitheater positioned their project as a canny contribution to state policy. These showmen's attempts to frame their project in the language of the reason of state is one of the more striking examples of how projects created an arena for political argumentation and jostling that, according to the very notion of the secrets of empire, should not have existed.

In 1620 King James issued a license for an amphitheater to John Cotton, John Williams, and Thomas Dixon, described in the license as "sergeants at arms" and all as royal servants; later, a Captain Robert Hasell came forward and claimed to be the original brains behind the project. Dixon might have been an employee of the projector Sir Giles Mompesson. Williams was a servant of Villiers.[54] In 1623, at the urging of Villiers and while the Spanish match for Charles I was under discussion, the King of Spain sent an elephant, Don Diego, and five camels as a gift to James I.[55] "John Williams, gentleman," was appointed "keeper of the elephant" by royal warrant in 1623 and paid 270 pounds a year for the animal's upkeep (the elephant drank a gallon of wine a day).[56] Henry Herbert, Master of the Revels, permitted Williams and others to put the elephant on display to a paying audience; later that year, they were also permitted to display a "live Beaver" and, fifteen years later, "an outlandish creature, called a Possum."[57] Williams may have gained access to these American specimens via Villiers's global collecting network, just as he had gained the care of the elephant through Villiers.

These later shows paled in comparison, however, to the scope of entertainment these men had proposed in 1620. They planned to build an amphitheater in London with space for "twelve thousand spectators at the least." They were also not bashful about the qualifications of the personnel they had secured. They had engaged "Cornelius the Dutchman [Cornelis Drebbel] the most admired man of Christendome for singuler Invention and Arte with divers others of our Nation, that will undertake for our Sea Fights, Prospectives, Nocturnalls, Driades, Naides, Fire and Water-workes."[58]

The prospectus for the amphitheater made the most of Drebbel's ability to stage eerily realistic, moving figures. Drebbel had often previously exhibited seemingly living figures in courtly optical displays and in the complexes of automatons he constructed in Eltham Palace outside London and in Prague. Drebbel was also secured to produce "Sea Fights with the resemblance of Shipps and Gallies in very Exquisit and Singuler order, worthy the view of the most Noble and Generous beholders"; "Masques of very Exquisit and Curious Inventions with the best Dauncers that can be, Mummeries allso, and Moriskors"; "Curious Prospectives in this Kingdome unnusuall of singuler rarietie, and high Invention"; and "Strange and unusuall Padgeants with very admirable and rare Inventions, never as yet brought forth to any Speculacon in theise Partes of the World, with all manner of Pleasures that may either delight the Eare, or content the Eye in them." There would also be seen "the lively Figures & pleasant demonstrations of the Driades, theire Pastimes, Natures, quallities and prime derivations"; "The nymble Niades in their proper Natures, and delightfull pleasures, in and about the Springes, Fountaines, and Water"; and "Nocturnalls of unexpressable Figures; Visions and Apparitions Figureing deepe Melancholly and unusuall Representations."[59] The deployment of philosophical terms such as the "Natures, qualities and prime derivations" in the display of wood nymphs illustrates the cultural blending and epistemic ambitions of the proposed show.

This proposal to bring courtly stage design to the city recalls Jonson's 1622 ridicule of Van Goose, a "projector of masques" who wished to bring city entertainment to court (discussed in chapter 2). In the amphitheater proposal, Drebbel would offer the city the Italianate perspective, stage machinery, and emotionally evocative set design that Inigo Jones had introduced to the English court some fifteen years previously.[60] If this theater had been built with the capacity to support Drebbel's proposed machined displays, its effects upon English scenic development might have been substantial, as no public theaters were built to accommodate such staging until the Restoration.[61]

This project mixed the highest and lowest forms of learning, entertainment, and cultural expression. While this prospectus assumed that "Noble and Generous beholders" would be among the spectators, the audience of twelve thousand it proposed would have been perforce socially diverse. This project thus mixed multifarious forms of knowledge before a varied audience. The displays of the amphitheater were to be "heroick and majestick." In addition to Latin and English "Tragedies, Comedies and Histories . . . full of high state and royall representments," there would also be "showes of great horse, and riche caparisons, gracefully prepared to entertaine Foraigne princes," and "the true use of all manner of armes and weapons for foote," including "pike, partizan, holbert, sword, rapier, muskett, [and] pistoll." Spanish pastimes, such as "Joco del Taurae" would be on view, and "fightings of wilde beasts," as well as "all possible exercises of the Olympiades, as wrestling in oyled skynnes . . . wrestling two or three against one, running, jumping, vaulteing, tumbling, daunceing on the ropes, gladiators," etc. From shows of "high state" to dancing on ropes, this proposal left nothing out in the range of its activities. It also left the door open to further possibilities, gesturing vaguely at "an infinite number of unexpressed properties of singuler order and composure."

The projectors had secured figures from the court to supply these entertainments. Whenever "fortie or fiftie great horse" might be needed, "a gentleman, His Majesties servant and Commaunder in his Highnes' stables, will be reddie for us." They had engaged a cast of "admirable daunce rs" and "great masters in musick . . . Mr. Alphonso [Ferrabosco II], Mr. Innocent Lanzire [Lanier], Mr. Bird [William Byrd?], Mr. [Robert] Johnson, and others." They had also secured "A Captaine of Foote, and his Officers of Excellent Experience," who would offer demonstrations of "many Excellent and Ingenious Experiences belonging to a Campe, Seidge or Garison," showing "the manly order and posture of a souldier."[62] According to the license, these military shows would include "other documentes and instruccions belonging to the honor and danger necessary for a Martiall Court with mathematicall & arithmeticall readings."[63] Such a mixture of a mathematical education with recreated battles and displays of martial drills would have brought the civic chivalry of the Artillery Garden to a large audience, and mixed it, in potentially disturbing ways, with courtly arts.

The figure of Endymion Porter looms large in this proposal. James I's Master of Horse, who could have given access to forty or fifty horses whenever needed from the king's stables, was George Villiers, in whose household Porter had served before he entered the service of Prince Charles.[64] The proposed Spanish entertainments recall Porter's early education as a page to Count Olivares in Spain. The references to the Olympiades currently being practiced point to the Cotswold "Olimpick Games" (figure 14) of Robert Dover (1582–1652). Dover, with the support of Porter, "between 1612 and 1642 staged the Cotswold Olimpick Games as annual celebrations of community sport from the highest

Figure 14. Michael Drayton, *Annalia Dubrensia. Vpon the yeerely celebration of Mr. Robert Douers Olimpick games vpon Cotswold-hills* (London: Raworth, 1636). Folger Library. Call #: STC 24954 Copy 3. Used by permission of the Folger Shakespeare Library under a Creative Commons Attribution-ShareAlike 4.0 International License

classes to the lowest."[65] Finally, the vision of the drilled honor guard recalls Porter's proposals for the reform of the Artillery Garden. Indeed, Porter would be revealed as a chief player in this project in a later iteration in 1626.

Objections to the project as proposed in 1620 soon emerged, and King James sent a letter to members of his Privy Council with some amendments to the license he had already issued. Among other concerns, he opposed the social leveling that such public education/entertainment might encourage. Tilting, tourneys, barriers "and such like reserved for Solempnities and Triumphs of Princes" were not "to be vilified dayly in the Eyes of the Vulgar for money offered." In general, James stressed that the amphitheater's shows would offer only a "Spectacle to the People, not pretending to make yt an Academy to instruct, or Teach the Nobilitie or Gentrie of this Kingdome a worke onely possible and fitt for Princes to Undertake, and not to be mixed with Mercenary or Mechanick Ends."[66] The martial honor that the amphitheater hoped to impart to its massive, paying public should properly belong to a princely academy.

The promoters of the amphitheater did not give up after this setback. A second license for the amphitheater was issued to John Williams and Thomas Dixon, with a note that it was "procured by Mr. Endymion Porter."[67] However, it was not realized, and in 1634 the project for the amphitheater was renewed once again. Yet another petition listed the original projector as John Williams, together with "Capten Robert Hasell," who was claimed as the "first Inventor and profeser of the busines of the Amphitheater." Hasell was a dubious character who had attempted to win a vast estate for Captain John Martin in Virginia and was known to the Virginia Company as "an Interpreter to a Polonian Lord of his own creatinge."[68] Williams and Hasell partnered with three courtiers, Sir Richard Young, Sir Richard Darley, and Henry Murray, Esquire; like Porter, Murray was a Groom of the Bedchamber to Charles I.[69]

This time, the projectors were far more intentional in placing their project within the context of overarching policy, possibly in response to the rival proposal for founding a noble academy advanced by Sir Francis Kynaston and discussed in the next section. Kynaston argued that the amphitheater projectors were only interested in their own profit, and not in public benefit. In response, the amphitheater projectors clarified how their proposal served the commonweal on several fronts. They claimed the amphitheater would provide profit to the Crown from the land and buildings, honor to the kingdom through its variable and delightful recreations, and benefit to the commonweal by replacing drunkenness and lasciviousness with the practice of manly sports. It would perfect the "exercises of the Olimpiads." The design of the theater would "accomodate a number of excellent propertyes, and inventions to entertaine Princes Embassadors Strangers, and honourable Natives and Subiects." Its ability to ward off dissent and rebellion through entertainment was a well-known political stratagem; it would "embusy diverse private consultacions and Conventickles, with matters of pleasant levity. . . . A pollecy oftentymes permitted in the beste, and most flowrishing Commonwealthes."[70]

Despite such arguments and repeated proposals, the amphitheater project came to naught, seemingly in deference to Kynaston's proposals. Kynaston would harshly condemn the presumption of the amphitheater projectors, who sought to arrogate forms of courtly and martial knowledge and prostitute them to the public gaze. In so doing, however, the amphitheater projectors were merely taking a cue from the court's own eagerness to mix diverse forms of knowledge, including noble and mechanical and manual arts, in ways that undercut older epistemic and social hierarchies. In this, the projectors were preceded by how Prince Henry and Prince Charles modeled their own training on the deeds of military engineers and the drills of London citizens. Kynaston too would make an extraordinarily wide survey of knowledge across London's social ranks the basis for his new experimental academy.

Kynaston's Academy

Kynaston's academy has often been seen as following on the heels of Bolton's initiatives; however, it in fact suggested a far different range of knowledge. It lay the greatest emphasis upon experimentation and the conjectural investigation of the secrets of nature, an arena of knowledge that Bolton avoided. Kynaston's academy bore more similarities to the project for a public amphitheater than to Bolton's proposals.

Like Nicholas Crispe, Kynaston was criticized in the period as one of its more extreme projectors.[71] Sir William Boswell complained that Kynaston "proposed impossible and impracticable things, a project with too many windings and too much ostentation."[72] Yet, Samuel Hartlib also recorded the view that he was "a good Scollar" with "an oeconomical contriving head."[73] When Hartlib was making notes on how best to form a London academy, he listed both Sir John Danvers, who became engaged in questions of pedagogy in part through his involvement with the founding of Henrico College in Virginia, and Danvers's friend Kynaston as offering models to follow.[74]

In one 1635 Latin work, *Musae Aulicae* (Courtly Muses), by the Scottish royal physician Arthur Johnston, with English translation by Kynaston, the "liberal arts" address Charles I.[75] As befitting a "courtly" rather than an academic grouping of the muses, this was a rather strange set of the liberal arts, including music, writing (*Graphica*), languages, astronomy, geometry, medicine, arithmetic, "Chymicke," fencing, dancing, and "Pictures Royall." Notably, the trivium (grammar, logic, and rhetoric) was missing, as were Bolton's preferred studies of antiquities and history. *Graphica* referred not to the study of letters but to the actual act of penmanship, whereas languages were specifically modern languages not usually taught in schools. Likewise, "Chymicke," interpretation of "Pictures Royall," dancing, and fencing were also not usually part of the academic curriculum at this time.

The arts of the *Musae Aulicae* flaunted expansive views of nature enlarging the power that such arts might lend kings. "Chymicke" offered an imperial rule over all the gods (referring to how metals were known under the names of gods). Geometry announced to Charles that "the earth's too little for thy throne. / And that Sage that did infinite worlds faine, / Would say thee worthy 'ore them all to raign." This view of Charles's potential extraterrestrial rule referred to Anaxarchus's theory of the infinite universe and its affects upon Alexander the Great, according to Plutarch, which Peter Heylyn had once described to Charles when he was prince. In a 1621 dedication to Charles, Heylyn depicted Alexander the Great's reaction to hearing about Anaxarchus's theory. Learning of all those worlds still to conquer, Alexander burst into a "violent irruption of teares."[76] This passionate response bore witness to Alexander's great desires and thus noble character. Pursuing such extreme goals as extraterrestrial conquest was the mark of a prince, since the "hearts of Princes are in a manner boundlesse, one world is not sufficient to terminate their desires."[77] Early Stuart promoters of colonial expansion knew to contrast this ambition to the limits that bookish schoolteachers wished to set upon their students. Promoters of colonial ambition spurned the pedantic teachings of schoolmasters. A preacher supporting Virginian colonization in 1609 ridiculed those who would advise princes that it was illegal "to inuade the Territories of other Princes, that are in quiet possession." He compared their "schoole" to the devil's hellish "cabinet of Abstruse Studies." They treated royalty like little children when they took it upon themselves "to nurture princes, as petties: telling them that they must not make offensiue warres."[78]

Verses in the *Musae Aulicae* also celebrated the court's revels and the ways their mastery of stage effects dramatized imperial rule of the elements. A dramatic ocean scene showing the waves first frozen by winter and then melting and storming with "fiery foaming surges" was achieved by "flouds of silke." These staged Charles's claim to "the whole right of seas, / Who mak'st and boundest

Oceans as thou please."[79] Sir Balthasar Gerbier described the high bar set for staging realistic seascapes and other natural phenomena in the period. "Seas must not only be seen to have a naturall motion, but heard to make a noise of breaking of their Waves on the shoar, and against the Rocks . . . Spectators may think those sights to be naturall operations."[80] Gerbier claimed to have put on such a show for Villiers in 1628 at the state-of-the art theater Villiers had built at his private property, York House, just before setting out for the naval campaign in France.[81] Thus, just as the public amphitheater projectors aimed to stage convincing illusions of nature and grand naval battles, Johnston and Kynaston linked the courtly muses to the abilities to symbolize control over the elements in the contemporary masque.

Kynaston saw his project for a noble academy as a rival to the public amphitheater. He anticipated that one objection to his academy might be that letters patent had already been issued "to erect an amphitheatre . . . to entertaine Masters that shall practice armes and shall exhibite showes and spectacles to the people." However, "the maine end therein being gaine and commoditie to the private purses of the patentees," the amphitheater could not compare to his more publicly spirited academy. His academy was "noe emulous projecte of some odd fellows who gave an outrecuydance [sic] of themselves," that is, who believe too much in themselves.[82] In contrast to the London amphitheater of presumptuous "odd fellows," Kynaston described his project as also operating in the limelight, but one which would bring glory to the nation. By being situated in London, he claimed his academicians would "best shine . . . as on a loaftie stage and publique amphitheatre."[83] By "public," Kynaston referred to the role of his institution in the body politic. Since the aim of the London amphitheater projectors was purely financial, he claimed, this made their proposal private, as it served their "private purses." His noble academy, while far more restrictive in terms of who could in fact be admitted into the space, was public because it served an important role, he claimed, for the state.

Kynaston argued that his academy was just like any other transplanted industry that augmented the powers of the state. It was modeled on the noble riding academies of France, where English gentlemen often sought training in riding the great horse (as had George Villiers at Angers in 1611).[84] London was flooded, Kynaston claimed in one manuscript defense of his work, with French masters of riding and dancing, an issue he analyzed through the lens of economic policy. He compared the education of English gentlemen in French academies to the importation of French wine; both lost the nation revenue that could be recouped by establishing the industry locally.[85] In his published account of the academy, he noted that other nations reap both "honour" and "gaine" when Englishmen repaired abroad for their training. Some of his subjects, like the "sciences of Navigation, Riding, Fortification, Architecture, Painting, and such like" were taught around London but "in dispersed places" and often "perfunctorily . . . for gain." This was especially the case with "diverse strangers" in the city who claim to teach "all the liberall arts and sciences" but upon trial were found "egregiously ignorant." Thus, his academy served "the publick good of the common wealth."[86] Kynaston's academy, as Hartlib noted, offered a prime example of how those "Inventions which a Man gaines in forrain partes are as well legal and are encouraged by the king's Patent as wel as those which a Man find's out by his owne industry."[87]

In fact, Kynaston's characterization of his academy as a French importation undersells its innovation. It offered not only training in riding and dancing, but forms of knowledge drawn from a wide social scale across London. It balanced the training of the body with the mind; each student in the academy (unless of extraordinary ability) should only work at any one time upon "two particular Science Arts or Qualities whereof one shall be Intellectual, the other Corporall."[88] Even the intellectual forms of knowledge, however, integrated manual skills through experiment and invention.

Kynaston designed novel infrastructures for teaching and investigating experimentally. As the prospectus for the academy demanded, the site would provide "a place haveing all mathematicall instruments and other occomodations for demostracon and accon [demonstration and action]."[89] He envisioned collections of books, charts, instruments, and rarities.[90] Faculty should teach through "Demonstration and Experiment." Furthermore, this was a research faculty, with specific requirements for faculty to produce and record new knowledge. Each faculty member was required to leave "in writing some memoriall of the most selected Points, Secrets, Experiments and Demonstrations which do belong to every of their Arts and Science."[91] They were to do so both annually at Christmas and at the end of their lives. In particular, the "Professour of Philosophe and Physick and his Assistant or Assistants shall from time to time . . . make experiments of naturall things, chiefly for medicinall use, and what they finde certain shall be recorded, and what they finde otherwise shall in a book by itself be noted, how and in what manner experiment was made, and how it failed." In a clear departure from contemporary recipe practice, both certain and doubtful knowledge were to be recorded, but specified and separated. Specifying failure would save time and trouble and alert others to claims of "Imposters."[92]

Knowledge would only be "received for the doctrine and learning" of the academy "which shall be found true, after sufficient experiment, or demonstration, and no other."[93] The "Bookes of *Secrets, Experiments,* and *Demonstrations*" collectively authored by the academy would only be communicated to faculty and students who had studied at the academy for more than seven years (unless a very great reason urged the regent and faculty to show the books to somebody else).[94] Hartlib considered having access to this resource as one of the main benefits of subscribing to the academy.[95]

Kynaston also offered inducements for experimental gentlemen at large to communicate their knowledge to the academy. These were awards of reputation rather than of money. Any "Gentleman" communicating any "naturall Experiment or Secret" to the academy, "and upon tryall it be found true and good his name and Experiment shall be recorded in *Libro Nobilium* [book of nobles], for a perpetuall honour unto him."[96] Practitioners of any form of learning were also encouraged to confer with the academy's faculty. If they imparted "any singular demonstration in Science or Secret in nature profitable," they would both be honored in *Libro Nobilium* and receive a financial honorarium.[97] Notably, gentlemanly social status did not verify knowledge claims; the experiments of gentlemen in particular required trial. Conversely, profitable secrets rendered by practitioners were not automatically suspect due to their base pecuniary motives, but also found a place in the book of nobles.

Kynaston's refocusing of noble training on demonstration and experiment earned him royal patronage. Charles I granted Kynaston's academy a royal license as well as 100 pounds from the treasury.[98] As a further reward and incentive, Charles also "offered him to aske many things which yet hee hase never done and so waites only for an Opportunity," or so suggested Johann Christoph Berger von Berg, a projector with whom Kynaston partnered.[99] The academy was established in the new development of Covent Garden and at first did well, although both failing finances and an epidemic of the plague soon scattered it, and it was never reconvened.

"Nothing upon Trust"

The best surviving record of Kynaston's natural investigations from the period of his academy are to be found in his manuscript commentary appended to his Latin translation of Chaucer's *Troilus and Criseyde*.[100] In his comments, we hear how he dug, planted, prodded, poked, burned, calcined, observed with telescopes and magnifying glasses, and collected phenomena from a wide social array of informants, from gentlemen to poor midwives. By casting his experimental activities as a commentary upon Chaucer's chivalric poem, Kynaston made clear the connection he was attempting to forge between these investigations and an English tradition of knighthood.

Kynaston is today best known as a so-called metaphysical poet who engaged many scientific themes in his poetry. What has not been recognized is the extent to which these were not merely tropes that he deployed but in fact reflections of his actual experiments and observations. Kynaston's commentary on Chaucer records his investigations of many of the phenomena that appear in his own poetry, from the sweetness of the "sugar of saturn" (lead acetate) to the "blue light" of the subterranean fairy world, to the movement of the sensitive or "coy shame-fac'd plant."[101]

Few topics escaped Kynaston's interest, from the weird sisters of Scotland to wraiths, topics he was able to investigate while accompanying Charles I on his progress to Scotland in 1633.[102] He discussed the theories of Paracelsus, Croll, Burggrav, Galileo, and Kepler, noting his own chymical experiments and microscopic and telescopic observations. Through the microscope, Kynaston observed veins, hair, a nettle, "West Indian" silk grass, and an astroites or star stone. He also described many objects he had observed in collections: the animal from whose horns Russian bows were made in the St. James's menagerie; the Poliphon of "my excellent friend Daniel Farrant" [a violist and the Poliphon's inventor] and a wonderful harpsichord of the king; the strange "houre flower [*Mirabilis jalapa*], which I saw in the gardens of my ever honor'd frend Sr Edward Sackville Knight marchall"; the marvelous "West Indian" sensitive plant, in the garden of his friend, Sir John Danvers; a half-inch square piece of parchment with all the Psalms on it, in the cabinet of King Charles, written by a blind man, "one Mr. Switzer dwelling in Shoe Lane"; a cup of peppercorns holding 365 turned and gilt ebony cups, in the collection of "my most excellent friend Mr. Wolfe Pombler [Johann Wolfgang Rumler]," the royal apothecary; an ancient rag of incombustible asbestos in the collection of the "noble & learned Mr. Gilbert North"; and a thunderstone the Earl of Bristol had had dug up and presented to the king, of which Kynaston received a two-pound piece.

Kynaston drew these phenomena from across a markedly wide social range. His survey of the London experimental scene likewise reaches far across social divides. He points out topics that should not be considered a "contemptable contemplacion in philosophie"; the seemingly mundane act of moderating an "all consuming fire" by throwing ash or turf upon it could produce wonders "past beliefe."[103] He reasoned about the nature of flame and of fuel from phenomena as diverse as the making of charcoal and a secret "imparted unto mee by an excellent Chymist" of how to age wood and draw out its oily parts, for the making of viols and lutes "as light & drie & as well seasoned as if it had been cut a hundred yeares."[104] While Kynaston offers many accounts attested by socially elevated observers such as himself, he was also keenly interested in feats performed by London's poorest residents. He wondered at the secret of curing thrush of one "poore woman": "let a young wolfe but sucke of a womans breast once or twice," and then let any child troubled with thrush, "of which many infants doe weekely die . . . sucke of that womans brest," and the child will be cured. Kynaston's interest included the potential for profit in such secrets; as he noted, "By which secret I have knowne a poore woman get her living & maintaine her selfe well."[105]

This social range appears in the two plants that Kynaston singled out as the most "admirable" things he had ever seen. The first was the sensitive plant in the Chelsea showcase garden of his friend Sir John Danvers (discussed in chapter 5).[106] The second was "a shrub then a plant which I got from a midwife in London & had a quarter of a yeare in my study which she called a Ladies Rose."[107] The twigs of this plant which grew up "close together" would open up "if it be brought into the chamber of any woman that is travailing of child just as the mouth opens for the birth," but "if the woman shall die in travaile the shrub will not open at all." Kynaston juxtaposed two plants drawn from opposite ends of the social spectrum and from the very different contexts of Danvers's garden and the chambers of laboring women around London. In so doing, he dramatized how he hunted down wonderful properties of nature all across town with an open, experimental attitude. In case these two plants or anything else in his commentary "shall seem strange to the reader," Kynaston avowed that he had "written nothing upon trust, but what himself hath been an eie witnes of & is found to be true & experimented."[108]

The ways that Kynaston blended gentlemanly curiosities and for-profit secrets of nature in his academy and experimental interests also apply to his own career as an inventor. As von Berg suggested, Charles's patronage of the academy included a promise to extend privileges and protections to Kynaston. The inventions that Kynaston developed in relation to the academy swiftly gained state support. Dr. Edward May, physician-in-ordinary to the queen and the professor of philosophy and physic in Kynaston's academy, wrote to Kynaston describing a "pendent furnace or fireplace for use on board ship," at the end of 1636.[109] Two weeks later, the Lords of the Admiralty were considering "Sir Francis Kynaston's proposition" for installing it.[110] The next month, the Earl of Berkshire, himself a self-proclaimed inventor of furnaces, was discussing Kynaston's proposal with the Lords of the Admiralty, and ensuring that the persons appointed to oversee building the furnace in the ship would "be at the appointment of the Earl and the assignees of Sir Nicholas Halse's invention," another socially elite furnace patentee.[111] Five months later, the Officers of the Navy reported to the Lords of the Admiralty, that "we have given our best assistance to Sir Francis Kenniston and Dr. May for fitting a hanging furnace for one of his Majesty's ships."[112] Kynaston was not the original inventor of this furnace; rather, the knowledge-collecting infrastructure of his royally supported academy functioned as a way to move this invention swiftly into his hands and then into naval application.

Kynaston's Interloping

Although Kynaston sharply differentiated his academy as serving public ends from the private interests of the amphitheater projectors, in fact Kynaston participated in a wide range of for-profit projects intersecting with his academy and appropriating knowledge from others, including from individuals associated with the amphitheater project, such as Drebbel. One of the figures around town whom Kynaston appears to have tracked closely was Drebbel; Kynaston seems to have been greatly interested in how he parlayed the presentation of wondrous technologies into court access. Kynaston criticized Drebbel's perpetual motion, while describing a curious globe of his own that he had presented to the king. The "Drebbeli perpetuum Mobile," might "moove or bee true in smal Models but not in great," claimed Kynaston. Kynaston presented Charles I with "a Curious Magnetical Globe for a new years gift at which the King took great delight, and which hase many great Rarities in it," which must surely have reminded onlookers of the famous spherical perpetual motion Drebbel presented to King James in 1607.[113]

Kynaston partnered with several individuals in developing inventions, including Samuel Hartlib and the Moravian nobleman Johann Christoph Berger von Berg, who, like Kynaston, competed with and imitated Drebbel in many inventions.[114] Von Berg signed one contract with Sir Francis Kynaston, and another in 1640 with Caspar Kalthoff the Elder (1606–64) in Hartlib's home; in the latter, von Berg and Kalthoff vowed to share all their secrets to one another and not to disclose them to further parties without permission.[115] Von Berg attempted to patent five inventions in England.[116]

In 1632 von Berg claimed to fellow Moravian Jan Amos Comenius that Drebbel was still sweating after the perpetual motion, but in vain, for all of his "boasts." What von Berg seems to mean here was that Drebbel was still failing to remake his small perpetual motion into a larger form that might be utilized practically, as Kynaston had complained. Von Berg reported that he heard Drebbel say with his own mouth that "either he would find it or nobody would."[117] In 1635 Hartlib described von Berg as "very intimate" with "Kuffler the successor of Drebbel."[118] Von Berg traveled frequently between the Netherlands and England, taking out a patent in the Netherlands for a perpetual motion in 1629, as well as in 1631 for an invention for lifting up shipwrecks from under the water.[119] In 1634 von Berg obtained a letter from Gresham College professor Henry Gellibrand and from John Wells, Keeper of the Naval Stores, who worked closely with Gellibrand on improving the study of magnetic variation for the benefit of navigation. Gellibrand and Wells testified that von Berg's "modells of instruments" for surveying, sounding the depths at sea, and lifting sunken vessels were "very vsefull for the Commonwealth."[120] Von Berg (whom Hartlib refers to as "Sir Christopher"), informed Hartlib in 1635 that the iron ovens previously built by Drebbel and the Küfflers were being imitated in Holland out of cheaper materials and suggested several ways he might improve them further.[121] From 1635 to 1638, von Berg wrote several letters to Samuel Hartlib from the Netherlands about the inventions he was developing there, including in Drebbel's hometown of Alkmaar.[122] In 1635 Hartlib noted von Berg's invention for cooking with a lamp, "which may bee exceeding profitable and vseful in Travel . . . and Plantations in Navigation and ships and the like."[123] This recalls both the portable ovens of Drebbel and the Küfflers and the pendant furnace May and Kynaston would install on a navy ship.

Belying Kynaston's differentiation from the amphitheater project, he, his associate von Berg, and Samuel Hartlib were all interested in developing for-profit and didactic shows. Such ideas first reached Hartlib through discussions with the Westminster schoolmaster John Brooke in 1634.[124] Brooke critiqued the 1633 pedagogical work of Comenius for insufficiently stressing the importance of sensual apprehension of material things. He proposed publishing an *Encyclopaedia sensualium*, that is, a collection of knowledge gained through the senses, an idea that Hartlib embraced. Brooke conceptualized this encyclopedia in terms of wondrous, staged displays. He advised that the "Learned Shewes must first bee performed by Candle-light in darkned Roomes" and "the Artifice of them must bee hidden."[125] As a model show for the *Encylopaedia sensualium*, Hartlib noted "Drebbels feate to shew the didactick of thundring and lightning" by appearing to create storms, about which Kynaston had informed him.[126] Noting how Kynaston's academy was provisioned with the best musical manuscripts, Hartlib observed that "Mechanical Musical shewes and others of all kindes" could serve didactic purposes and there were also "shews for all sorts of Berg-werch [mining technology]."[127] He also noted that Kynaston's academy would have "a Latin Theater."[128] The Academy offered "publick Musick" each Tuesday afternoon, as well as "Publick Shewes" and "Presentations," which required tickets for admittance.[129] Based on Kynaston's 1635 published academic masque, the *Corona Minervae*, the theater of his academy allowed for "significant scenic effects," and seems "to have used a proscenium stage with a curtain in front, in imitation of Whitehall," thus fusing together "academy and court."[130]

In 1635–36, von Berg, the associate of both Hartlib and Kynaston, was much taken with this idea of didactic, for-profit shows. He noted one "Projector," Richard Bradley, who could "Invent most Curious things for shew's that will draw very much the People et bring in much gaine."[131] The inventions of Richard Bradley included massive spinning devices that could "set fortie or three score Children on worck."[132] Hartlib also recorded von Berg's interest in the idea of an inventor named Rous "to Invent profitable shewes especially some incredible and notable Effects of Nature and Art and by others to execute them in faires and Tearme-times. This way much and speedily may bee gotten."[133] Von Berg argued that his own invention of a mill could be used "by Way of Shews . . . both privatly and publikly without any staine of disreputation And the rather because it will be as well profitable and pleasurable and pose the best wits which shall not bee able to finde out the secrets of this Inventtion . . . or dive into the ground of them seeing both the Mathematicall and Mechnical principles are newly found out by the Author and have been heretofore unknowen."[134] Thus, inventions could produce profits in two ways, first for the purpose for which they were intended, and second as mechanical marvels for a paying public that could develop mental astuteness by attempting to pierce the marvel's mysteries. Testing the public's intellect with new knowledge, von Berg argued, elevated public, for-profit shows above the possibility of disrepute.

Ultimately, von Berg fell out of favor with Hartlib and his friends. Observers considered von Berg's inventions derivative, "for the most part nothing else but new Applications vpon old Principles or Inventions."[135] As John Dury wrote point-blank to Hartlib in 1643, "he hath no Conscience, but doth intend to defraud yow."[136] Von Berg, fitting many of our common stereotypes of projectors as foreign, fly-by-night swindlers, was exceptionally fraudulent in his practices. Appropriation of knowledge, however, was par for the course in this setting, producing much churn, enmity, and failed relationships, as the careers of Hildebrand Prusen (discussed in chapter 2), Thomas Russell (discussed in chapter 3), or Richard Steele (discussed in chapter 4) have already illustrated.

Interloping by social elites among practitioners of manual knowledge was often nothing to be ashamed of. It was, in fact, what won Kynaston's academy royal support, in contrast to Bolton's many learned proposals.

Conclusion: Knowledge and Power

A derisive attitude toward traditional bookish scholarship in the early Stuart period queries some assumptions about the ways that period artisans and scholars collaborated by joining together the hand and the pen. Francis Bacon, in particular, is often seen as canonizing a fusion of artisanal experience and scholarly literacy in such forms of knowledge as his *experientia literata* (literate experience), which sought to record experiential knowledge in textual form, much as Kynaston's books of experiments aimed to do.[137] Yet, this assumed collaboration does not account for the denigrated status of the liberal arts in the Caroline court nor for interloping practices of appropriation. Forms of physically demanding knowledge excluded from the liberal arts due to their socially inferior status were more successfully integrated into new noble academies in early Stuart England than were forms of textual erudition promoted by Bolton. Although Bolton repeatedly attempted to show that ancient history and antiquarianism could prove useful to a political career, to large-scale projects, and to other contemporary profitable enterprises, he failed to prove convincing. Framers of noble education, from the princes in their versions of the Artillery Garden to Kynaston in his establishment, wished to train interlopers. They wanted the elite student to become inventive, ambitious, manually dexterous, martially vigorous, experimental, and skilled at appropriating useful and profitable knowledge from across society and around the world. They did not wish to act like a "stupide pedant," in the words of King James,[138] telling princes not to make offensive wars nor to take what belonged to others.

Rather than collaboration between two clearly defined spheres of the hand and the pen, we find in these early Stuart educational projects interloping in all directions. The princes themselves imitated the martial and mathematical training of civic militias. That was also imitated by the amphitheater projectors, who combined it with a wide array of forms of knowledge, from juggling, to courtly masques, to natural knowledge, to both Latin and English "Tragedies, Comedies and Histories . . . full of high state." Kynaston's academy in turn competed with the amphitheater. While Kynaston condemned the amphitheater as a for-profit and self-centered initiative, his own academy integrated spectacular forms of education and intersected with his profitable projects, which themselves appropriated knowledge from others.

The vision of knowledge that Kynaston aimed to offer in his academy was encapsulated by a period commonplace: *arte et marte,* through art and war, or knowledge and power. Kynaston named his academy the *Musaeum Minervae* in order to call on such notions of fused *ars* and *mars* personified by armed Minerva. In a masque at the academy celebrated in the presence of the young princes and princess, Kynaston dramatized the figure of Minerva as a fusion of "Powe'r and Wisdome" who would "advance" the "use . . . of Armes and Arts."[139]

At one point in the masque, it seemed as though Minerva dared to offer her royal guests a pedantic form of erudition by distributing several books among them. This elicited the shocked query, "How dare you use the Prince thus, schoole-mistris? Are th'armes, and arts you promis'd to his view To be pick't out of bookes?[140] As it turned out, Minerva's books were in fact foods cleverly shaped into the form of books, complete with many an ingenious pun, which, along with the food, might be enjoyed with gusto. "Aulus Gellius," for example, on closer examination proved to be, "All gellies."[141] The simulated books offered a means to stage a deliciously flippant attitude toward traditional scholarship. If at first the young royal audience might have thought that Kynaston's academy wished to spoon-feed them the "sober" learning of history and Latin letters that Bolton promoted, Kynaston cleverly demonstrated to them that the *studia humanitatis* would not be on the menu.

CHAPTER 7

Unlimited Invention

The arena of the undisciplining of knowledge was a chaotic one. Projects often jostled against one another. They generated criticism and social unrest. Applicants for patents framed their proposals as interventions in the state in ways that contravened the crown's desire to maintain control over the secrets of empire. Interlopers' magpie-like tendency to steal and imitate pieces of knowledge also undermined the secrecy so important in military engineering especially, but also for the purposes of interstate commercial rivalry. As free agents, interlopers might also sell intelligence or a project to a rival power.

Charles I established a new state institution for invention at Vauxhall in Lambeth that drew on the patent system while minimizing some of its drawbacks. Vauxhall differed from the technical workshops founded at other courts by allowing its investigators to pursue self-directed, unbounded projects pushing the limits of possibility, as in many period projects. It differed from the patent system by moving invention away from intervention by investors. By situating several investigators alongside one another and by including a repository of models and previous attempts, it extended the time frame and intellectual abilities allocated to investigating difficult goals. It also made many specialized and innovative instruments, engines, and furnaces available for this research. This new institution grew from the valorization of hands-on learning discussed in the previous chapter, from Charles's long-term engagement with projects as in the great design discussed in chapter 4, and from efforts to recruit international engineering talent into the Ordnance office.

Recruiting Talent

Ben Jonson's 1625 play *The Staple of News* staged terrifying rumors of new enemy technologies. One rumor had it that "Spinola is made Generall of the Jesuits," the which society is now the "onely Enginers of Christendom." The Jesuits presented Spinola with an "Engine" to "winde himselfe with, up, into the Moone." Spinola also devised another "new Proiect: To bring an army ouer in corke-shooes" with "fourescore pieces of ordinance, Mounted upon cork-carriages, with bladders, In stead of wheeles." Spinola had an array of other wonderful inventions at his fingertips, including a burning glass found in "Galileos study" that could "fire any fleet that's out at sea."[1]

The fearmongering of Jonson's *Staple of News* drew on the sinister powers of innovative offensive technology, such as the hellburner or fireship pioneered at the Siege of Antwerp (1584–85), which was the weapon of mass destruction of its age. Fireships appeared to be regular ships but were in fact laden with bombs designed to explode on contact with the enemy. This was also an exciting time for pyrotechnics.[2] Seventeenth-century round shot was "travelling at similar velocities to bullets from modern handguns," and sometimes even at supersonic speeds.[3]

One credulous character in Jonson's play, fearing that Spinola's "strengths will be unresistible," asked if there was any "news against him, on the contrary?" He was relieved to hear of the "invisible eel" of "Cornelius-Son" for sinking shipping, a reference to Cornelis Drebbel's newly invented submarine, and a piece of prized "newes of State."[4] Cornelius-Son's boat was described as "an Automa" that "runnes vnderwater, / With a snug nose, and has a nimble taile / Made like an *auger*, with which taile she wriggles / Betwixt the coasts [*sic*] of a Ship, and sinkes it streight."[5] This wriggly automaton indicated a terrifying ability to insinuate and destroy, twisting itself into a ship's interior. Beginning with his 1622 project as prince, Charles I was eager to develop this technology of great subterfuge and power (explored in chapter 4). As Jonson suggested in his play, the submarine symbolized English resistance to the new-fangled military engineering dramatically deployed in continental war.

With the accession of James I, and the retreat from the Elizabethan confrontation with Spain, royal investment in ordnance and military engineering had been scaled back.[6] The youthful engagement of Charles in military engineering, "out of the way mathematics," and inventions led him to expand this investment already as a prince. Charles oversaw, tested, and even invented instruments. He also began to recruit foreign military talent even before he succeeded to the throne. A petition of February 27, 1624, from Arnold Rotispen (a.k.a. Rotsypen and Rotsipen) to Prince Charles asked to "be sworn his servant" and to be granted a patent for diverse "rare invencions and manufactures wherewith he hopes to serve your highness, and profitt the Common wealth."[7] In due course, James granted Rotispen a privilege according to the terms Rotispen had requested from Prince Charles. This was not only for his "new Invention of making Gunns," for which Rotispen had already shown the prince a proof, but also for "some other rare inventions and manufactures to be perfected by him."[8] According to the text of the privilege that was issued, the prince "had made triall of this invention and doth approve thereof."[9]

In 1626, Charles granted Rotispen letters of denization, a pension, and another patent for making guns a new way, illustrating Charles's willingness "to cherish such ingenious, usefull and profitable Inventions, to the Incouragement of others in the like laudable Endeavours."[10] He also ordered the development of Drebbel's underwater technology that year, without, at that point, any particular naval engagement in mind. Sir William Heydon was charged with overseeing "the making of divers water Mynes, water Pettards, forged cases to be shott with Fireworkes and boates to goe under water and the like."[11] This was a large-scale endeavor that Charles veiled in secrecy. Heydon turned to Villiers to request a warrant for 360 "forged iron cases with fierworkes," 50 "water mynes," 290 "water pettards," and 2 "Boats to conduct these fier Ingens under water."[12] The very next week, Heydon was ordered to provide Drebbel and his colleague Rotispen with workspace in the Ordnance Office. For "the better advancement" of their inventions, they were to be given space in the Minories in the "custody" of Heydon, "that they may be there at hand and fitted upon all occasions for the services abovesaid."[13] Charles also ordered secure storehouses "for the safe and private keeping of all such Artifices, Engines, Municions and habilements as from tyme to tyme shalbe by them contrived and performed."[14]

In their confinement in the Minories, Drebbel, and Rotispen set to work on their underwater technologies. In July 1626, Heydon was paid 1,525 pounds (more than $400,000 today) for "certain forged iron cases, with fireworks, water mines, and water petards, with boats to conduct fire-engines under water, appointed to be made ready for his Majesty's present service."[15] Rotispen and Drebbel were paid 100 lb. (about $30,000 dollars in today's money) each for their efforts.[16]

An opportunity to deploy such inventions soon arose at the battle of the Isle of Ré, which "developed into a war involving new-fangled special ships" on both sides.[17] The English faced as their enemy the famed engineer Pompeo Targone, who had previously been employed by Ambrogio Spinola, the Genoese commander of the Spanish forces at the siege of Ostend in 1604, where Targone's marvelous inventions gave rise to much reportage.[18] Thus, the battle realized the scene Jonson had imagined in the face-off between Spinola's engineers and Drebbel.

Villiers, the leader of the expedition, placed great confidence in Drebbel's underwater boats, floating bombs, and torpedos, which he endeavored to shroud in the greatest secrecy. He instructed John Coke, the Secretary of State, to make sure that "Cornelis the Dutch Ingenier be sent" along with more supplies. The greatest need of the Army was "able men of his profession, & lett him bring such fyrworkes with him as were left behind & meanes & materialls to make more heare needefull for the besieging of a Fort."[19] "Cornelius Drubble" received "thirty last of powder and 200 land Granados and 200 morter Granados" to deliver to Villiers in France.[20] Along with Drebbel sailed John Heydon as a replacement for his brother Sir William Heydon, who had drowned almost immediately upon arriving at Ré. Secretary Coke reported to Charles that in "the dispatch of Mr. Heydon and Cornelius we have used such expedition that this day they purpose to set sail, and that secrecie that Mr. Burrell [the shipwright, William Burrell], who with great diligence hath in two dayes prepared the foure barques, knoweth nothing of the design."[21] Privately, Heydon complained to Coke that "every day I find more instruments imployed which may make the business more publique then I could wish it who have not acquainted any soule with the least particuler more than Mr. Cornelius."[22]

The battle made manifest the extreme lack of engineering talent on the English side, and Villiers, assisted by Heydon, quickly began recruiting. A corps of six international engineers was recruited to the Navy, several from the Dutch, at the salary of 100 pounds per year for life, with an additional 4 shillings per day while employed on active service (compare that to the salary of 12 pounds per year of Ordnance Office artificers).[23] Given his prior career serving in the army of the Prince of Orange, John Heydon played a pivotal role in these efforts and was asked to review the quality of all the potential new recruits.[24] After the war, Heydon also recruited swordsmiths in the Netherlands. In 1672 two German swordsmiths working at the Hounslow sword manufactory recalled Heydon's efforts "to bring the Manufactorie of swordblademaking" from Holland to England in 1629. These German swordsmiths had been displaced by the

Thirty Years' War. In order to encourage them to move from Holland to England, King Charles "caused severall Mills to be erected at Hounslowheath for there use, where they made swordblades for his Ma[jes]ties stores and for the Gentryes wearing as good and as sharp as any in the world."[25]

Heydon faced difficulties recruiting talent from beyond Holland. Villiers asked him to find a French engineer since Villiers specifically wanted to fight the French with a Frenchman, according to a courtly logic of one-upmanship (and even though Targone was Italian). With some difficulty, Heydon identified as a candidate "Mons[ieu]r La Roche, who indeede is a frenchman of Sedan and a principall Petardier, but otherwise no professed Ingeneir, nor any of that nomber." Villiers accepted this idea.[26] This was Bartholomew de Créqui, lord of la Roche, who had served as the captain of the petardiers in the Dutch Army, and who would eventually be hired as captain of the fireworks in England and knighted by Charles I in 1645.[27]

Drebbel's inventions (as well as Targone's) did not stand up to the rough weather and choppy seas. Nevertheless, Villiers continued to rely heavily on them. Drebbel and his assistant and son-in-law, Abraham Küffler, prepared for Villiers's return trip to Ré, which never took place due to Villiers's assassination in 1628. Küffler made a list for the new engagement that indicated a planned massive scale of deployment, impressing twenty-four barges "to be fitted for our purpose" along with their watermen, thirty barrels of powder, forty "floters," and two hundred musketeers.[28] Heydon sounded hard-pressed in a hastily written letter to the Secretary to the Admiralty, Sir Edward Nicholas, describing how he had "received commandment from my Lord Duke for the furnishinge of 20 or 30 boats with a great quantity of Engins to be sodaynly prepared by Mr. Cornelius and his sonn."[29]

Thus, the failure of Drebbel's engines at their first engagement in France did not spell the end of his career. Villiers planned to redeploy Drebbel's engines on an even grander scale for his return to battle. Drebbel continued to serve John Heydon as an engineer in the Ordnance Office for the rest of his life.[30] The partnership between Drebbel, the Heydons, and the Crown that dated back to the 1622 project for deploying the submarine in the Indian Ocean thus persisted. The battle of the Isle of Ré offered an opportunity to utilize the underwater technology that Drebbel was developing in the Ordnance Office. Rather than retreating from investment in high-fangled military technology, after the battle Charles continued to develop the Ordnance Office, with the help of Heydon, as a site for experimentation.

Defending "Vulgar Teachers"

In 1633, the year of Drebbel's death, Heydon established Richard Delamain in the Minories, possibly in the space vacated by Drebbel. Delamain was charged with "framing and contriving of warlike engines . . . and forts, raising batteries and entrenchments for both assault and defence."[31] Beforehand, Heydon had employed Delamain, in place of "sundry aliens" hired as engineers in 1627, in a surveying "expedition for measuring forts and castles." Through the mediation of Heydon, Delamain requested to be admitted to the open position of one of the engineers who had been recruited in 1627. He also suggested that he might pen "sundry mathematical tractates" suitable for teaching the young prince.[32] In 1630 Delamain gave a pasteboard version of an instrument, called the "mathematical ring" or circles of proportion, as a New Year's gift to King Charles, together with a manuscript book explaining it.[33] He was awarded a patent of ten years, and he published his book, dedicated to Charles I. In 1633 he was awarded a second patent for an improved instrument for 14 years that also authorized him "to teach the use of the said instrument, or any other mathematicall arts."[34] He re-issued his book. He was also appointed as a royal engineer.[35]

Meanwhile, Delamain engaged in a very public priority dispute with the clergyman and mathematics tutor William Oughtred (1574–1660) of Surrey over both the circles of proportion and another instrument, the horizontal quadrant. Much has been written about this controversy, often from the perspective of who first designed these instruments.[36] This was certainly the view that Oughtred took of the argument. Katherine Hill has also framed the dispute as a disagreement over the appropriate social role of the mathematical practitioner. Here we look at how this dispute additionally involved the royal interest in these instruments, the suitability of instruments for educating princes, and the ways that inventions were considered in patents.

In his dedication to King Charles of his 1630 work on the circles of proportion, Delamain pointed out that "Magnanimous Princes and heroicall spirits" sought the "quickest and easiest waies in the operation" of the mathematical arts.[37] In a publication on the horizontal quadrant, Delamain likewise noted that "Nobilitie, Gentrie or others" would find it "a recreative Instrument," with "copious use" and "great facilitie, and expedition in operation."[38] Oughtred criticized this noble taste for easy instruments. According to Oughtred, he had first designed the horizontal quadrant when the instrument maker Elias Allen wished to give Charles a New Year's gift in 1627. Oughtred had observed the delight Charles took in the "great concave Dyall at White-hall" made by Edward

Gunter. Gunter published a description of the dial on the command of Prince Charles.[39] Seeking to please Charles with "some pretty Instrument," Oughtred designed a horizontal version, although neither he nor Allen bothered to make it. According to Oughtred, when Delamain saw that Oughtred did not "make any great account of it," he appropriated it as his own.[40]

Oughtred's student William Forster defended Oughtred's priority in a 1632 work of Oughtred's that Forster translated from Latin to English. There he cited Oughtred as explaining that he had not published on his instruments before Delamain because Oughtred did not much value instruments. He did not believe in teaching mathematics "by instruments, but by demonstration." Only "vulgar teachers," Oughtred had said, began "with instruments, and not with the sciences . . . to make their scholars only doers of tricks, and as it were jugglers."[41]

In his response, Delamain argued that what mattered was making an invention available for public use. He criticized Oughtred's own admission that he had possessed the circles of proportion "for more then twelve yeares past, but put it not to use."[42] Delamain, "by opening the Cabenet," shewed its treasure to the world. Furthermore, in developing the practical uses of the instrument, he had encountered many difficulties and expended much effort. He reasoned that those who did not realize their inventions in a timely fashion could not legally restrain others from making them available for public use. It was "too malepart, & too rigid, to ty the liberties of others in their actions, who desiring the good of others in not concealing things, if they shall by their industrious labours, shew the excellent use of such or such a thing, to a publicke view."[43] To that, Oughtred retorted that it was "malapert sawsinesse" for someone of Delamain's status to comment on legal and political issues.[44] Delamain's reasoning, however, was sound from the perspective of patents. As we have seen, it was also not unusual for figures from across the social spectrum to debate precisely these questions as part of an application for a patent. Unlike Delamain, Oughtred did not engage in patenting instruments.

In responding to Oughtred's comments about his teaching style, Delamain conceded that it was accurate for Oughtred to call him a "vulgar teacher," but not a "juggler."[45] Oughtred's remarks plainly insulted "many noble personnages, with too grosse . . . an attribute, in tearming them doers of tricks" when they made use of practical instruments because "their more weighty negotiations" did not allow them the time for more abstract studies. Oughtred's self-reported snubbing of the New Year's gift he had designed to please Charles also could be seen as insulting to the royal taste in instruments. Delamain argued that instruments were not just a quick means to mathematics, "but even the glory of it." It was better to teach through instruments that made ideas obvious to the senses than by textual means alone.[46]

Delamain's acceptance and defense of the label of "vulgar teacher" might seem odd for someone receiving royal patronage. It spoke to how Charles sought learning from across the social spectrum while spurning scholarly forms of education, as discussed in chapter 6. Delamain's claims about the public good likewise agreed with the contemporary patent system. Patents were not rights awarded to first inventors, but grants made to those who put inventions to public use or who developed any resource lying waste. Oughtred claimed that Delamain saw an opportunity and seized it. For projectors, such an action was worthy of reward. Delamain not only won a second patent and an official post, but as described below, continued with the help of Heydon to create instruments highly valued by Charles.

John Heydon's Gallery of Inventions in the Minories

Heydon sought knowledge across a wide social domain, both from the conventionally learned as well as from those who acquired their knowledge by less erudite means. As he wrote to Sir Thomas Browne in 1652, "I have ben abundantly satisfied by the testimony of very knowinge men, as well literat as illiterate." For, as he continued (in Latin), it is not "he who knows Greek, or who proclaims Latin words, who is learned or wise, but he who finds out truths and can distinguish them from falsehoods" (Non qui graeca scit, aut verba calet Latina doctus est, aut sapiens; sed qui vera videt; & falsis secernere novit). Heydon presented himself as an informed skeptic concerning traditional learning. He admired how Browne did not subscribe to the "traditions" of either philosophy or medicine, "beyond what effects of Nature, his owne Experimentale practice, or demonstrable Reason shall manifest to be really true."[47] Heydon's house in the Ordnance Office in the Minories, marked by today's Haydon Square and Haydon Street, offered a venue where he and a cast of inventors, including Cornelis Drebbel, Arnold Rotispen, and Richard Delamain, could experiment in a setting equipped with many new-fangled instruments and models of inventions.

According to one informant in 1634, Heydon "had a hundert Models et Rarities. Amongst others to read an Inscription in the night halfe a mile of which hee kept very secret."[48] Drebbel's son-in-law, Dr. Johann Sibbert Küffler, informed Samuel Hartlib that at the end of his life, Drebbel was pursuing the art of flight and was also busy with optical inventions that remained in the possession of Sir John Heydon.[49] He was working on a telescope made of liquors of diverse colors, as well as a new "kind of Circle or segment or section," but "he dyed in the pursuit of that Invention. Sir Iohn Heyden interrupting him always when hee was about to reveale it."[50] According to Küffler, the "Model or Instrument" of Drebbel's last telescope was "left amongst his other many works in a Gallery in Sir. I. Heydens house."[51] Using the thermostatic, self-regulating oven invented by Drebbel, Heydon and Sir Kenelm Digby experimented in the Ordnance Office with hatching chicken eggs and investigating their development.[52] Heydon also knew a technique "of turning salt water into fresh so that ships will bee suplied sufficiently hereafter in this kind."[53] As noted in chapter 2, in Heydon's own philosophical writings, he cited Drebbel's philosophical opinions, illustrating how he made use not just of Drebbel's devices, but also his more theoretical ideas.[54] Others reported that Heydon's brother-in-law, Sir Christopher Gardiner, "got all Drebbel MS. and Arcana."[55]

Drebbel's collaborator in the Ordnance Office, Arnold Rotispen, seems to have continued several of Drebbel's inventions, such as his automated mill for grinding optical lenses described in chapter 2. In 1634 Charles granted Rotispen a patent for twenty-one inventions, many of them sawing, cutting, milling, boring, pressing, printing, polishing, and turning engines for accomplishing the work of craftsmen "with less paynes" and "more exactly and in farre shorter space then can be done by hand." Several of these engines produced parts, such as wheels or "all manner of opticall sections," that were required in other instruments and engines. Charles specified in the patent that Rotispen's engines or instruments had not been "exposed to the viewe of anie, but only of Our Selfe and of such Wee have especially deputed hereunto." He required that Rotispen's inventions stay "framed and sett up at Our storehowse in the minorites, there to remayne for Our peculiar service."[56] These would have joined the gallery of models and instruments in Heydon's house in the Ordnance Office.

Vauxhall

This gallery seems to have informed a new institution for Caroline invention established with Heydon's help. One of the most significant and currently largely ignored technological developments of Charles's reign was the center for experimentation known at the time as Vauxhall, after the manor whose grounds it occupied. During its Interregnum resuscitation, Vauxhall would be termed a "college of artisans."[57] In 1667 Edward Somerset, 2nd Marquis of Worcester (1602–67), would dub the "house called Vauxhall" an "operatory."[58] Today it is sometimes called the "Vauxhall Ordnance Factory" or the "Vauxhall Operatory" but to avoid both anachronism and predetermination concerning what this institution was, I will refer to it simply as "Vauxhall," as the sources of the period do.

The concept behind Vauxhall was the creation of a well-endowed site for innovation in military and industrial engineering that allowed for collaboration between different investigators. According to a 1645 inventory following Parliament's seizure of the site, Vauxhall was a substantial institution for the time. It held nine forges, thirteen vices, seven furnaces, and ten workbenches. It contained many engines for filing, boring, and turning wheels for firelocks that may have been the work of Rotispen.[59] Most striking was its collection of models. The survey noted many models built of lead, copper, iron, tin, brass, leather, horn, paper, wood, and pasteboard. These included "Two perpetuall mocions, one of tinne plate, another of paistboard," and "seven great wheeles made for a perpetuall mocion" in addition to horseless wagons, boats that navigated against the tide, engines for cutting tobacco and mills for grinding corn, among many other inventions.[60]

Vauxhall was designed to be a space that Charles could personally visit. It boasted a new, large two-story brick building and a smaller, older building containing workshops as well as a mill for boring guns, and on the upper floor, "one faire roome called the King's dining roome." There was a separate brass foundry, a purpose-built brick structure for the manufacture of leather guns (invented by Robert Scott, the first employee at Vauxhall), stabling for six horses, a garden, and an orchard.[61] It stretched over eight acres and was ideally situated beside the river.[62]

Due to a later historiographic association between experiment, innovative technology, and hot Protestant rebels, the Caroline origins of this institution have long been overlooked, although one recent revisionist account argues that it may have been most productive before 1640.[63] Historians have tended to associate Vauxhall instead with Edward Somerset, 2nd Marquis of Worcester, who had some interactions with the royal servants at Vauxhall already during the reign of Charles I and who later revived the institution.[64] Many of the investigations at Vauxhall, such as the development of steam power, have been attributed to Somerset's patronage of this technology, rather than to Charles I's.[65] Somerset himself later claimed to have spent 50,000 pounds on experiments at Vauxhall.[66]

Vauxhall, however, was not Somerset's initiative, but the result of longer-term Caroline investments in innovation. When Charles purchased the site in 1629, he already had in mind to settle there the gunmaker Captain Robert Scott, who arrived in 1630. A Scottish soldier, Scott invented his lightweight leather guns while serving in the Swedish army. He was recruited by the Danes for his invention, and then by England. In 1629/30 Scott, described as "the King's Servant, one of the Gentlemen of the Privy Chamber," received a grant of denization, as did his wife, two children, and brother, as well as his nephew and fellow inventor, James Wemyss (a.k.a. Wimes or Weemes) (1610?–67).[67]

Vauxhall offered a new space for gunnery trials and training, in place of the contested Artillery Garden (discussed in chapter 6). In 1634 Charles issued Heydon a warrant "for making such a butt and platform at Foxhall [Vauxhall] as is at the artillery garden."[68] Simultaneously, Heydon was expanding workshop space in the Tower. In 1633 Heydon composed a memorandum, "On the necessity of increasing the number of foundries for brass ordnance," as well as "on setting a part a place in the Tower a workshop for Jasper Calthorpe, and erecting a forge there."[69] This was Caspar Kalthoff the Elder (1606–64), best known for his work on repeating and multifire guns, as well as various pneumatically driven engines. His son, also Caspar Kalthoff, lived in Dordrecht and worked as a lens grinder.[70] One year later, brickwork for the forge of "Gaspar the engineer" in the Tower was paid for.[71] In addition to his workspace in the Tower, Kalthoff also had a forge and his own "little working house or shop" built at Vauxhall.[72]

At the same time, a workshop was dedicated in the Tower to Arnold Rotispen, who previously worked alongside Drebbel in the Minories.[73] Rotispen and Kalthoff were likely collaborating, as Rotispen and Drebbel had previously.[74] Kalthoff also set up a perpetual motion in the Tower, shown to a visiting John Pell in 1639 and discussed by Theodore Haack and Samuel Hartlib.[75] Many others were interested in Kalthoff's perpetual motion, including Sir John Evelyn, Sir Endymion Porter, and William Frizell, an art agent who worked for the Earl of Arundel.[76] As described in chapter 6, in 1640 in the home of Samuel Hartlib Kalthoff signed a contract with Johann Christoph Berger von Berg to share all their inventions in common.

Heydon's memoranda in support of Kalthoff agreed with his valorization of "illiterate" people, which usually meant in this context those who lacked a knowledge of Latin. Joachim Hübner noted to Hartlib in 1640 that "Kalthoffs several Inventions are not brought to paper."[77] Another correspondent lamented that Kalthoff "doth not care for writing or cannot write."[78] Nevertheless, Benjamin Worsley characterized Kalthoff as a fascinating interlocutor "about Glasses Telescopes Things in Natural Philosophy and Chimistry." Worsley visited Kalthoff the Elder's home in Dordrecht, to where Kalthoff had retreated during the Interregnum. Worsley found it always full of gentlemen dropping in "to speake with him."[79] We might imagine Kalthoff equally engaged in such conversations with his fellow workers in Vauxhall. The design of Vauxhall was to allow such conversations by situating many inventors together, and at the same time, to secure these conversations out of the public eye. By working in Vauxhall rather than in their own homes, these inventors would not have gentlemen dropping by and potentially appropriating the knowledge they found.

In addition to Scott, Wemyss, and Kalthoff, further Vauxhall employees included the gunfounder William Lambert, the gunmaker Harman Barne, the engineer and modelmaker, William Joulden, and a Captain Ghest.[80] Later, John Bishop joined them to serve as "Engineer & Overseer of all the Instruments of Warre, made, moulded & contrived in Fauxhall."[81] Samuel Hartlib described how, after Bishop had left Vauxhall and moved to Southwark, he had "incomparable Optike and Acoustiques besides Burning-Glasses," but would "admit no living soule to see his tooles and workings."[82] John Lanyon, a major projector during the Interregnum, likely also worked at Vauxhall. He first approached Charles I in 1633 with a petition for multiple endeavors, including proposals for roofing, for dyeing using "diverse sorts of Indian colours," and for water pipes.[83] He received a privilege for both covering of houses and "freyming of clothes in the Indian manner" in 1635, a technique he was still working on in the 1650s.[84] Lanyon was soon recruited to work on gunnery, and he was made proof-master of arms in 1638.[85] He showed Hartlib a large collection of "his inventions and experiments," including "Military ingenuities."[86] Lanyon also made new tools, such as augurs and saws, as well as a model of the watermill used in Brussels.[87]

Charles I granted inventors at Vauxhall many resources for experimentation. For instance, Kalthoff recalled to Worsley how, "striving to put nature to extremity," he attempted ever larger, stronger, and more expensive vessels to withstand the steam pressure he was building up to power his attempt at the perpetual motion. His last trial was six feet in every direction, and "of Copper, and so thick the very metal came to 6[00] or 700 lb. sterling which vessel was . . . at the kings Charge and . . . left at fox-hall [Vauxhall]." To this "great Vessel" he affixed a pipe full of water, and by applying fire to the bottom, he could "make it run out with a constant spout 60 foot high. . . . All which hee had compleatly finished before our troubles, and expected a very good reward from the king."[88]

The approach Kalthoff describes here, in which he constantly searches for more power through attempts in varying shapes and substances demonstrates an open-ended manner of investigation that Fokko Jan Dijksterhuis has called "'artefactual': getting conceptual grip on matters by tinkering with and reflecting upon artefacts like thermoscopes and stoves."[89] This also represents the central approach of making and knowing identified by Pamela Smith as a "cycle of trial, failure, replication, and a responsive, adaptive approach to unexpected outcomes."[90] Finances and material infrastructure were limiting factors to investigation in this manner, especially for difficult to achieve, ambitious goals, which might require long-term material tinkering with expensive materials. In a typical project, investors did not grant inventors much time to experiment. We have seen, for example, in chapter 3, how Thomas Russell rushed to try out one form of pan and then another in his alum works, complaining that he was not granted the "patience or grace" of time. A place such as Vauxhall placed an interval between the development of models and the rush to application. In its gallery of models, it also usefully stored a showcase of previous attempts.

After the Commonwealth government gained control of the Vauxhall property, Hartlib and his interlocutors worked to salvage the royal institution for Commonwealth use. John Dury composed a 1649 memorandum on Vauxhall, commending it as a site for "all manner of Ingenuities Rare Models and Engines," for making "Experiments and Trials of profitable Inventions," and as "a place of Resort whereunto Artists and Ingeners from abroad and at home may repaire to meete with one another, to conferre togetther." It was for these reasons that King Charles had founded Vauxhall, Dury noted. If the Commonwealth government did not follow through on the king's plan, "it will bee a disparagement vnto vs" that "wee should bee lesse mindful of the Publick then hee did seeme to bee."[91] Dury and others were very cognizant of Charles's role in establishing and funding Vauxhall, which served as a model for many of their further plans in the Interregnum.

Dury also compared Charles's Vauxhall Operatory to the "kunst-kameren that is Chambers of Artifices" found on the Continent. There was crossover between the large collection of models held in Vauxhall and the collecting of models in the *Kunstkammer* or curiosity cabinet. The *Kunstkammer* was often also associated with workshops bringing together innovative craftsmen who could blend a variety of types of crafts and knowledge in order to produce the boundary-defying objects found in chambers of wonder. Both *Kunstkammer*-affiliated workshops, therefore, and Vauxhall encouraged the mixture of a variety of forms of practical knowledge to produce innovative works that could not be defined according to the traditional parameters of a single craft. However, there were some important differences. Unlike the one-off curiosities produced in workshops associated with the *Kunstkammer*, at Vauxhall technologies were developed for large-scale agricultural, manufacturing, and military application.

Continental princely collections, such as the Medici Galleria degli Uffizi or the Rudolfine collections in Prague, functioned as a form of cultural politics displaying the power of the prince to master nature.[92] Charles, by contrast, designed Vauxhall to keep its inventions out of the public eye.

Vauxhall might also be compared to the galleries of the Louvre, which housed ten instrument makers between circa 1600 and circa 1660.[93] There were several differences, here too, however. The Louvre galleries housed a wide variety of specialized craftsmen, from painters to sculptors to theater designers and clock-makers. Vauxhall held individuals who tended to work across fields. Moreover, whereas the Louvre instrument makers often functioned in close collaboration with particular requests from the state or other figures, the Vauxhall inventors appear to have been granted latitude and funds to pursue difficult, long-term goals they proposed. In this way, Vauxhall grew from practices of projecting that combined together many experimental enterprises rather than drawing on specialized craft expertise.

Vauxhall was located merely three-quarters of a mile from Tradescant's Ark, a collection that did put early Stuart unbounded knowledge of nature and art upon display to the public gaze. However, Tradescant's Ark also speaks to the distinctiveness of the early Stuart culture of interloping. Rather than the possession of the prince and the work of his servants, the Ark was the idea and work of two upward-interloping gardeners. Its collections drew upon much individual donation from across a wide social spectrum, blended with royal and courtly participation. These donors ranged from Charles and chief courtiers, to global merchants, to a number of artists and engineers, including the stone sculptor "Mr. [Nicholas] Stone," the designer "Mr. Francis Cline" (recruited to the Mortlake tapestry works), and the painter "Mr. Rowland Bucket." At the bottom of Tradescant's list of donors were inventors, such as the follower of Bacon, "Mr. [Thomas] Bushell" who had also moved to Lambeth, as well as "Mr. [John] Lanyon," "Mr. Gasper Calthoofe," and "Mr. William Lambert," all of nearby Vauxhall.[94]

Sovereign of the Seas

In 1630 Charles commissioned the talented medalist and alchemist Nicholas Briot (1579–1646) to strike a medal proclaiming his dominion of the seas, shown in Figure 15 in a later version owned by the king himself and given to William Juxon. Charles's ship of state set sail into the open ocean. The motto around the rim suggests how Charles saw his power extending further beyond: "NEC META MIHI QVAE TERMINVS ORBI" (Nor is that a limit to me, which is

Figure 15. Nicholas Briot, *The Juxon Medal: Charles I, 1600–1649, King of England, 1625* (obverse); *The Dominion of the Seas* (reverse), 1639. National Gallery of Art, Washington, DC. Gift of Drs. Yvonne and A. Peter Weiss, 2019.117.1.a & b. Public domain

a boundary to the world). Much as Holy Roman Emperor Charles V (1500–1558) chose the two columns of the Pillars of Hercules and the motto *Plus Ultra* (further beyond) as his motto, Charles I linked the image of his sovereignty to ideas of unboundedness.[95] His ship of state did not remain within safe, known waters, but ranged where others dared not go.

Such ideas would soon be embodied in an actual vessel, when in 1634 he commanded the building of the enormous ship, the *Sovereign of the Seas*. The *Sovereign*, launched in 1637, was the most heavily armed ship in the world, with 102 pieces of ordnance consuming 145 tons of bronze.[96] Like the large ship that Charles wished to send in the 1622 "great design" for Asia, the size of the ship alone was a statement about the English Crown's claim to glory.

In case such claims were not already transparent, the ship was also covered in gilt emblematic images whose interpretation was published in print. These included an image of King Edgar trampling seven kings, since Charles was making a claim to oceanic rule dating back to Edgar's time in the tenth century.[97] James I had already begun to address the issue of rights over water by prohibiting foreigners from fishing along British and Irish coasts. Charles took claims to oceanic sovereignty much further, claiming English control over all the waters (hazily defined) stretching to the coasts of the continent.[98] Sir Francis Kynaston titled him "great Charles, King of Great Britain, France and Scotland and sole prince of the Ocean."[99] This title reflected the ancient English style of "oceani dominus & imperator" that Kynaston's friend John Selden traced back to Edgar. In the second edition of his *Titles of Honor*, Selden expanded upon this (adding a citation to John Dee's account of Edgar's imperial rule of the sea), arguing that Edgar's usage proved that "the King of England or Great Britain so long since wrote himselfe, and was Emperor and Lord of the British Sea; as the expression is that of Oceani Dominus & Imperator."[100] Of course, the extent of ocean where empire was a possibility in Edgar's time differed drastically from Charles's. In the motto, "Nor is that a boundary to me which is a limit to the world," Charles seems to be making a much wider claim.

Charles expressed another meaning to the idea of sovereignty of the seas in his personal development of nautical instruments. At the time of the launch of the *Sovereign of the Seas*, according to Richard Delamain, Charles I himself invented an instrument the identity of which is unknown today. One of Delamain's many petitions specified that the king ordered Delamain to make up "sundrie new Instruments in silver for your Majesties particular use, one of which being your Majesties invented by your Majestie about the time of the Launching of the Soveraigne." To make all these instruments in silver, Delamain would need thirty-six pounds of silver delivered to him "at your Majesty's house at the Minories, at Sir John Heydon's."[101] If Oughtred claimed that Delamain acted as a "jugler" when he entertained social elites through instruments, Charles's invention of his own instrument alongside Delamain seems designed to challenge the notion that princes did not understand the mathematical toys with which they played.

Delamain hired several individuals both to build instruments and to test ideas experimentally, as in 1637 he again petitioned Charles I to pay additional workmen "for the confirmation of some new propositions, and instruments to be made for his Majesty's bedchamber and for the Great Ship [that is, the *Sovereign of the Seas*]."[102] Producing the instruments that the king had ordered required the work of both "sundry artificers" as well as "sundry gentlemen, amongst whom Sir John Heydon has not been the least, and an eye-witness of the many labourers daily employed in your Majesty's house at the Minories."[103] Charles cherished the instruments made by this group. Shortly before his execution, he ordered that "his large Ring Sun-Dial of Silver, a Jewel his Majesty much valu'd . . . invented and made by Mr. delamaine," be given to his son.[104]

The Philosophy of the Voyage

In 1634 a cryptic note by Samuel Hartlib suggested that Delamain's instruments related to the intertwining of voyages of discovery with new attitudes toward knowledge. Hartlib wrote, "De la Mainne in Chancery-lane [Richard Delamain] is to know a certain Noble-Man who is about a great worke in that kinde which Wats hase written in his Appendix." Hartlib considered this text, which he also called Watts's "Appendix on Natural Philosophy [Appendix de Physica]," of great importance.[105] This appendix likely refers to a text by Watts published alongside the 1633 edition of Captain Thomas James's narrative of his search for the Northwest Passage in the Arctic.[106]

Charles's pursuit of the title of monarch of the unbounded ocean perhaps motivated his sponsorship of several voyages of investigation to search for a northwest passage, as well as the 1635 search for a northeast passage past Japan by the Courten-Porter association.[107] Thomas James described how after he and his collaborators approached King Charles with their plan to search for the Northwest Passage, Charles "graciously accepted of the offer" for a voyage to the Arctic (figure 16).

Once Charles accepted his offer, James "with all speed" contrived in his mind "the best modell I could; wherby I might effect my design."[108] He gave careful thought to the "quality and abilitie" of the men he should hire. One

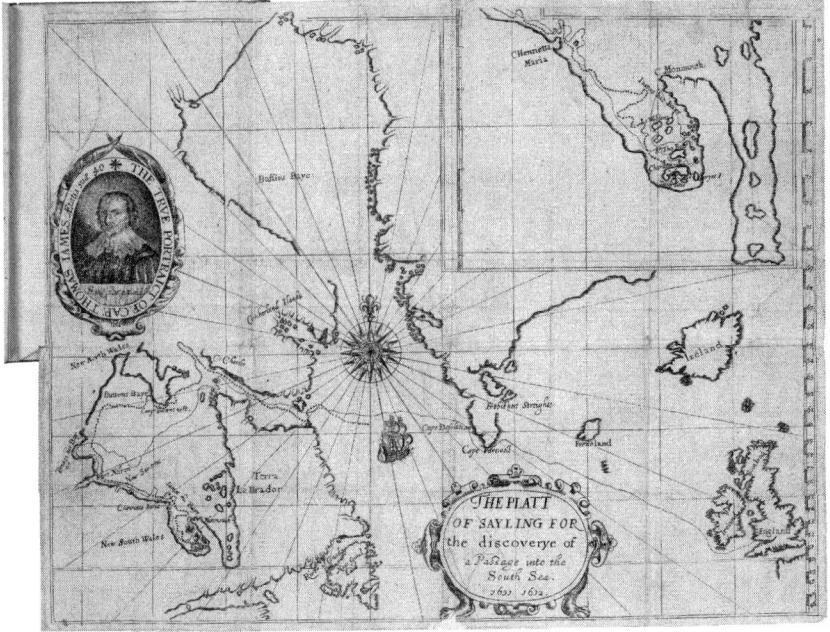

Figure 16. Thomas James, *The Strange and Dangerous Voyage of Captaine Thomas Iames, in his Intended Discouery of the Northwest Passage into the South Sea* (London: Legatt, 1633). Folger Library. Call #: STC 14444 Copy 1. Used by permission of the Folger Shakespeare Library under a Creative Commons Attribution-ShareAlike 4.0 International License

counterintuitive decision he made was to reject all who approached him who "had bin the like voyage, or adventures." He preferred to keep "the power in my owne hands" and make "all the men to acknowledge immediate dependence upon my selfe alone." In comparison to another Arctic explorer Charles I also sponsored, Luke Foxe, "James was the theoretical explorer, Foxe the practical and 'very skilful seaman.'"[109]

James studied "Journals, Plots, Discourses, or what-ever else might helpe my understanding," and set "skilfull workmen" to make instruments.[110] This attention to textual study and new-fangled tools was unusual in England at the time.[111] James also arranged to observe the lunar eclipse that would take place during his voyage while Henry Gellibrand was observing it at Gresham College in London, so they might better position the point of longitude where James made landfall.[112]

According to the full title of James's travel account, it described "the Rarities observed, both Philosophicall and Mathematicall" and included "An

Appendix concerning Longitude, by Master Henry Gellibrand" as well as "an Advise concerning the Philosophy of these late Discovereyes" by Watts. In his own appendix to his narrative, James listed thirty-five instruments taken on the voyage, many of them specially constructed for it, such as "many small [hour] glasses"; "Two paire of curious Globes: made purposely: the workeman being earnestly affected to this Voyage"; and "four speciall Needles (which my good friends Master [Elias] Allen and Master [John] Marre gave me) . . . toucht curiously, with the best Loade-stone in England." Additionally, James carried with him "a Chest full of the best and choisest Mathematicall bookes," and "Study Instruments, of all sorts."[113]

James's use of instruments by several well-known instrument makers has been read as a sign of the valorization of expertise on this journey.[114] Yet, as in the case of projectors like Thomas Russell, James's voyage as a whole was not about developing a particular expertise—Luke Foxe, instead, embodied the experienced seaman. James's account of the voyage, which he claims Charles I asked him to write, serves as a model of how to plan discovery while remaining open to contingency.[115] His description of his ship and its technical accoutrements makes it appear as though it were a research vessel avant la lettre. In this, it was the floating equivalent of Vauxhall; both were designed to equip voyages into the unknown.

It was such characteristics of James's ship that inspired the third appendix concerning the "philosophy" of the voyage by William Watts, the same figure who was an early user of the term "experimental philosophy" (as discussed in chapter 3) and who opposed the use of "Systematical Method" (as discussed in the introduction). The Anglican clergyman William Watts was made a prebendary of Wells in 1633 and would become royal chaplain to Charles I in 1639.[116] He addressed his appendix to the students of divinity at the University of Cambridge. It was an extraordinary move to not only suggest that a sea captain had a "Philosophy" but that it was one that should serve as a model for students of divinity, queen of the sciences, at Cambridge.

James did not succeed in achieving his target of the Northwest Passage, and unlike the experienced seaman Foxe, he returned home with almost no other discoveries or accomplishments other than surviving an Arctic winter.[117] In fact, Watts suggested that the idea of the Northwest Passage might be a Spanish plot to distract English shipping away from "their golden Indyes." Watts was not interested in the "finding" of the Northwest Passage, but in the "Searche." It was in the journey and not in the fulfillment of the stated goal that much could be learned. "For mine owne part, I suppose that the *Philosophers stone* is in the *North-West Passage*," wrote Watts, "For that theres so much *Philosophy* in the way to it."

Watts contrasted the philosophy that could be learned in the pursuit of ambitious (and in his view, fictitious) goals to knowledge bound by the teachings of a master.[118] He addressed two "propositions": "Whether those Rules of Aristotles Philosophy be to be allowed so Vniuersall, that they hold all the world ouer" and "Whether they ought to be so magisteriall, as to prescribe against all other examinations." Global journeys of discovery showed that no rules could be universal, not only Aristotle's: "the great and infinite *Creator* hath so disposed and varied euery thing, that it is impossible for mans reason and obseruation to conclude him: and therefore, though vulgar and receiued *Philosophie,* may giue a man a *generall* hint, all the world ouer; yet no *Vniuersall* and vnfayling certainty."[119] Nor could anybody's authority be so great as to preclude investigation by others. Watts admire Aristotle. Aristotle's greatness required acknowledgment, much as every other ship encountering "the Kings Ships in the Narrow Seas" was required to strike its sails "in acknowlegement of a Soueraignty." Yet "no humane dictates" could be so "Magisteriall" as to prevent all others.[120]

By arguing against the existence of universal rules and against obedience to authority, Watts intended to "encourage and countenance future undertakings." Only by demonstrating that much natural philosophy remained merely at the level of "hints" to be pursued by anyone could "Physicks" advance as had "our Geography, our Mathematicks, and our Mechanicks."[121] Watts cited from the Book of Daniel, "Many shall passe to and fro, and knowledge shall be encreased."[122] No human's authority was so great as to contravene further examination and exploration by others; knowledge increased in the unpredictable journeys of many.

Conclusion

It is perhaps counterintuitive that Charles I should simultaneously support such views contravening authority and aim to establish his own position as sole monarch of the seas. In this, he drew upon the culture of projects, which attempted to yoke the initiative and energy of ambitious interlopers trying out new ways of doing things to the augmentation of royal power. Vauxhall embodied (and attempted to resolve) some of the internal contradictions generated by the patent system as a branch of state power. The patent system relied on individual projectors and their own abilities to arrange investment, materials, personnel, and space. Thus, it could not ensure secrecy even concerning initiatives that were conceptualized as belonging to the secrets of state. Vauxhall, by supplying all the needs of inventors, was designed to enable a degree of state secrecy that the patent system had not secured. It thus closed the door into reasonings of state that the system of projects had opened for individual projectors.

Vauxhall also addressed material needs of artifactual ways of knowing. The tendency to extrapolate from individual curiosities to widespread utilitarian application often failed, as we have seen. Although models remained an important part of practice at Vauxhall, inventions were conceptualized as utilitarian from the get-go, and were sometimes built in a rather large size, as in Kalthoff's six-foot by six-foot steam engine. Vauxhall also extended the timeframe of experimentation, both by linking investigations to previous attempts through the stored gallery of models and accumulated experience of several artificers, and by separating the investigations by inventors from the pressing needs of investors to raise immediate capital.

For Vauxhall, Charles did not recruit experts or craftsmen in particular fields. There were certainly other arenas of his patronage, such as the Mortlake tapestry factory or the Hounslow sword manufactory, where he encouraged the transplantation of a specific, established craft from elsewhere to England. The investigations in Vauxhall were of another nature, however, far more conjectural, ambitious, and hybrid. These were attempts to push, in Kalthoff's words, nature "to an extremity." Each of these might not succeed, but they were pursued in an environment of probability. Watts advised that philosophy was more to be found in the "Searche," even those that did not succeed in their original aims, than in adhering to discipline and the authority of a master.

As discussed in chapter 1, James I had identified the increase of knowledge "of all Artes and Trades" as a "principall part" of "soveraigntie"; the furthering of "plantations, increase of Science, and works of industrie" also advanced royal honor.[123] Such a political viewpoint on the advancement of knowledge set the stakes of the search for knowledge high. As discussed in chapter 5, early Stuart thinking concerning colonial projects suggested that states could not afford to rely upon seemingly assured certainties. Not just for their advancement, but even for their survival in a context where other states were continually fomenting projects and pursuing hidden interests, states needed to take risks to expand their power to the uttermost degree possible. In the gallery at Heydon's house in the Minories and later in Vauxhall, inventors had space and resources to pursue highly desirable and difficult quests—perpetual motion, horseless carriages, submarines, flight, telescopes, and the like. These were technologies that pushed the limits of possibility and thereby appeared to expand human power and the power of the state.

Conclusion

If "desire is the travel, knowledge is the inn," wrote Donne, but in this uncertain world, where "we are ignorant of more things than we know," rarely do we reach that safe harbor.[1] Even princes who glory "*in Arcanis*," that is, in secrets of empire unknown to other men, find themselves suddenly hapless when "thoughts and purposes indigested fall upon them and surprise them."[2] Projects were supposed to offer crowned heads the secret keys for unlocking power. Yet, we might imagine them viewing the constant crowd of competing, jockeying projects with more than a touch of indigestion.

In this contingent, incoherent world, we are all often taken by surprise. By undisciplining knowledge, interlopers cut away the strict epistemic procedures and guardrails that in the past had offered distinct methods and types of proof for varied forms of knowledge. Instead, careening wildly between one form of knowledge and the next, interlopers seized upon distinct phenomena and linked them together through loose forms of reasoning and calculation. Such undisciplining of knowledge produced confusion, doubt, and great social disruption. And still there is no inn of knowledge awaiting us at the end of the road where we might finally get some rest.

Knowledge could not afford to be at rest. The culture of interloping infused knowledge with ever greater desires and fears. Projects promised great profit and power should risks be taken. They painted a picture of lost opportunities and even existential danger should they be avoided. They urged quick action; "Occasion is pretious, but when it is occasion."[3] The most advantage possible needed to be extracted from the entire world. Projects extrapolated wildly, took action hastily, appropriated knowledge widely, and arrogated power both at home and abroad.

This book has delved into this knowledge culture by tracing numerous schemes spanning Europe, Asia, Africa, America, and the Arctic. A wide cast of characters from across the social scale pursued these schemes—and these represent only a fraction of the projects and projectors of the time. This investigation has revealed many arenas of contention, internal contradiction, and unintended consequences, but also shared commitments, arguments, and intellectual practices. The enterprises investigated here have stretched far beyond the boundaries of any one discipline. In so doing, they questioned the value itself of disciplines, borders, and mastery, abandoning knowledge safeguards in favor of risk-taking at the edges of the possible.

This portrait of an unrestrained knowledge culture has aimed to question still influential accounts that insist upon discipline as the defining trait of European modernity. Such accounts, contrasting discipline with passion, and the civilizing process with violence, situate Europe and European men of science in particular, as distinctively civilized, dispassionate, and regulated. They thus recapitulate a colonial contrast between European civilization and ungoverned embodiment elsewhere. Historically, this contrast was deployed to justify conquest as in the best interests of the uncivilized. In short, the claim to a pacified, dispassionate, self-limiting, and civilized Europe has acted as a cover for what was an insatiable hunger for power. The knowledge practices that served this hunger underwent processes of undisciplining, rather than civilizing. They rejected the categories and strictures of traditional disciplines and social and epistemic divides. By valorizing unrestrained knowledge-making, they hoped to open epistemic possibilities foreclosed by more sober, stable, and systematic knowledge structures.

Antisystematic accounts that leaped wildly both into future projections and laterally across knowledge domains not only allowed interlopers to make many arguments that would not have passed muster according to prior standards of *scientia* (science). They also permitted them to pass over and remain silent about many consequences and effects of their proposals. Despite the vast intellectual and financial resources invested in projects and the enormous consequences that often resulted for both people and the environment, projects were not forms of knowledge that were easily held to account. They quickly spun out plans taking advantage of the moment's opportunity and the concatenation of many factors. In this hothouse intellectual environment we have observed many knowledge hybrids springing up, such as experimental philosophy, garden philosophy, and the philosophy of the voyage, whose loosely defined and probabilistic nature would hardly have warranted them the title of philosophy in more disciplined settings. Thus, the speed, looseness, contingency, and distance built into projects was generative. Yet, due to those same qualities, much could be overlooked, such as counterevidence or harms wrought. In this way, the undisciplining of knowledge both created and destroyed.

This concluding chapter points to the future of undisciplined knowledge in the legacies it bequeathed to modern science and technology. It begins with the ways that Bacon took advantage of the holes punched in the structure of knowledge. This account offers a critique of how Bacon has previously been situated in relation to his early Stuart contemporaries. The term "Baconian" has often been applied to Bacon's contemporaries and successors as though everyone was following his well-charted program, when this was not the case. On the other hand, by thus situating Bacon among his contemporaries, I do not mean to argue that Bacon's works had no effect upon a wider culture of interloping. In addition to critiquing the view that nearly every early Stuart naturalist or experimentalist drew inspiration from Bacon, this view also questions the alternative historiographical claim that Bacon did nothing but copy the achievements of others. Bacon has been seen as merely codifying prior artisanal practices.[4] Scholars have identified the source of much of *Sylva sylvarum* in the works of Giambattista della Porta and Hugh Plat.[5] Deborah Harkness accused Bacon of stealing Elizabethan empiricism.[6] As a participant in the wider world of early Stuart projectors, inventors, and experimenters, it would not be surprising if Bacon appropriated knowledge from elsewhere. This was the intellectual modus operandi of his setting. Yet, I argue that Bacon also made his own contributions to the further undisciplining of knowledge.

During Bacon's lifetime, the forms of loose knowledge Bacon and other interlopers promoted did not enjoy the backing of a "civic epistemology," or widely shared agreement concerning what counted as authoritative knowledge. It was Bacon's work (above all his approachable *New Atlantis* rather than his more cerebral texts) that gradually helped make the values of interlopers part of a civic epistemology. His lush rhetoric and evocative imagery did much to sketch a vision of a future, perfected knowledge that continues to distract us from past failures and present unfulfillment. He located the safe harbor of knowledge on a fantasy island toward which we still voyage.

Francis Bacon and His Contemporaries

Many were the individuals in early Stuart England who proposed fusions of power in its many interlinked guises (technological, military, financial) with knowledge. Yet, the widespread commonplace linking knowledge and power has in modern times often been associated with the solo figure of Sir Francis Bacon.[7] Indeed, almost any early Stuart figure with an interest in exploring nature, it seems, has been called a follower of Bacon, from gentlemen such as Sir Francis Kynaston, to artillerymen like Robert Norton, to surveyors like John Norden, to clerics like William Watts.[8] Charles I's institution of Vauxhall has been interpreted as a direct attempt to realize Bacon's *New Atlantis.*[9]

Numerous early Stuart individuals and enterprises have been called Baconian on the basis of little evidence. Vauxhall had more to do with long-standing dynamics between Charles, the Heydon brothers, and the recruitment of foreign talent dating back to the 1622 Southeast Asian project than with Bacon's posthumous publication of the *New Atlantis* in 1627. Neither Kynaston nor Norton nor Watts ever cited Bacon. In the manuscript defense of his academy, Kynaston described many English models of learning: Roger Bacon, John Scotus, John Dee, William Gilbert, John Napier, and Henry Briggs.[10] Notably, he did not mention Francis Bacon. Samuel Hartlib noted in 1635: "Hee [Kynaston] slighted much Verulam though hee had not read them. Idem rejected Galen, Aristotel, prized only Rational and Experimental Studies."[11] Kynaston cited thirty-eight post-classical authorities in his commentary on Chaucer, but not Bacon.[12] Watts also never cites Bacon in his 1633 appendix to *The Strange and Dangerous Voyage,* although he does cite other figures such as Pierre Gassendi. Nevertheless, this text has been called "very Baconian."[13] In fact, in all his conversations with Samuel Hartlib, Watts seems not to have discussed Bacon, even when the conversation would have seemed to naturally tend in that direction (as in a conversation concerning broken forms of knowledge such as aphorisms, or the need for laboratories in universities, or the ways that mariners could collect knowledge). Watts appears to be far more taken with other figures, such as Thomas Russell, Richard Delamain, Henry Gellibrand, and Edward Herbert, Lord Cherbury.[14]

Claims have also been made that Cornelis Drebbel's investigations were "part of a unique assembly, an early collaborative laboratory," including Bacon and Drebbel.[15] Rosalie Colie saw Drebbel's "laboratory" as "essentially a microcosm of Bacon's."[16] She pointed out the many similarities between the investigations of the fellows of Bensalem in Bacon's *New Atlantis* and those pursued by Drebbel in the early Stuart Court.[17] Interestingly, Drebbel's purported "Baconianism" has appeared as two different claims: that his activities influenced Bacon and that, at the same time, his activities show evidence of Bacon's influence. Charles Webster, for instance, picked up on Colie's views of Drebbel's influence on Bacon, while also situating Drebbel as a follower of Bacon.[18] Despite the two-way directionality claimed for such influence, agency remains subsumed under Bacon. Bacon, for example, is never described as a Drebbelian in modern scholarship.

In their own lifetimes, Bacon and Drebbel were compared, not least by Drebbel's friend Constantijn Huygens Sr., who cast Bacon as the sun and Drebbel as the moon.[19] Despite his lunar metaphor, Huygens did not mean to imply that Drebbel merely reflected Bacon, but that Drebbel was second to Bacon, in Huygens's estimation, for natural knowledge in London at the time. Others also considered Bacon and Drebbel as the top two naturalists in the area. The Herborn philosopher Johann Bisterfeld complained to fellow Herborn academic Johann Heinrich Alsted that, when visiting London in 1625, he couldn't visit Drebbel because of the plague, Bacon was leading a "private life" and he couldn't see him, and there was nobody else in London worth talking to.[20] While Bacon and Drebbel might have been comparable, they were certainly not friends nor collaborators. Drebbel never mentions Bacon, and Bacon's mentions of Drebbel are either offensive or dismissive. The important takeaway here, however, is that period comparisons of Bacon and Drebbel do not place the former in the category of philosopher and the latter in the category of craftsman, but set the two more or less on a par. Drebbel considered himself, and was considered by many, to be a philosopher in his own right. He had no need to partner with Bacon. His unbounded, expansive form of knowledge-making meant that he would never have accepted a limited role in a program prescribed by Bacon. Having Sir John Heydon looking over his shoulder was bad enough.

Similarities appear between Bacon and his contemporaries because they all drew from a shared cultural reservoir of intellectual practice, attitudes, and commonplace imagery and arguments. In his discussion of "Garden Philosophy," for instance, John Ferrar might appear to be taking from Bacon the idea of advancing "*Plus Ultra*" past Hercules's columns of knowledge. William Watts, in his account of the philosophy inspired by sea journeys, might also seem to reflect Bacon, who also cited from the Book of Daniel on the title page of his

Novum Organum. Bacon did not have a monopoly on such images and citations, and all these authors might be drawing upon widely spread ideas. However, even in the event that Ferrar and Watts were cribbing from their copies of Bacon's texts without choosing to cite him, then they, as Bacon frequently did, were acting as interlopers, appropriating and claiming such knowledge as their own, and not as Bacon's.

The idea that Bacon rallied his contemporaries to follow his lead is based upon a still powerful historiographical conception originally developed by mid-twentieth-century Marxist historians: the "Baconian program." The notion that democratic social and epistemic planning undergirded modern society was persuasive at a time when large-scale programming like the Soviet space program achieved epochal landmarks, or rather, spacewalks. Benjamin Farrington, in his 1949 *Francis Bacon, Philosopher of Industrial Science*, wrote about the importance for Bacon of developing a "program," a contention he extended in his 1963 *Francis Bacon: Pioneer of Planned Science*.[21] Paolo Rossi, discussing the role of crafts in Bacon's science, used the term "programma baconiano" often.[22] The prominent Marxist historian Christopher Hill, argued that men "like Recorde, Dee, Digges, Hood, Gilbert, Briggs had practised new methods and glimpsed some of their possibilities. But Bacon gave men a noble and all-embracing programme of co-operative action in which the humblest craftsman had a part to play."[23] In current historiography, the notion of a seventeenth-century culture of improvement or growth has replaced what mid-twentieth-century historians like Benjamin Farrington would have called progress, but the notion of a Baconian program looms as large for Paul Slack in *The Invention of Improvement*, for Joel Mokyr in *A Culture of Growth*, and for many other current scholars as it did for Farrington and Hill.[24]

If any such program existed, surely it should have been known to John Dury and Samuel Hartlib, chief proponents of the Baconian program according to accounts such as Webster's *Great Instauration*. Yet, in 1646, while praising the project of Pierre Le Pruvost (which did offer a program for social coordination) to Hartlib, Dury lamented how no such plan previously existed. The kernel of Le Pruvost's proposal, Dury reported, lay in the "subordinacion of mens endeavors . . . to bring them to act orderly within themselues." In the absence of such coordination, "all will turne to Confusion and nothing will be safely enioyed by any, because euery one shifting for himselfe alone in his privat wayes of advantage will crosse another, and strive to pull out of anothers hand that which hee is managing." This was because men acted like children, laying "hands on that which is next unto them, to make it their owne by grappling for it."[25]

In several writings, Bacon does seem to be describing a course of action with differentiated roles for various types of people. Calling himself a trailblazer in the *Novum Organum*, Bacon asked his audience to

> look at what you must expect from men following my directions (*indicia*) with plenty of spare time, from the labours of many working together, and from the passage of time—especially following a path which is (unlike that of the rationalists) open not only to single travelers but one in which men's work and labour (especially for the gathering in of experience) can best be shared out and then brought together. For only then will men begin to know their own strength, when instead of countless men doing the same thing, some will be responsible for some things, others for other things.[26]

Yet, characterizing Bacon's remarks here as the outlining of a program would be a mistake. A program (from Greek, "to write before") offers literally a script to be copied. It was a term used in the early modern period for forms that could be copied by schoolchildren to practice their handwriting. It could also mean a printed schedule of events.[27] As a way of organizing knowledge, therefore, it suggests that such knowledge had already been defined by the writer of the program. In order to reach inexorably the knowledge goals set out by the program, each member of it need only follow the given script. A program might seem to be exactly what Bacon intends when he referred to "men following my directions" in the quotation above. However, these directions or *indicia*, were not fully written out instructions for how to proceed, as a program implies. Rather, an *indicium* is a signpost, indicating a way but not leading the traveler there by the hand. Bacon did not designate precisely what could be found at the end of this journey or exactly how the traveler should proceed.

A more appropriate period analogy for this type of social arrangement would be not a program, but a project. Projects were assemblies of varied talents undertaking risky ventures and trying out endeavors that had no assurance of success. The relationships between the various individuals on a project were far messier and, as we have observed many times in this book, at times antagonistic. Programs disciplined knowledge. In order to map out a path of action that could be followed, they required a basis of already conquered knowledge. Projects undisciplined knowledge. They offered a quick-changing tool for ventures into the unknown.

Bacon and the Gaps of Knowledge

Rather than in the authorship of a predictable program, Bacon's contributions lay in the way he furthered practices of loose reasoning, suggestion, and deferral of proof that often appear in the texts of projects. Thus, in place of the positive "knowledge is power" with which Bacon is so tightly associated, we might notice how he cleverly located power in forms of disknowledge—gaps, silences, unrealized proposals, never completed works, and partial truths. One of his greatest legacies lies in the disjuncture between his advertisements for the advancement of learning and what he actually accomplished. In 1608 he exhorted himself in private memoranda to take on "a greater confidence and Authority in discourses of this nature, *tanquam sui certus et de alto despiciens* (as of one sure of himself and looking down from above)."[28] Evidently, Bacon did not feel such confidence on the interior, but he succeeded marvelously in overcoming his impostor syndrome and presenting himself in authoritative fashion, even for forms of knowledge that he had not yet discovered.

The grand architectural facades Bacon sketched for knowledge concealed the fact that he left his work unfinished. Bacon never discovered a new logical instrument, or organon, to replace Aristotle; nevertheless, he characterized his work as just such a *New Organon* (*Novum Organum*). Rather than a completed work, this title represented a wish for the future, a genre in which Bacon excelled. He had "no universal or systematic theory to put forward," and he did not expect to live to complete his work.[29]

Bacon's goal has never been reached, a fact better remembered in ages closer to Bacon's than our own. As Comenius noted, "Truly, Verulamius [Bacon] is by some . . . shrewdly lasht, and ill-spoken of because he promised a great repairing of Arts, but did not performe it."[30] As the canny Joachim Hübner observed to Samuel Hartlib, humans would never come to know the underlying principles of nature at which Bacon "aimes so much," unless "God himselfe" reveals them.[31] Likewise, Marin Mersenne opined in 1625, "we will never be able to make our intellect equal to the nature of things, which is why I believe Bacon's design is impossible."[32]

As many have pointed out, Bacon never invented the scientific method, and indeed, there is no such thing as a unitary scientific method.[33] Nevertheless, with his wishes taken as accomplished deeds, Bacon is still known popularly as the inventor of the scientific method. In this way, he helped to set up a powerful disjuncture between science as a realm of inventive, probabilistic, even fringe thinking, and a popular conceptualization of that realm as produced by following method, that is, a rote sequence of steps, divorced from human subjectivity and producing certain knowledge.

Instead of a new method, Bacon offered many stopgaps and strategies that gestured at a future completion of an aim. One of his greatest abilities lay in further embellishing the forms of rhetoric deployed by projectors in order to gain support and investment for their projects. Despite offering no particular program that told any individual what to do, Bacon did write gorgeously about epistemic risk-taking and the possibilities of future advantage in ways that could and did tempt many to join in the pursuit of seemingly unlikely forms of knowledge. In this way, Bacon's writings helped translate the attitudes of the interlopers, which during Bacon's life were shared by only a subset of the population and contemned by most, to a wider, civic epistemology.

The section where Bacon described how others might follow his directions, or *indicia*, was part of his argument for why we should hope that forward movement is possible. Bacon's other arguments for expanding our hopes for the future included the imagination of how impossible current knowledge must have seemed before it was acquired. "This fact too can elicit hope, namely that some things already discovered are of a kind that before their discovery the least suspicion of them would scarcely have crossed anyone's mind, but a man would simply have dismissed them as impossible."[34] Guns, silk, and the compass were all inventions that Bacon gave as examples; if, before their discovery, these were proposed by anyone, he would have been considered a laughingstock. Likewise, many things stood "well off the beaten track of fancy and are still undiscovered."[35] Bacon's signposts (*indicia*) aimed to point out such unbeaten paths.

Concluding his many reasons for eliciting hopes, Bacon offered a cost-benefit analysis, describing the future potential for the discovery of knowledge as a risk-taking voyage toward an unknown land, whose potential profits from discovery outweighed the costs of the venture.

> Lastly, even if the breath of hope blowing from that new continent were much weaker and less perceptible, yet I have decided that (unless we evidently wish to be mean of soul) we must make the attempt. For not to try and not to succeed are quite different risks, for by not trying we cast aside an immense good but by not succeeding we lose a little human labour. But from what I have said, and also from what I have not, it seems to me that we have a great deal of hope, not only to persuade a keen man to have a go [*homini strenuo ad experiendum*], but also to make a wise and moderate [*prudenti & sobrio*] man believe in it [*credendum*].[36]

Bacon sought to expand hopes for seemingly impossible, off-the-beaten-track ideas, pursued by groups endowed with diverse skills. Here, he aimed to persuade individuals of two sorts. There were those who were already active, keen individuals, prone to experimentation, who, by what he had said (and quite mysteriously, also by what he left out), might be induced to try Bacon's proposal. Then there was another sort, a more sober type of individual not generally so eager to innovate. Bacon here admits that he might never be able to persuade this kind of person to be involved hands-on, but such an individual might be led to find such a proposal at least credible.

He flattered those who were willing to try out such gambles as "keen," far different from those "mean of soul" opposed to all risks. Bacon's arguments here reek of the defenses deployed by period projectors for new, improbable undertakings. The voyage of Columbus was an especially frequently invoked example, as a proposal that had been presented to the English Crown and rejected as improbable, only to be realized to the great advantage of England's enemy, Spain. The alchemist and agricultural experimentalist Francis Segar, around 1600, for instance, criticized two sorts of people who often opposed any new proposal. These included both the "vulgar (slothful of spirit and dispisers of all good and New Inventions)" as well as scholars, who "when any new matter is propounded of which they have not heard or read, presently they are sturred with disdayne to condempt it, to dispute agaynst it, or with some Jest to put it out of counttenance."[37] The latter were like "unskillfull Cosmographers before Americus tyme . . . affermying that Americus, Columbus, and Magellanus did but Imagyn cassells in the aire."[38]

The argument that all seemingly reasonable inventions and successful navigations seemed like utterly improbable projects when they were first proposed was such a common feature of projectors' rhetoric that it was called out by Thomas Brugis in *A Discovery of the Projector*. He criticized how projectors

> dare most impudently to affirm that all Arts, Trades, Crafts, Sciences, Mysteries, Occupations, Professions, Devises, and slights whatsoever, were merely Projects, in respect of the Authors that devised them, whom they will needs have . . . to be Projectors . . . likewise they will tell you that the first Discoveries of the West Indies, by Columbus, Magellane, and Drake were Projects, seeking thereby to bring into their ranke of Projectors, three such Ingenious, Noble, and Venturous persons.[39]

Such arguments for encouraging hope left many things out, as Bacon hinted. Through selective memory, they feature discoveries and inventions that, although improbable when first proposed, in the end panned out. Vast mountains of failed projects were passed over. Amnesia concealed the huge resources invested in such projects. Silence also stifled the memory of multitudes of people who suffered a loss of livelihood, sovereignty, and life in the name of these endeavors. The diverse knowledge cultures of pre-Columbian populations died unmourned deaths, and unmentioned went the toll inflicted upon the natural environment.

The Epistemology of the Hint and Experimental Philosophy

Bacon's works suggested future directions in epistemic risk-taking without decreeing a program for specific actions leading to assured results. This view of his work counters both the old idea that Bacon invented the scientific method, and newer and more sophisticated arguments, such as William Eamon's account of Bacon's venatic (hunting) epistemology. Eamon based his argument on two articles published by Carlo Ginzburg in 1979 and 1980, where Ginzburg identified clues as the "roots of a scientific paradigm."[40] Ginzburg suggested that the roots of clues were ancient, dating back to hunters and diviners who "learnt to sniff, to observe, to give meaning and context to the slightest trace."[41] Ginzburg saw the clue as part of the pursuit of an almost mechanically followed method. With their noses to the ground, hunters could follow trivial details leading ineluctably to the end of their quest.

Inspired by Ginzburg, William Eamon associated the clue with a more particular time and place: Francis Bacon and his successors in the Restoration Royal Society. According to Eamon, following clues represented Bacon's new epistemology of the hunt. While Bacon, according to Eamon, may not have developed the scientific method, he did offer a signal turn away from previous understanding of the pursuit of knowledge, such as the reading of the book of nature. He and his successors followed an experimental methodology of "the hunter, who follows clues that lead to an unseen quarry."[42] Other scholars have argued that the Royal Society developed literary forms to direct such hunting on a global scale, using for example, questionnaires as a tool of "disciplining travel," in service to the "disciplining of speech and writing" that accompanied the "institutionalization of natural philosophic knowledge with the community of the Royal Society."[43] Yet, many of the techniques for restraining knowledge in this way, from clues to questionnaires, were far from new in the seventeenth century.[44] Further scrutiny of the conceptual history of clues questions Eamon's association between the clue and a novel epistemology in the seventeenth century.

The clue was quite old (as Ginzburg would agree) in Bacon's time. From the ninth century, "clew" meant a ball, and soon after, a ball of yarn. Since the fourteenth century, a "clew of thread" often referred to the guide leading a figure in a labyrinth to the exit. By the late seventeenth century the clue indicated a guide or piece of evidence to a solution or discovery, with its previous meaning as a thread beginning to be obscured.[45] Clues did not particularly apply to hunting.

There was another term, however, new to the seventeenth century, that did relate directly to hunting and that almost immediately far outstripped the clue in popularity. This was the hint. The *Oxford English Dictionary* dates the first use of the noun "hint" to 1616 and the verb "hint" to 1648, from "hent," to grab or get ahold of, and related to "hunt."[46] While at first glance the hint and the clue might appear to serve more or less the same function, they in fact differ drastically in terms of the levels of certainty and guidance they offer. The hint suggested that the seeker of knowledge had grabbed onto a piece of a larger, un-known body, and passed on that handhold for another to use when hunting down the prey. The hint, a grasping in the dark, is hazy. While a hint giver passed something along, whether it was anything of value remained to be seen by future investigation. Hunters had no assurance that, after giving chase, they would capture their unknown, lurking prey. Hints were often described as "brief" or "short"; they gestured vaguely at something larger but cut off before pursuing the topic at any further length. The clue was rarely further qualified with additional adjectives, since in the early modern period it already connoted a slender thread leading precisely to a predetermined end.

The clue might be easily equated with the sense of method, which also of-fered "a path through" (Gr. μετ' + ὁδός) to a predetermined target. Thomas Hobbes referred to the "Clue of Reason" in order to stress the importance of beginning inquiries at precisely the correct point leading infallibly to certain knowledge.[47] According to Hobbes, the clue of reason led from the darkness of doubts "as 'twere by the hand into the clearest light."[48] The claims clues made about precision, certainty, and eventual fulfillment could not differ more from hints. The clue, leading out of a labyrinth toward a specific answer, and assur-ing those who followed it of certainty, was medieval. It offered a method. The hint, a loose, hazy grasping in the dark, was novel in the early seventeenth century, and was antimethodical. It gestured at a vague suggestion that might or might not lead to results. Often, hints were couched in the language of prob-ability. A hint might show that something had "possibility," or more than "bare probability," or even "fesibility."[49]

In an era of epistemic risk-taking that rebelled against method, the hint very quickly outstripped the older clue in popularity. Despite the fact that the word "hint" did not really take off until around 1640, it was used about six times more frequently in printed texts published between 1600 and 1700 than was "clue."[50] William Watts, for example, both took a hint and gave a hint in his 1633 appendix to Thomas James's *A Strange and Dangerous Voyage.* The content of the travel narrative and the observations it contained, he claimed, had given him the "hint" for his own philosophical ruminations in his appendix. Those ruminations, in turn, included the idea that varied nature showed that philosophy could at best give a "generall hint" concerning how things were in the world, "yet no *Vniuersall* and vnfayling certainty."[51] In *Virgo Triumphans,* John Ferrar "hinted" at the benefits of transplantation; he had no certain explanation of why it worked, but it has been found "with an experimented happinesse" that transplantation improved plants, to the extent that Ferrar considered transplantation a tenet of "Garden Philosophy."[52]

Hints were intimate. They were emotional. One intimated a hint about a feeling that knowledge was somewhere in the vicinity. As a form of transmission, hints were social, but they did not transmit equally to everyone. Hints implied the selective transmission of partial knowledge to those with the ability to follow up on them. Whereas following a clue was supposed to be a matter of simply hanging on to a guiding thread, making something of a vague hint required its own form of knowledge, personal initiative, and a congenial attitude. Sir Isaac Newton thus described a letter of his that was printed in the *Philosophical Transactions* as communicating "onely to those that know how to improve upon hints of things."[53]

Hints served as a litmus test for how industrious and eager other individuals were to follow up on a new proposal or not. Following Ferrar's hint in *Virgo Triumphans,* five years later, Edward Digges in Virginia recalled how Ferrar had "hinted" about the possibility of a "double Silk harvest," something Digges had not yet achieved but remained hopeful about.[54] By contrast, others were depicted as obdurately unexcited by hints. A promoter of the colonizing of Madagascar, recalls how, endeavoring "to doe my King and Country service, as well as to benefit my self," he had given the East India Company "some hint" of the opportunities in Madagascar for plantation, "which they contemned."[55]

The vague and merely probabilistic qualities of the hint did not make it any less appealing to naturalists of the time. Early modern scientific genres highlighted doubts, and scientific knowledge remains probabilistic today.[56] By calling received knowledge into question, hints served as tools for advancing knowledge into the future. Bacon himself never used the term "hint," which had not yet risen in popularity, but he came very close to it in one of his "Instances with Special Powers," the "Intimating Instances (*Innuentes*)." These were suggestions that gestured in a direction of possible future research, but made no claims that such suggestions could be realized.[57] As one scholar has suggested, Bacon's works did not give a method for discovering depersonalized, objective knowledge, but rather intimate "hints," that is, "an awakened alertness to the feeling of being 'certaine.'"[58]

Many others who have been described as following the Baconian venatic epistemology of clues in fact embraced hints. Hartlib's associate, John Dury, explained that he wrote "imperfectly" in the form of "hints" on purpose in order to spur others "to undertake for their owne advantage any thing."[59] The young William Petty (future fellow of the Royal Society) advised Hartlib that, in order to advance learning, books should be examined to "Re-experiment the Experiments contained in them, and withall to give hints of New Enquiries."[60] Celebrated gentlemen philosophers of the Restoration Royal Society deployed the hint often and prominently in order to highlight their knowledge as fragmentary and open to doubt.[61] It was the hint, rather than the clue, that would typify the loose, undisciplined forms of reasoning and knowledge appropriation that characterized emergent experimental philosophy.

Hints allowed for gestures at explanations and proof, rather than actual explanations and proof. Robert Hooke used the word "hint" forty-six times in his landmark 1665 *Micrographia*. By comparison, he used the word "clue" in the sense of a guide out of a labyrinth only once, when he opened his text by admitting that he possessed no certain clue.[62] Hooke used "hint" another twenty-eight times in his 1679 Cutlerian Lectures. John Evelyn employed it seventeen times in his 1670 *Sylva, or A Discourse of Forest-Trees*. Robert Boyle used it seventeen times in his 1663 *Some Considerations Touching the Usefulnesse of Experimental Naturall Philosophy*, twenty-seven times in his 1664 *Experiments and Considerations Touching Colours,* and twenty-four times in his 1665 *New Experiments and Observations Touching Cold*.[63] Evelyn and Boyle did not use the term "clue" at all. Rather than the narrow, methodical epistemology of the clue, it was the vague, loose reasoning of the hint that the early Stuart period bequeathed to subsequent generations.

The hint allowed experimental philosophers to draw connections and inferences and conjecture about the future in ways that experimental evidence could not conclusively support. In contrast to what they claimed was the specious certainty of scholastic deductions leading to axioms, Boyle and other experimental essayists flaunted epistemic modesty. They eschewed method and embraced happenstance. Boyle favorably contrasted "accidental Hints" with "accurate Enquiries."[64] The power of the former lay partially in their divine origins. It was God who caused the experimental mishaps in the laboratory that conveyed "accidental Hints" and led to powerful discoveries, far beyond what "accurate Enquiries" might achieve.[65]

Hints allowed for loose gesturing at future knowledge that was not supported by knowledge's present state. A case in point is the claim of Walter Charleton, FRS, that the College of Physicians had realized Bacon's "Solomons House" from *New Atlantis*.[66] Most of his evidence came in the form of unrealized hopes and hints. Charleton had "some reason to put you in hope, that ere long you may see a Collection of most of the Anatomical Experiments these Men have made," which would provide a "Method" to "perfect that Comparative Anatomy, whose defect the Lord St. Alban [Bacon] so much complained of, in our Art."[67] Such a work did not appear. Other members of the society, according to Charleton, were pursuing "the hint" given by a German scholar about how to explain muscular motion through mechanics, an "enterprise of great difficulty, and long desiderated."[68] These physicians projected a future state of knowledge, greatly desired but of no assured fulfillment. They also interloped laterally into the knowledge of others, thus acting, according to Charleton, as Bacon's "Merchants of Light." Physicians not only came to know "American druggs," but also acquired such knowledge of subterranean phenomena that "Lapidaries and Miners come to learn of them." Some physicians "enquire into the mysteries of Refiners, Belfounders, and all others that deal in Metals," others research the "adulterating of Wines," and still others the "hurtfull arts of Brewers, Bakers, Butcher, Poulterers and Cooks."[69] Charleton framed this invasion of other arts as having a medicinal purpose, namely, being able to advise against threats to health that these other practitioners posed. However, as the former assistant to royal physician Sir Theodore de Mayerne, Charleton would have known that members of the College of Physicians also interloped into other trades for reasons of profit.[70]

Charleton suggests that both this quest for knowledge from other domains and the projection of a grand design for the future of knowledge was stimulated by the noble passion of rivalry. The "noble Emulation that hath equally enflamed their ingenious breasts, makes them unanimous in cooperating toward the Common design, the erecting an intire and durable Fabrick of Solid Science; such as posterity may not only admire, but set up their rest in."[71] In great contrast to the view that the "Baconian program" required dispassionate obedience, Charleton argued that it was passion itself that could be trusted to build up a future, solid, and durable science which never need be doubted nor changed by later generations, and where future knowledge might find "rest."

In fact, thankfully, medicine has continued to move on from the seventeenth century. What has endured are practices of directing attention to the fulfillment of all hopes and desires at some point in the future. This move serves to distract from the disconnect between the current state of a science, in which many topics are extremely difficult to investigate and lack a clear method, and the authority that science enjoys. With Bacon's help, Charleton imagines a future science constructed as an integrated, ordered, "intire and durable Fabrick." We have been imagining that future science ever since.

Of course, experimental natural philosophy did not go unchanged over the course of the seventeenth century, but continually convulsed with major transformations. Academics of the late seventeenth and early eighteenth centuries worked to reframe undisciplined knowledge, including experimental philosophy, into academic forms, as I am continuing to explore in a related book project, *Curating the Enlightenment.* Throughout this process of attempting to reinsert undisciplined knowledge back into disciplines, knowledge never lost the dynamism endowed by early Stuart interloping. Rather, it was precisely its wild inheritance that revamped a notion of discipline designed for the stable transmission of knowledge into a research discipline, that is, an enterprise aimed at transforming knowledge. In research, the greatest intellectual rewards go not to the maintainers of knowledge, but to those daring minds who upend their field or even create new ones.

The importation of interloping, which aimed at infinite advancement in all directions, into the very core of an inherited disciplinary structure created an unstable amalgam of two contradictory tendencies. One is the undisciplined, aggressively competitive advancement of knowledge. The other is the disciplinary and corporate (that is, faculty-based) structure of the university. These two opposing forces of innovation and structure pull against each other, creating internal tensions in the structure of research disciplines that have never been fully resolved, but which continually generate ad hoc knowledge innovation, upheaval, and churn.

Shipwrecked Knowledge

Science never achieved the solid fabric that Charleton promised. Still today it resembles more the figure of Curiosity than that of Science that we contemplated in the Introduction. At the outset of this book, we considered how odd it was that despite never truly changing her habits, Curiosity managed to assume the authority of Science. Curiosity continually roved the world, hungrily grabbing particular phenomena that could be linked together through suggestion and inference, rather than systematic theorizing. These connections, however, were only probable and therefore often only temporary. They continually were dissolved, undone, and contradicted by the next phenomenon Curiosity seized upon. How then did science come for many to embody epistemic authority? Why do we feel like certainty exists somewhere in this world, if our most prestigious forms of knowledge are still always open to being overturned?

The undisciplining of knowledge provided both the disease and its medicine. Curiosity's saving grace was her amnesia and perpetual forward motion. In the ability to forget what was rejected, and to not even categorize it anymore as science, lay science's claim to being uniquely cumulative, and thus more forward moving than other forms of knowledge. As old, wild-eyed ideas were disproven and drifted forgotten to the ground, Curiosity distracted attention from them by drawing on those forms of persuasion developed by projectors and honed by Francis Bacon: always holding out hope for that next thing, the next connection, the next discovery. Antisystematic, loose forms of reasoning allowed for suggestion and hand waving that could defer proof indefinitely into the future. Curiosity, flying about the world of becoming rather than inhabiting the world of being, is always on the move and in the making. She inherited Science's aura not in her own guise, but as a blurred, deferred holographic projection of her promised future.

Interlopers bequeathed us alluring messages about future possibilities, about aiming high, and about being all that we can be. Innovation, the malleability of identity, and hope for the future are now beloved as ideals. They form a central part of our civic epistemology concerning the place of science and technology in the commonweal and its future promise. We have consumed these messages in various guises, from elementary-school textbooks to superhero comics. It takes some effort to look beyond them and notice what they leave unsaid. Our tendency to chart one achievement after the next, according to a notion of eternally onward-marching progress, overlooks the many ways knowledge has buckled, bent, and broken down. The contingent history of projects, including unsavory consequences, violent behaviors, and failure, offers a much-needed correction to this selective gaze.

The image of Merchants of Light heading off to the distant horizons that adorned Bacon's title pages has often been viewed through such retrospection.

Given how science and technology have transformed our world, it has been possible to imagine these voyages as self-assured journeys following a map that Bacon laid out to our scientifically and technologically endowed present. Yet Bacon himself encouraged us to forget what we know about already conquered knowledge goals and to reconsider them when they were still unlikely, seemingly impossible proposals. From this viewpoint, we can look at his famous imagery again and perhaps see something new. What we see is not a map charting a way forward, but an invitation to an open-ended riskscape, with no landmarks to guide us.

Bacon's frontispiece to his 1620 *Novum Organum* represents an unbuilding of expected architecture. Innumerable early modern books came preceded with architectural imagery that set the book into a structure of knowledge, often loaded with allegorical personifications and historical figures that gave knowledge a hierarchy, an order, and a sense of tradition in which to place the printed work. Rubens designed such a typical frontispiece, for example, for Justus Lipsius's 1615 edition of the works of Seneca printed by the Plantin Press of Antwerp (figure 17). Two floors of severely smooth columns order the frame, on which are positioned serried ranks, with gods and heroes (Hercules, Pallas, Ulysses) up above, classical Greek Stoic philosophers Zeno and Cleanthes flanking the doorway, and backing the base, their later Roman and Greek followers, Seneca and Epictetus.

On Bacon's title page (figure 18), we see two columns, just as rigid, impervious and stony as those holding up the facade of the Senecan edition. But these hold nothing up, and there is nobody there. The expected architecture has been removed, and we find ourselves outside, unsheltered and undirected, at the edge of an open sea. Rather than providing the building blocks of knowledge in which Bacon's work can be set, these columns represent the Pillars of Hercules, or the two precipitous cliffs at the Strait of Gibraltar. They symbolize the very edge of safety and that which was already known. Journeying beyond them meant intentionally leaving all structures behind and venturing into chaotic, dangerous waters.

As many ran to and fro, knowledge increased. Yet, this was no well-charted voyage sailing into a secure harbor. These ships were manned by interlopers, shifting course as they hunted advantage wherever it could be found in unknown seas. In their holds, they chained up many people forced to labor on their behalf. If they had the ability to do so, the interlopers "swam between two waters," sneaking up on others unexpectedly, blasting them to bits and making

Figure 17. Theodore Galle, after designs for Peter Paul Rubens, frontispiece to Justus Lipsius, ed., *L. Annaei Senecae Philosophi Opera* (Antwerp: Plantin, 1615). Rijksmuseum, RP-P-OB-6890. Public domain

Figure 18. Simon de Passe, frontispiece to Francis Bacon, *Novum Organum* (London: Bill, 1620). Matteo Omied / Alamy Stock Photo

away gleefully with their prizes. The choppy waters they prowled swam with hidden shoals, monsters, and pirates. If Bacon's imagery is meant to tempt us forward with the exciting prospects of what lies ahead, we should never forget that in these unbounded waters many adventurers foundered and sank to the ocean floor, taking many of the unwilling with them.

Abbreviations

Add. MS.	Additional Manuscripts
BL	British Library
CSPC	*Calendar of State Papers Colonial*
CSPD	*Calendar of State Papers Domestic*
CSPV	*Calendar of State Papers Relating to English Affairs in the Archives of Venice*
HP	M. Greengrass, M. Leslie, and M. Hannon, *The Hartlib Papers.* Published by The Digital Humanities Institute, University of Sheffield. Available at: https://www.dhi.ac.uk/hartlib. © 2013 The University of Sheffield	Version 3.0
IOR	India Office Records
ODNB	*Oxford Dictionary of National Biography*
RVC	*Records of the Virginia Company of London*, ed. Susan Myra Kingsbury, 4 vols. Washington, DC: Government Printing Office, 1906–1935
SP	State Papers Online. https://www.gale.com/primary-sources/state-papers-online
VCA	Virginia Company Archives. https://www.amdigital.co.uk/primary-sources/virginia-company-archives

Introduction

1. Ripa, *Iconologia*, 67. The *Iconologia* was first published in 1593. For an introduction to the meaning of *scientia,* see Sorell, Rogers, and Kraye, "Introduction." On the history of curiosity as a vice and its recuperation in the early modern period, see Kenny, *The Uses of Curiosity.*

2. Ripa, *Iconologia*, 20.

3. Weisheipl, "Classifications of the Sciences"; Ovitt Jr., "The Status of the Mechanical Arts"; Whitney, "Paradise Restored"; Reisch, *Natural Philosophy Epitomised,* xxxv.

4. Arthur Lovejoy's grand 1936 overview of the idea of intersecting hierarchies in *The Great Chain of Being* has been criticized in many particulars. Nevertheless, a notion of hierarchy similar to what he traced did obtain in metaphysical discussions from the thirteenth to the sixteenth centuries; Mahoney, "Lovejoy and the Hierarchy of Being,"

220. Although for some philosophers, these notions of hierarchy may have been metaphorical rather than a representation of a literal understanding of cosmic structure, popular culture does indicate such hierarchies obtaining in common understandings of space: Mahoney, "Metaphysical Foundations of the Hierarchy of Being"; Grant, *Planets, Stars, and Orbs*; Tillyard, *The Elizabethan World Picture*. On constraint, see Monagle, *The Scholastic Project*.

5. Lindsay, *The Three Estates*.

6. Marshall, "Forgery and Miracles."

7. Millstone, "Seeing like a Statesman"; Loe, *The Kings Shoe Made*.

8. Burckhardt, *Die Cultur der Renaissance*.

9. Koyré, *From the Closed World*; Wolfe, *Freedom's Laboratory*.

10. C. Hill, *The Intellectual Origins of the English Revolution*; Webster, *The Great Instauration*; V. Keller, "Deprogramming Baconianism."

11. Park and Daston, "Introduction"; Roodenburg, Muchembled, and Monter, *Forging European Identities*, which is the final volume of the four-volume series *Cultural Exchange in Early Modern Europe*; Burke and Hsia, *Cultural Translation*; Smith and Schmidt, *Making Knowledge*. The new focus on labor and craft revives some perspectives from earlier Marxist work: Freudenthal and McLaughlin, *The Social and Economic Roots of the Scientific Revolution*; Zilsel, *The Social Origins of Modern Science*. For a critique of genius and a study of the more temporally accurate notion of ingenuity, see Marr, Garrod, Marcaida, *Logodaedalus*.

12. Portuondo, *Spanish Disquiet*, 8.

13. Portuondo, *Spanish Disquiet,* 9. Goldgar, *Tulipmania*.

14. Mulsow and Daston, "History of Knowledge."

15. Stark, "Emergence"; Bauer and Norton, "Introduction."

16. Proctor and Schiebinger, *Agnotology*.

17. HP 29/3/13A.

18. Daston and Park, *Wonders and the Order of Nature*.

19. For an attempt at mapping rapidly evolving interdisciplinary research areas, see Okamura, "Interdisciplinarity Revisited."

20. Thurs, "Myth 26"; Woodcock, "'The Scientific Method' as Myth and Ideal."

21. For example: Secord, "Knowledge in Transit"; P. Smith, "Science on the Move"; P. Smith, "Itineraries of Materials and Knowledge"; Solomon, *Objectivity in the Making*, 107; Heesen, "Accounting for the Natural World"; Harkness, "Accounting for Science"; H. Cook, *Matters of Exchange*; Östling et al., *Circulation of Knowledge*; Heringman, *Sciences of Antiquity*. Marchitello and Tribble's *The Palgrave Handbook of Early Modern Literature and Science* includes a section entitled "Pre-disciplinary Knowledge." For a review of recent discussions of hybrid knowledge, see: Winterbottom, *Hybrid Knowledge*; Long, "Multi-tasking 'Pre-professional' Architect/Engineers"; Biagioli, *Galileo, Courtier*; Burke, "The Renaissance Translator as Go-Between"; D. Turnbull, *Masons, Tricksters and Cartographers*; Schaffer et al., *The Brokered World*.

22. For example, "trading zones" in Long, *Openness, Secrecy, Authorship*, 15, 211, 234, 243, and 246. There is one episode of theft in Long's account, at 91–92. See also Long, "Trading Zones in Early Modern Europe."

23. Bod et al., "The Flow of Cognitive Goods," 488.

24. Rieppel, Lean, and Deringer, "Introduction," 22.

25. Cotgrave, *A Dictionarie*, [Sssvi^v].

26. As discussed in Keller and McCormick, "History of Projects." Analogous terms proliferated across Europe, such as *arbitrista* in Spain and *donneur d'avis* in France, but each of these was flavored by the local situation and was not necessarily interchangeable.

27. Knowler, *The Earl of Strafforde's Letters and Dispatches*, 186.

28. Monson, *Naval Tracts*, 453.

29. For example, Novak, *The Age of Projects*. Defoe, *An Essay upon Projects*.

30. Cited in Keller and McCormick, "Towards a History of Projects," 437. Samuel Johnson, *The Works*, 223.

31. Peck, *Consuming Splendor*; Ash, *The Draining of the Fens*; Cavert, *The Smoke of London*; McCormick, *William Petty*; Pastorino, "Weighing Experience"; Yamamoto, *Taming Capitalism*.

32. Working, *The Making of an Imperial Polity*.

33. Many themes discussed in this book recur in their Restoration manifestation in Mulry, *An Empire Transformed*.

34. "Interlopers: Leapers or runners between; it is usually applied to those that intercept the Trade or Traffick of a Company, and are not legally authorized": Blount, *Glossographia*, [X8^v].

35. The financial projector Gerard Malynes accused the English Merchant Adventurers of coining the term (Malynes, *The Center of the Circle*, 98–99). On the Company's attempts to deal with their opportunistic, "disorderly brethren," see Leng, "Interlopers and Disorderly Brethren."

36. Gerbier, *Subsidium peregrinantibus*, 116. For critical views of interlopers as spies, see Gerbier, *Subsidium peregrinantibus*, 112; Hall, *Quo vadis?*, 5.

37. Markley, "Riches, Power, Trade, and Religion," 498.

38. In this I differ slightly from Clucas. I would argue that Donne here contrasts internally differentiated structures with structureless chaos; Clucas, "Poetic Atomism," 329.

39. "An Anatomy of the World" in Donne, *Poems* (1633), 242.

40. Elegy 14: "Julia," in Donne, *Poems* (1639), 97.

41. Norton, *The Gunner*; Parkinson, *Paradisi in Sole Paradisus Terrestris*.

42. Malynes, *Consuetudo, vel Lex Mercatoria*, 262.

43. Malynes, *Consuetudo, vel Lex Mercatoria*, 182–84.

44. Compare Malynes, *Consuetudo, vel Lex Mercatoria*, 62 and Finkelstein, *Harmony and the Balance*, 33. For Malynes's eclectic mix, see Wennerlind, *Casualties of Credit*, 48.

45. Müller-Mahn, Everts, and Stephan, "Riskscapes Revisited," 207.

46. A. Wood, "Custom, Identity and Resistance."

47. Kelley, "The Problem of Knowledge," 15.

48. Speer, "Schüler und Meister."

49. Working, *The Making of an Imperial Polity*, 34–35.

50. Kew, UK National Archives, PROB 11/125/121.

51. Chamberlain, *The Letters of John Chamberlain*, 377.

52. Renn, *The Evolution of Knowledge*.

53. Mokyr, *A Culture of Growth*.

54. V. Keller, "Accounting for Invention."

55. Montaigne, *Essais*, 37.

56. Bacon, "Of Seditions and Troubles," *Essays*, 84.

57. Donne, *Works*, 536.

58. V. Keller, "Into the Unknown."

59. Muraca, "Décroissance," 154.

60. Heyd, *Be Sober and Reasonable*.

61. Shapin, *The Scientific Revolution*, 90–94.

62. Bauer, "Zur entwicklung des 'homo oeconomicus,'" 65.

63. Forman, "On the Historical Forms," 62–63.

64. Forman, "On the Historical Forms," 62.

65. P. Miller, "Nazis and Neo-Stoics," 145.

66. Elias, *The Civilizing Process*, vols. 1 and 2.

67. Linklater and Mennell, "Norbert Elias," 385.

68. Linklater and Mennell, "Norbert Elias," 404. On the death of Elias's mother, see Jitschin, "Family Background of Norbert Elias."

69. Working, *The Making of an Imperial Polity*, and the literature cited there on p. 17.

70. Oestreich, *Geist und Gestalt*; Oestreich, *Neostoicism and the Early Modern State*.

71. Oestreich, *Geist und Gestalt*, 19.

72. Oestreich, *Antiker Geist und moderner Staat*, 104.

73. Oestreich, *Antiker Geist und moderner Staat*, 164.

74. Oestreich, *Antiker Geist und moderner Staat*, 268.

75. Oestreich, *Antiker Geist und moderner Staat*, 269–70.

76. Raeff, *The Well-Ordered Police State*. Wakefield, *The Disordered Police State*, 157, notes the similarities between Oestreich and Raeff.

77. Srbik, *Wilhelm von Schröder*; Stollberg-Rilinger, *Der Staat als Maschine*; Raeff, *The Well-Ordered Police State*.

78. P. Smith, *The Business of Alchemy*; Sandl, "Development as Possibility"; V. Keller, "Happiness and Projects between London and Vienna."

79. Wakefield, *The Disordered Police State*.

80. P. Miller, "Nazis and Neo-Stoics"; Kelly, "The Politics of Intellectual History."

81. Borkenau, "The Sociology of the Mechanistic World-Picture," 117; a translation from Borkenau, "Zur Soziologie des mechanistischen Weltbildes." A fuller discussion is in Borkenau, *Der Übergang*.

82. Borkenau, "The Sociology of the Mechanistic World-Picture," 116–17.

83. Foucault, *Surveiller et punir*. For Foucault and Lipsian Neostoicism, Brooke, *Philosophic Pride*, 33–35.

84. Dear, "Mysteries of State," 214.

85. Koenigsberger, *Politicians and Virtuosi*, 260.

86. Dear, *Discipline and Experience*; Dear notes that Simon Schaffer brought Borkenau to his attention, but did not note the tensions between Borkenau's views and those of Oestreich ("A Mechanical Microcosm," 61n31). Shapin, *A Social History of Truth*, lists Oestreich in his bibliography, but not Borkenau. Borkenau's main follower in the Anglophone history of science is Richard Hadden: Hadden, "Social Relations"; Hadden, *On the Shoulders of Merchants*.

87. Biagioli, "Etiquette, Interdependence, and Sociability."

88. Shapin, *A Social History of Truth*, 162–65.

89. Shapin, "A Scholar and a Gentleman," 283.

90. Guillory, "The Bachelor State," 267n10.

91. Gaukroger, *Francis Bacon*, 13–14.

92. V. Keller, "Deprogramming Baconianism."

93. Gaukroger, *Francis Bacon*, 74.

94. Gaukroger, *Francis Bacon*, 10. Gaukroger notes Julian Martin, *Francis Bacon*, and Leary Jr., *Francis Bacon*, as emphasizing these points.

95. Gaukroger, *Francis Bacon*, 12.

96. Jalobeanu, *The Art of Experimental Natural History*, 74.

97. Dear, "Mysteries of State," 214.

98. Shapin, *Never Pure.*

99. Lorraine Daston has criticized the many accounts of legitimizing knowledge found in the sociology and history of science of the late 1980s and mid 1990s in "The Nature of Nature in Early Modern Europe."

100. Shapin, "The Invisible Technician."

101. Govier, "The Royal Society, Slavery and the Island of Jamaica"; Winterbottom, *Hybrid Knowledge*, 18.

102. Robert Boyle, *Some Considerations*, 29; V. Keller, *Knowledge and the Public Interest*, 232. This passage been discussed fairly recently without broaching the issue of slavery; see, e.g., Irving, *Natural Science*, 80–81.

103. Otremba, "Inventing Ingenios."

104. Malcolmson, *Studies of Skin Color*, 21.

105. Boyle, *Of the reconcileableness of specifick medicines*, 26.

106. L. Stewart, "Global Pillage."

107. Hirschman, *The Passions and the Interests*, 12.

108. Zabel, "Introduction."

109. Hirschman, *The Passions and the Interests*, 44.

110. V. Keller, *Knowledge and the Public Interest.*

111. "Disknowledge" was coined in Eggert, *Disknowledge.*

112. As in Hirschman, *The Passions and the Interests*, 62na. For example, Gerbier, *A Sommary Description*, [B]: "theire Slavish condition proves to them a verry great blisse."

113. Hirschman, *The Passions and the Interests*, 62.

114. Hirschman, *The Passions and the Interests*, 129.

115. Hirschman, *The Passions and the Interests*, 132.

116. It is a privilege to imagine, as Connor does, that those pointing out privilege reap an "abundant harvest of pleasure" in so doing, rather than feelings of exhaustion, disgust, and impotency; Connor, *The Madness of Knowledge*, 32. For a volume studying emotions in the marketplace of ideas, see Leemans and Goldgar, *Early Modern Knowledge Societies.*

117. Johns, *Piracy*, 19.

118. Andrews, *The Spanish Caribbean*, 252.

119. Forster, *The Debates on the Grand Remonstrance*, 228, 230–31.

120. Jowitt, *The Culture of Piracy*, 35.

121. Hanna, *Pirate Nests.*

122. Hebb, *Piracy and the English Government*, 2, 20.

123. McDonald, *Pirates, Merchants, Settlers, and Slaves.*

124. Hebb, *Piracy and the English Government*, 248.

125. Craven, "The Earl of Warwick"; Murphy, "Merchants, Nations and Free-Agency"; Matar, *British Captives*, 75–91.
126. Bacon, *New Atlantis*, 42–44.
127. Harkness, *The Jewel House*, 241–53.
128. Merchant, *The Death of Nature*.
129. Werrett, *Thrifty Science*.

Chapter 1 · The Political Economy of Projects

1. Wrightson, "The Politics of the Parish," 22–25.
2. Yamamoto, *Taming Capitalism*.
3. Herbert, *Newes out of Islington*, 3–4.
4. "What is a *Proiector?*": Jonson, *The Devil is an Ass* [1616] in *Workes*, 110.
5. Brugis, *The Discovery of a Projector*, 19.
6. Scot, *The Projector*, 6.
7. Ash, *The Draining of the Fens*, 13.
8. Hindle, *The State and Social Change*, 9; Braddick and Walter, *Negotiating Power*; Yamamoto, *Taming Capitalism*, 13.
9. Hindle, *The State and Social Change*, 21.
10. Botero, *Della ragion di stato*, 1.
11. von Friedeburg and Morrill, *Monarchy Transformed*.
12. Melton, *A sixe-folde politician*, 93–94.
13. Loe, *The Kings Shoe*, 17.
14. Donne, *Pseudo-martyr*, 47; M. Price, "Recovering Donne's Critique."
15. Descendre, *L'État du Monde*; V. Keller, *Knowledge and the Public Interests*.
16. Donne, *Pseudo-martyr*, 47.
17. Cotgrave, *A Dictionarie*, [Hhii^r].
18. Cotgrave, *A Dictionarie*, [Pp^v].
19. Cotgrave, *A Dictionarie*, [Dddiiii^r].
20. Bullokar, *An English Expositor*, [O2^r].
21. SP 14/180/90.
22. Yamamoto, *Taming Capitalism*, 112.
23. Gorges, *A True Transcript and Publication* (1611), [C2v].
24. Clegg, *Press Censorship in Jacobean England*.
25. Thanks to Richard Serjeantson for this suggestion about the possible connection to the suppression of political publications.
26. Botero, *Della ragion di stato*, 1.
27. Peck, *Consuming Splendor*; Kupperman, "The Love-Hate Relationship with Experts."
28. Botero, *Relations, Of the Most Famous* (1611), 328, 329.
29. Palmer, *An Essay*, 23.
30. *Trésor Politique*, 143.
31. Testa, "From the 'Bibliographical Nightmare,'" 7.
32. Brendecke, *Empirical Empire*.
33. Cressy, *Dangerous Talk*, 11.
34. Melton, *A sixe-folde politician*, 93–94.
35. Melton, *A sixe-folde politician*, 95.
36. Melton, *A sixe-folde politician*, 97.

37. Notestein, *The Journal of Sir Simonds D'Ewes*, 36. Another account identifies the projector as Sir John Marley; both Marley and Melton were involved with coal. See, e.g., *CSPD*, 1638–39, 397.

38. Cited in Cuttica, "Sir Francis Kynaston," 137.

39. Loe, *The Kings Shoe*, 31.

40. Gorges, *A True Transcript and Publication* (1611), [F2r]. Gorges wished to have the office established on the exchange called Britain's Burse, where his account of his office was published. *CSPD*, 1611–18, 17.

41. Gorges, *A True Transcript and Publication* (1611), [C2v].

42. Gorges, *A True Transcript and Publication* (1612), [E3].

43. Gorges, *A True Transcript and Publication* (1612), [E3v].

44. Cressy, *Dangerous Talk*, 103.

45. Ferrar, *Memoirs*, 115–16; Sutto, *Loyal Protestants and Dangerous Papists*, 25.

46. Ferrar, *Memoirs*, 116–17.

47. Kostylo, "From Gunpowder to Print"; Adam Mossof, "Rethinking the Development of Patents," 1259.

48. An example of where the two are related is Amundsen, "Thinking Metallurgically."

49. Macleod, *Inventing the Industrial Revolution*, 10; Yamamoto, *Taming Capitalism*, 50–51.

50. Edmunds, *Law and Practice*, 106–8; Rooke, "Report of the Clerk."

51. Macleod, *Inventing the Industrial Revolution*, 12.

52. Macleod, *Inventing the Industrial Revolution*, 12; Ash, *Power, Knowledge, and Expertise.*

53. Macleod, *Inventing the Industrial Revolution* 14.

54. Mossof, "Rethinking the Development of Patents," 1261; Woodcroft, *Titles of Patents of Invention.* For some of the issues with drawing long-term comparisons based on Woodcroft's numbers, see Yamamoto, *Taming Capitalism*, 29. Among other issues, Woodcroft only considers inventions, rather than other sorts of monopoly patents.

55. Yamamoto, *Taming Capitalism*, 32.

56. James I, *A Declaration of His Maiesties Royall pleasure,* 21.

57. Edmunds, *Law and Practice*, 118.

58. Biagioli, "Patent Republic."

59. Gray, *A Good Speed to Virginia.*

60. Scott, *A Collection of Scare and Valuable Tracts*, 424.

61. Buning, "Inventing Scientific Method"; van Cruyningen, "Dealing with Drainage."

62. R. Johnson, *Nova Britannia,* [Dv].

63. St. John, "Norden's Project"; discussed by Richard Hoyle, "Disafforestation and Drainage." See also, Owen, *Calendar of the Manuscripts,* 368–89.

64. Malynes, *Lex Mercatoria*, 216.

65. *CSPD*, 1639–40, 236–37.

66. St. John, "Norden's Project," 289–94.

67. "Intégrer les inventions dans la forme du projet économique"; Hilaire-Pérez, "Patents et privileges," 147.

68. Yamamoto, *Taming Capitalism*, 62.

69. Ruellet, *La Maison de Salomon.*

70. Wrightson, *English Society*; Heal and Holmes, *The Gentry in England and Wales*; French, *The Middle Sort of People*; Kamen, *Early Modern European Society*, 76.

71. Higton, "Portrait of an Instrument-Maker," 151.

72. Higton, "Portrait of an Instrument-Maker," 152.

73. [Bolton], *The Cities Advocate*, [A2v].

74. [Bolton], *The Cities Advocate*, 19.

75. Stone, *The Crisis of the Aristocracy*, 376–77.

76. Gough, *The Rise of the Entrepreneur*, 12.

77. Thirsk, *Economic Policy and Projects*, 24–50.

78. Cramsie, *Kingship and Crown Finance*, 35–36.

79. Slack, *The Invention of Improvement*, 60.

80. See, e.g., Cust, *Charles I and the Aristocracy*, 34, 41.

81. Ash, *Power, Knowledge, and Expertise*; B. Jardine, "Instruments of Statecraft."

82. Buning, "Inventing Scientific Method."

83. Pumfrey and Dawbarn, "Science and Patronage in England."

84. Stone, *The Crisis of the Aristocracy*, 363.

85. Stone, *The Crisis of the Aristocracy*, 364.

86. Manning, *Swordsmen*, 17–19.

87. Nerlich, *Ideology of Adventure*, 1:4.

88. Nerlich, *Ideology of Adventure*, 1:12–14; Truitt, *Medieval Robots*.

89. Muchmore, "Gerrard de Malynes," 348; BL Add. MS 10038, 23.

90. Cochran-Patrick, *Early Records*, 129–38.

91. Malynes, *Lex Mercatoria*, 265.

92. *Instructions for the increasing of Mulberie trees*.

93. Shaw, *Letters of Denization*, 42.

94. Shaw, *Letters of Denization*, 42, 44. One block was for nine denizations, another was for forty naturalizations. On von Wulffen, see Kirchner, *Das Schloss Boytzenburg*, 321; and *CSPD*, 1637–38, 50, 108; *CSPD*, Addenda 1625–49, 179.

95. Shaw, *Letters of Denization*, 45; Wijnman, "Vernatti, Sir Philibert (1)."

96. For example, Todt, "A Venture of Her Own."

97. Bacon, *Works*, 44; Steen, *The Letters of Lady Arabella Stuart*, 59.

98. Patterson, "Projects," in *A Catalogue*, 5, 34.

99. Cf. Clark, *Working Life of Women*, 28.

100. Leong, *Recipes and Everyday Knowledge*.

101. Pal, *Republic of Women*; DiMeo, *Lady Ranelagh*.

102. Hartlib, *A rare and new discovery*;

103. Gerbier, *A Sea-Cabbin Dialogue*, 33; Van den Heuvel, *Women and Entrepreneurship*.

104. Oestreich, *Neostoicism*, 268.

105. D. Lindley, *Court Masques*, 277. On Howard's art collecting, Wilks, "Art Collecting at the English Court," 34, 40.

106. "Appendix to the Third Report," in *Third Report of the Royal Commission*, 15.

107. Thrush and Ferris, "Howard, Sir Thomas"; *A Publication of Guiana's Plantation*.

108. *CSPD*, 1637–38, 79, 128, 236; A Proclamation concerning certain Kilnes.

109. Cited in Page, *The Victoria History*, 262.

110. Whitelocke, *Memorials*, 24.

III. Halse, "Great Britains Treasure," BL Egerton MS 1140, f. 44; Cavert, *The Smoke of London*, 58.

112. Halse, *Patent of Sir Nicholas Halse*, 1.

113. *CSPD*, 1636–37, 369–70.

114. *CSPD*, 1638–39, 251.

115. Woodcroft, "A Commission Directed to Sir Richard Wynne."

116. *CSPD*, 1637–38, 79.

117. T. Porter, "Thin Description."

118. BL Egerton 1140.

119. BL Stowe 325, 151.

Chapter 2 · Cast of Characters

1. Ratcliff, "Art to Cheat the Common-Weale."

2. Macintyre, "Buckingham the Masquer," 65–66.

3. Sharpe, "Virtues, Passions, and Politics," 778, 790.

4. Knowles, "Jonson's *Entertainment at Britain's Burse*." I first identified the figure of Van Goose as Drebbel in the *Masque of the Augurs* in my dissertation (V. Keller, "Cornelis Drebbel"), and this has since become an accepted interpretation. See Speake, "The Wrong Kind of Wonder."

5. For example: Webster, "William Harvey," 20; Coffey, "As in a Theatre"; A. Keller, "The Age of Projectors," 469; Davids, "Amsterdam as a Centre of Learning," 307.

6. Jonson, *The Masque of Augures*, [A3ᵛ].

7. Dacos, *La Découverte de la Domus Aurea*.

8. For example, Babington, *Pyrotechnia*, 38.

9. Jonson criticized Drebbel's automatons in similar ways in *Entertainment at Britain's Burse*.

10. Jonson, *The Masque of Augures*, [A3ᵛ].

11. Gregg, *King Charles I*, 219.

12. Malcolm Smuts describes Charles I as dabbling in inventions; Smuts, *Court Culture*, 153.

13. Harkness, *The Jewel House*, 180. On Henry's court, see Cormack, "Twisting the Lion's Tail."

14. Hill, *The Intellectual Origins*; Webster, *The Great Instauration*. See also, e.g., Newell, *From Dependency to Independence*.

15. Cust, "Charles I's Noble Academy," 340.

16. Bibliothèque Municipale Inguimbertine, Carpentras, Ms. 1776, 411v.

17. Major, *Dissertatio epistolica de cancris*, 63.

18. Hubert, *A Catalogue*, 60, 41.

19. Harkness, *The Jewel House*, 142–180.

20. Ash, *Power, Knowledge, and Expertise*.

21. Pastorino, *Weighing Experience*, 51, 65; Cramsie, *Kingship and Crown Finance*, 86–87, 129.

22. Sharpe, "The Earl of Arundel," 209–44; Brenner, *Merchants and Revolution*, 289. In 1635 Samuel Hartlib noted that Arundel gained a patent for a maltkiln and thus suppressed Küffler's; HP 29/3/44B. With the exception of his promotion of the colonization of

Madagascar, he seems to have been more of an investor than an active participant. On Arundel's investments, see Stone, *Crisis of the Aristocracy*, 376.

23. Lockyer, *Buckingham*, 93. Throughout, I refer to him as Villiers rather than Buckingham.

24. Bellany, "Singing Libel," 185.

25. Hille, *Visions of the Courtly Body*," 99.

26. Macintyre, "Buckingham the Masquer," 60.

27. Hille, *Visions of the Courtly Body*, 60.

28. Hille, *Visions of the Courtly Body*, 116.

29. Hille, *Visions of the Courtly Body*, 117.

30. Cited in Bellany, "Naught But Illusion?" 156.

31. "Jolis Romans." *Le Surveillant de Charenton*, 3.

32. Cogswell, "The Symptomes and Vapors," 327, 333.

33. Cogswell, "The Symptomes and Vapors," 335.

34. Gough, *Sir Hugh Myddelton*; Ash, *Draining the Fens*.

35. Winfield, *British Warships in the Age of Sail*, 148.

36. "Den overledenen hertoch van Buckingham heeft onder zijnne rariteijten bewaert een vertroude descriptie van een rijcke goud mijne . . . die den vermaerde see helt Sir Walther Ralleij opgedaan en gevonden hadde." Cited in Sellin, "Michel le Blon and England," 105 (my translation). V. Keller, "The 'Framing of a New World'"; Gerbier, *Sea-Cabbin Dialogue*, 21.

37. Townshend, *Life and Letters*, 13.

38. Cited in Jowitt, "To Sleep, Perchance to Dream," 258.

39. Jowitt, "To Sleep, Perchance to Dream," 257.

40. Jenkins, "Paper-Making in England," 585; *CSPD*, 1637–38, 395.

41. *CSPD*, 1629–31, 131.

42. *CSPD*, 1637–38, 371.

43. T. Osborne, *Dynasty and Diplomacy*, 72; Keblusek and Noldus, *Double Agents*.

44. Gerbier, *Baltazar Gerbier Knight to all Men that Loves Truth*, 3.

45. For example, Gerbier, *The Second Part*, 29. Gerbier claimed authorship of this anonymous pamphlet in a letter to Bulstrode Whitelocke: *The Whitelocke Papers*, f. 36.

46. Feingold, "Experimental Philosophy"; V. Keller, "A wild swing to phantsy."

47. Stevenson, "Occasional Architecture."

48. V. Keller, "*Pennetrek*."

49. H. Williamson, *Four Stuart Portraits*, 26; J. Wood, "Gerbier."

50. BL Add. MS 16371 h; BL King's Topographical CXVII, 98; V. Keller, "*Pennetrek*."

51. Feingold, *Mathematician's Apprenticeship*, 144.

52. Aylmer, *The King's Servants*, 288.

53. Aylmer, *The King's Servants*, 289.

54. *CSPD*, 1633–34, 323. *CSPD*, 1625–26, 494.

55. Aylmer, *The King's Servants*, 288; McGowen, *The Royal Navy*, 265–66. Rymer, *Foedera*, 7:388–93.

56. Trim, "English Military Émigrés," 237–60. Bull, *"The Furie of the Ordnance,"* 52–53. In a letter ca. 1630 to Charles, Heydon recalls how he went to serve in the Netherlands with Charles's recommendation in 1618, and returned to assume his brother's

position in the Ordnance Office in 1627, with favorable recommendations from the Dutch army: SP/16/179/90.

57. SP 16/30 f.86 and SP 16/31 f.20.

58. McGowen, "The Royal Navy," 266.

59. Feingold, *Mathematician's Apprenticeship*, 208.

60. *CSPD*, 1627–28, March 6, 1627, 83.

61. BL Harleian 1581, 118.

62. BL Harleian 1581, 118.

63. Bodleian Library, Ashmole MS 1446, 166v.

64. Morton, *New English Canaan*, 182.

65. See, for example, the notes and sketches of equipment added to a report Gardiner sent Heydon of his progress in an alchemical process of 1637: SP 16/374 f. 109. For Gardiner's biography and an overview of his alchemical correspondence with Heydon, see Saco, "Sir Christopher Gardyner"; V. Keller, "The Authority of Practice."

66. Digby, *Two treatises*, 220.

67. SP 16/510/76.

68. *CSPD*, 1634, 435.

69. SP Privy Council 2/36 f. 240. Digby also coauthored a report on the behavior of saltpeter men: SP 16/165/76. On the Evelyn family saltpeter business, see Darley, *John Evelyn*, 3–4.

70. SP 16/179/100. Evelyn and Officers of the Ordnance had previously disagreed over saltpeter supplies; Hodgetts, *The Rise and Progess*, 234.

71. The astrologer John Heydon possessed manuscripts of both "Sir Christopher Heydon" and "Sir John Heydon," confiscated with the rest of his possessions in 1668; SP 29/238/118.

72. Ruellet, *La Maison du Salomon*, 267. "Stoves or Furnaces for Drying and Heating" in Woodcroft, *Appendix to Reference Index*, 16.

73. PROB 11/164 and PROB 11/152.

74. PROB 11/164 and PROB 11/159; M. Kennedy, "Charles I and Local Government"; Thomas et al., *The propositions of Sir Anthony Thomas*.

75. Ash, *Power, Knowledge and Expertise*.

76. V. Keller, "The Authority of Practice."

77. In his autobiography, Huygens referred to both John and William Heydon ("Heidonios hic nempe duos, par nobile fratrum"): C. Huygens, *Gedichten*. Colie, *"Some thankfulnesse to Constantine"*, 103.

78. On Killigrew's alchemical interests and medical cures, see Trevor-Roper, *Europe's Physician*, 182–83, 186, 356.

79. C. Huygens, *Gedichten*, 209–10.

80. County of Middlesex, *Calendar to the Sessions Records*, 305. "Hellibrand Spruson of Tower Wharf, sailmaker": *CSPD* 1627–28, 529.

81. Gover, *The Place-Names of Middlesex*, 151.

82. Hildebrand Prusen married Anna ap William, who in turn was related to Trevor; Palmer, "A History of the Old Parish of Gresford," 270. "Original will of John Trevor of Trevalyn," GLY/738, East Sussex Record Office. Thrush, "The Navy Under Charles I," 360–62.

83. McGowan, *The Jacobean Commissions of Enquiry*, 95, 98, 241.

84. McGowan, *The Jacobean Commissions of Enquiry*, 42–43.

85. *CSPD*, 1625–26, 494.

86. For Prusen's misadventures as "Merchant of the East Indies," see *CSPC East Indies, China and Japan*, vol. 4, 1622–24, 22–26, 176–77, 232, 248, 258–59, 275; *CSPD* 1611–18, 613.

87. *CSPD*, 1603–10, 640.

88. McGowan, *The Jacobean Commissions of Enquiry*, 42.

89. BL East India Company Court Book, IOR/B/7, 363.

90. *CSPC*, East Indies, China and Japan, vol. 4, 1622–24, 259.

91. Birdwood, *The Register of Letters*, 490–91.

92. Virginia Company, *A Declaration of the State*, 35; Lefroy, *Memorials of the Discovery*; "Hiddlebrand Preiwsen," *RVC*, 1:609.

93. *RVC*, 1:401–2, 414.

94. T. Hardy, *Annual Report*, 98; Kirk and Kirk, *Returns of Aliens*, 194. This is apparently the same Abraham Baker involved in smalt production; SP 16/91 f. 46. Tait, "Southwark (Alias Lambeth) Delftware."

95. Coldham, *English Adventurers*, 174. Hildebrand Pruson v. Don diego Sermento, legate of Spain, Kew, UK National Archives, High Court of Admiralty, 23/6/494; see also SP 14/104/203.

96. SP 16/158/46.

97. For Heydon's complaints about Burrell, see Add. MS 64893, 66.

98. SP 16/158/46.

99. SP 16/ 267/25. See also SP 16/339/36, for the division of rents between Kirke, Killigrew, Dawes, Long, Prusen, and Heydon, and SP 16/251/8, Nov. 26, 1633.

100. *CSPD* 1633–34, 35; *The Case of the heir of Sir Anthony Thomas*.

101. "En croissant d'aage il aloit tousiours croissant d'inventions, qui procedoient de la vivacité d son esprit, sans ayde ny lecture de livres qu'il a tousiours mesprisé, tenant pour maxime que la verite et l'excellence des sciences consiste en la cognoissance des secrets de la nature dans laquelle elles sont tout comp[osées]." Peiresc, *Relation*, 408v.

102. "Il vit tout à faict en filosofe, ne se soucie que de ses observations, et mesprise toutes les choses du monde et les grands et saluera plustost un pauvre homme, qu'un grand seigneur. Il vit selon les loix de la nature et ne croit à rien." Peiresc, *Relation*, 410r.

103. HP 30/4/35A.

104. "Magne Senex," "fronte Batavum Agrigolam, sermone Sophum Samiumque referret et Siculum." C. Huygens, "Vita Propria," in *Gedichten*, 203–4. Perhaps Huygens refers to Pythagorus, Aristarchus, or Euripides of Samos and Archimedes of Sicily.

105. Czech National Archives, Národní archiv 1, M. Horákové 133, Prague 6, Ms. A 215.

106. Cornelis Drebbel to Ijsbrandt van Rietwijck, in The Hague, Koninklijke Bibliotheek, KA 47, Constantijn Huygens, *Verzameling musica, medica, etc.*, 207r–207v.

107. Drebbel, *Wonder-vondt*, [B^r].

108. Drebbel, *van de Natuere*, 42; Drebbel, *Wonder-vondt*, [B^v].

109. Cornelis Drebbel to James I, National Library of Scotland, Adv. MS. 33.1.1.

110. "Nicht allein der König in Engeland / sondern noch viel 1000 Menschen die Prob gesehen." Harsdörffer, *Delitiae mathematicae et physicae*, 400.

111. "Postquam ei curiosam manum admovisset Regina, ita ut omnes actiones cessarent." Drebbel to James I, National Library of Scotland, Adv. MS. 33.1.1.

112. "Dieser Motus Dreppelianus ist hernacher in unterschiedlicher Teutscher Fürsten Kunstkammer kommen." Saledinus, *De perpetuo mobile*, 25. "Il avoit inventé 17 ou 18 sortes d'instrumens, qui monstroient le flux et le reflux de la mer." Peiresc, *Relation*, 411ᵛ.

113. Doering, *Des Augsburger Patriciers*, 167. "Perpetuum mobile, welches in ainem gläserinen Ring ascendiert vnd descendiert." Zeiller, *Handbuch*, 490. "Tagebuch Christian II. von Anhalt-Bernburg, Jahr 1623." Worm, *Epistolae*, 1085–86.

114. Gabbey, "The Mechanical Philosophy," 9–84. Recent reinterpretations of Drebbel's instruments and natural philosophy can be found in: Borrelli, "The Weatherglass and its Observers"; V. Keller, "Drebbel's Living Instruments"; Dijksterhuis, "Magi from the North."

115. "Ghedenckende / dat rechtveerdicheyt niet wil / dat straf de misdaet sal overwegen / maer liever dat straf door beweghelijcke barmhorticheyt soude verlicht worden / op dat alle Menschen souden smaken die aenghename vrucht van de wyse Regenten / en in plaets van 't wreede bittere Oorlogh / haer vermaken met de soeticheyt van de Consten." Drebbel, *Wonder-vondt*, [Biiʳ].

116. HP 29/3/48B.

117. "Instrumentum quo, tempore necessitatis aut belli, litterae distantia unius aut duorum milliarum leguntur; item ut de nocte in milliare unum atque alterum visum nostrum extendamus." National Library of Scotland, Adv. MS. 33.1.1. See also, "In gratiam Regiae Majestatis atque Illustrissimi Principis diversa perfeci Instrumenta. Nempe Perpetuum Mobile, Perspicilla quae Longinqua in angustum contrahunt." BL Harleian MS 7011, f. 56.

118. Drebbel to James I in Beeckman, *Journal tenu par Isaac Beeckman*, 3:440.

119. "Il se promet de faire une lunette de longue veüe cappable de distinguer de sept lieües jusques à de l'escriture; de faire un miroir pour brusler de demy lieüe loing; de multiplier la lumiere d'une estoile en sorte qu'elle puisse faire lisre une lettre de nuict, et esclairer une espace trente pas de diametre." Bibliothèque Municipale Inguimbertine, Carpentras, Ms. 1774, 407r.

120. "Rubenius ante annos aliquod ad Peireskum scripsit, Heymum Pictorem perspexisse apud Drebbelium Opticum Tubum, diametri palmaris: quo liceret in disco Lunae discernere Campos, Sylvas, Aedificia & Munimenta locorum, nostratibus non absimilia." Gassendi, *De vita Peireskii*, 216, 303. "Dr Kuffler confessed that his Father in Law Drebbel was vpon the Invention of bringing the Moone so neare to ones face, as to see things in it. Hee did labor with great earnestnes in it." HP 29/8/12B.

121. HP 29/5/73A.

122. MacGregor, "The Tradescants," 21.

123. McEvansoneya, "A Note on Cornelius Drebbel."

124. Leith-Ross, *The John Tradescants*, 97.

125. "Cavillati aliqui cum Jacobo Rege sunt, vix operae quicquam edidisse perpetuum inventorem, cuius utilitate impensa rependeretur." C. Huygens, "Fragment Eener Autobiographie," 116.

126. HP 29/3/47B.

127. Czech National Archives, Ms. A 215.

128. Bäcklund, "In the Footsteps of Edward Kelley."

129. HP 8/27/1A.

130. HP 29/5/100B.

131. HP 15/2/26A.

132. "En sa maison auprès de Londres." "Et cela est grandement aisé, car on n'a que mettre le verre sur l'outil et après faire aller le moulin par un petit garçon. On peut aller se pourmener et n'y regarder de trois et quatres heures plus ou moins, cela ne manque point." Bibliothèque Municipale Inguimbertine, Carpentras, 1776, 412v–413r.

133. Brunsman, *The Evil Necessity*, 23.

134. Richard Bradley, in 1635–36: HP 29/3/58B, HP 29/3/59A.

135. Shirley, *Triumph of Peace*, 7–8.

136. Whitelocke, *Memorials*, 20.

137. Yamamoto, *Taming Capitalism*, 97.

138. Knowler, *Earl of Strafforde's Letters*, 71.

139. Peck, *Consuming Splendor*, 76.

Chapter 3 · *"Projectors are commonly the best Naturalists"*

1. HP 29/2/11A.

2. HP 29/2/11A.

3. Feingold, "Experimental Philosophy," 3. I will discuss the transformation of experimental philosophy into an academic discipline in universities in German-speaking lands in my next book project.

4. Leslie Hotson performed an exhaustive search for all Thomas Russells with the status of esquire at the time and found two: the London metallurgist mentioned by Bacon and the overseer of Shakespeare's will: L. Hotson, *I, William Shakespeare*, 16.

5. Aubrey, *Brief Lives*, 130–31.

6. Sturtevant, *Metallica*, [A]. Sturtevant was termed a gentleman in the letter patent he published in *Metallica*, 5.

7. He excuses his coinage of new "scholasticall tearmes." Sturtevant, *Metallica*, 89.

8. *CSPD*, 1603–10, 474. Pastorino, "The Mine and the Furnace," 654, 658–60.

9. Pastorino, "The Mine and the Furnace," 650, 652.

10. Pastorino, "The Mine and the Furnace," 656.

11. For instance, the patent to Russell quoted in Pastorino ("The Mine and the Furnace," 649) was made out to "Tho:s Russell gent."

12. Bacon, *Works*, 63.

13. Bacon, *Works*, 63.

14. Malynes, *Consuetudo, vel Lex Mercatoria*, 263.

15. Malynes, *Consuetudo, vel Lex Mercatoria*, 264.

16. BL Add. MS. 19402.

17. Pastorino, "The Philosopher and the Craftsman," 760–61.

18. *CSPD*, 1611–18, 250.

19. For arguments for and against deploying a concept of expertise in the early modern period, see Ash, "By Any Other Name."

20. Sturtevant, *Metallica*, 5, 30; Sherman, "Patents and Prisons," 243.

21. Brugis, *Discovery*, 2.

22. For example, Bacon, *Works*, 52.

23. Thomas Russell to the Earl of Salisbury, 1609, in Owen, *Calendar of the Manuscripts*, 74.

24. Bacon, *Works*, 63; Feingold, *Mathematician's Apprenticeship*, 208; Simon Healy, "CHALONER, Sir Thomas."

25. Sir Thomas Chaloner to Sir Thomas Lake, 3 March 1608, cited in Sherman, "Patents and Prisons," 248.

26. Camden, *Britain*, 721.

27. W. Price, *English Patents of Monopoly*, 97.

28. W. Price, *English Patents of Monopoly*, 83.

29. W. Price, *English Patents of Monopoly*, 88.

30. W. Price, *English Patents of Monopoly*, 91; West Yorkshire Archive Service, Leeds, 100/PO/8/I/19.

31. Davidson and Sgroi, "Brooke, Sir John."

32. Turton, *The Alum Farm*, 124.

33. Clow and Clow, "The Natural and Economic History of Kelp," 305.

34. Atkinson, *Acts of the Privy Council*, 641.

35. Upton, *Sir Arthur Ingram*, 122; Bickley, "The Manuscripts," 12.

36. Upton, *Sir Arthur Ingram*, 123.

37. Upton, *Sir Arthur Ingram*, 123.

38. West Yorkshire Archive Service, Leeds, WYL100/PO/8/I/3.

39. Upton, *Sir Arthur Ingram*, 124.

40. Upton, *Sir Arthur Ingram*, 124–25.

41. Bickley, "The Manuscripts," 15.

42. Bickley, "The Manuscripts," 15.

43. Bickley, "The Manuscripts," 16.

44. Bickley, "The Manuscripts," 16.

45. Thomas Russell to the Earl of Salisbury, 1609, in Owen, *Calendar of the Manuscripts*, 74.

46. Bickley, "The Manuscripts," 17.

47. Bickley, "The Manuscripts," 16.

48. Bickley, "The Manuscripts," 18.

49. Bickley, "The Manuscripts," 19.

50. *CSPD*, 1623–25, 15, 42, 83, 86, 125, 126, 128, 154, 230, 234, 249, 256, 277, 299, 303, 314, 318, 320, 330, 336, 375, 378, 380, 403, 468, 471.

51. Owen, *Calendar of the Manuscripts*, 74.

52. Owen, *Calendar of the Manuscripts*, 200.

53. Pastorino, "The Philosopher and the Craftsman," 761.

54. *CSPD*, 1611–18, 468.

55. Hammersley, *Daniel Hechstetter the Younger*, 38.

56. Owen, *Calendar of the Manuscripts*, 202–3.

57. Thomas Russell to the Earl of Salisbury, in Owen, *Calendar of the Manuscripts*, 74.

58. "Salisbury to Undertakers of the Copper Works in Cornwall," "wishes John Milward to give occasional information" on the works. *CSPD*, 1611–18, 79.

59. *CSPD*, 1611–18, 111. Milward, sending Cecil a treatise on the coin of the realm, wrote to Cecil from Truro in Cornwall that he could "do little service at the copper works, another being placed there." *CSPD*, 1611–18, 113. Milward would later be granted the office of the Controller of the Mint in the Tower. *CSPD*, 1611–18, 230.

60. Hammersley, *Daniel Hechstetter the Younger*, 304.

61. Hammersley, *Daniel Hechstetter the Younger*, 38.

62. Hunneyball, "Russell, Thomas." There he is identified tentatively with the "Thomas Russell of Lambeth," who died in 1635; however, the latter's will does not identify him as a gentleman. Kew, Public Record Office, 11/169/412.

63. "Petitions to the Westminster Quarter Sessions," WJ/SR/NS/011/19.

64. *CSPD*, 1638–39, 253.

65. *CSPD*, 1623–55, 75.

66. *CSPD*, 1619–23, 269.

67. *CSPD*, 1623–25, 289–90.

68. Bickley, "The Manuscripts," 16.

69. Cressy, "Saltpetre, State Security and Vexation," 104.

70. Cressy, "Saltpetre, State Security and Vexation," 106; Wijnman, "Vernatti, Sir Philibert (1)."

71. Cressy, *Saltpeter*, 12.

72. Cressy, *Saltpeter*, 3.

73. Russell, *To the Kings most Excellent Maiestie*.

74. Cressy, *Saltpeter,* 82.

75. Russell, *To the Kings most Excellent Maiestie*.

76. Russell, *To the Kings most Excellent Maiestie*.

77. A copy of the proffer is at Parliamentary Archives, House of Lords, Parliament Office, Journal Office, Main Papers 1509–1700, 10/1/30.

78. House of Lords. *Journal of the House of Lords*, 541.

79. Tullock, *The Rise and Progess*, 242.

80. "A Proclamation for the better making of Saltpeter within this Kingdome," 2 Jan. 1626/7, in Larkin, *Stuart Royal Proclamations*, 116–20. Cressy, "Saltpetre," 105.

81. "At the Court at Whitehall the 19th of December, 1627," SP, Privy Council 2/36 f. 240; Lyle, *Acts of the Privy Council of England, 1627–28*, 192–93.

82. "At the Court at Whitehall the 19th of December, 1627," SP, Privy Council 2/36 f. 240; Lyle, *Acts of the Privy Council of England, 1627–28*, 192–93.

83. *CSPD*, 1627–28, 303.

84. Bull, "Pearls from the Dungheap"; Robertson, "Reworking Seventeenth-Century Saltpetre."

85. Russell seems to have erred on the side of caution. As far as modern scholars have been able to ascertain, the mean household size in the period 1564–1649 may have been 5.073 people: Laslett, "Size and Structure of the Household," 210. Malynes estimated English households of six members: Malynes, *Consuetudo, vel Lex Mercatoria*, 235.

86. Deringer, *Calculated Values*.

87. Slack, *Invention of Improvement*, 48–51.

88. Deringer, *Calculated Values*, 7.

89. Jasanoff, "Citizens at Risk."

90. Keeler, Cole, and Bidwell, *Commons Debates 1628*, 396.

91. Keeler, Cole, and Bidwell, *Commons Debates 1628*, 397.

92. *RVC*, 1:403.

93. *RVC*, 1:485, 1:494.

94. McCormick, *William Petty*.

95. McCormick, "Food, Population, and Empire."

96. "Mr. Russell's Project Touching Artificial Wine in Virginia, July, 1620," in *RVC*, 3:366–67.

97. *RVC*, 3:624–628. *CSPC*, America and West Indies, 1675–1676, also Addenda, 1574–1674, 72.

98. H. Porter, "Alexander Whitaker," 332; *RVC*, 1:232.

99. Brock, *Abstract of the Proceedings*, 166, 174, 185, 274.

100. Maydom, "New World Drugs," 20.

101. Maydom, "New World Drugs," 124, 125; cited from W. M. *The Queens Closet*, 281.

102. Parkinson, *Theatrum Botanicum*, 1558.

103. George Thorpe, "A Letter to John Smyth, December 19, 1620," in *RVC*, 3:417.

104. Cowen, "Boom and Bust," 9.

105. Buckinghamshire Archives D-C/3/18 holds an agreement between Sir William Russell, Bt., Sir Basil Brooke, Kt., Sir Nicholas Fortescue, Kt., Sir William Foord, Kt., Thomas Jones, Esq., and Thomas Russell, Esq., for the establishment of a soap works for the manufacture of soap under royal Letters Patent granted to the Society of Soapmakers; Yamamoto, *Taming Capitalism*, 95–97.

106. Woodcroft, *Patents of Inventions*, 5. The patent is reprinted in W. Price, *The English Patents*, 206–213.

107. W. Price, *The English Patents*, 209.

108. W. Price, *The English Patents*, 208, 212.

109. Clow and Clow, "The Natural and Economic History of Kelp," 304.

110. Andrew Palmer to the Earl of Salisbury, 1608, Owen, *Calendar of the Manuscripts*, 239.

111. *CSPD*, 1623–25, 256.

112. 27 May 1624, Northamptonshire Archives, Cockayne collection, C 3023. In 1630 it was claimed that Thomas Jones, Thomas Russell, and William Turner owed Cockayne's estate 1,300 pounds: Northamptonshire Archives, Cockayne collection, C 3086.

113. W. H. Overall and H. C. Overall, eds., *Analytical Index,* 224–25. See also *CSPD*, 1631–33, 321, which seems to refer to this 1624 trial rather than the 1632 investigation.

114. Rymer, *Foedera*, 8, Part 2:206–210.

115. Rymer, *Foedera*, 8, Part 2:240. W. Price, *The English Patents*, 119.

116. *CSPD*, 1631–33, 213.

117. "Note of the true state of the new soap settled by patent. The advantages of the new soap as made altogether from home materials are briefly set forth, with a computation of the profit to be anticipated from its sale." *CSPD*, 1631–33, 263.

118. Rymer, *Foedera*, 8, Part 2: 240–42.

119. *CSPD*, 1634–35, 394.

120. Birch, *The Court and Times of Charles*, 229.

121. Birch, *The Court and Times of Charles*, 230.

122. Birch, *The Court and Times of Charles*, 231.

123. "A Proclamation concerning Soap and Soap-makers," in Larkin, *Stuart Royal Proclamations*, 405. *CSPD*, 1633–34, 316, 336–38.

124. "A Proclamation concerning Soap and Soap-makers," in Larkin, *Stuart Royal Proclamations*, 405.

125. Dear, "The Meaning of Experience."

126. Shapiro, "The Concept 'Fact.'"

127. Shapin and Schaffer, *Leviathan and the Air Pump*.

128. Yamamoto, *Taming Capitalism*, 125, although Yamamoto dates these practices to the 1640s and '50s.

129. Matthews, *Proceedings Minutes and Enrolments*, 6.

130. *CSPD*, 1637, 285.

131. *A Short and True Relation Concerning the Soap-busines*, 5.

132. *A Short and True Relation Concerning the Soap-busines*, [D2].

133. *A Short and True Relation Concerning the Soap-busines*, [D2v].

134. *A Short and True Relation Concerning the Soap-busines*, 6. See also Rushworth, "A Decree Concerning Soap-Boilers, 1633."

135. *A Short and True Relation Concerning the Soap-busines*, 10. *CSPD*, 1633–34, 43, 310, 444, 461 and 515.

136. *A Short and True Relation Concerning the Soap-busines*, 10.

137. *A Short and True Relation Concerning the Soap-busines*, 11.

138. *A Short and True Relation Concerning the Soap-busines*, 12.

139. *A Short and True Relation Concerning the Soap-busines*, 12.

140. House of Commons, *Journals of the House of Commons*, 2: 260.

141. Aaron, *Global Economics*, 161.

142. *The Proiectors Down-Fall*, 1–2.

143. T. Miller, "Pleasure, Honor, And Profit," 43.

144. Feingold, "Experimental Philosophy."

145. Feingold, "Experimental Philosophy," 4.

146. HP 29/2/19B.

147. HP 29/3/37B.

148. Mason, *Serving God and Mammon*.

149. HP 10/2/27/2A. For more on Gerbier's understanding of experimental natural philosophy, see V. Keller "A wild swing to phantsy."

150. HP 28/1/15B.

151. North, *The Life of the Honourable Sir Dudley North*, 286.

152. Aubrey, *Brief Lives*, 1:134.

Chapter 4 · Statecraft

1. "Quae in aëre homines factitamus." C. Huygens, "Fragment Eener Autobiographie," 117.

2. Bourne, *Inuentions or Deuises*, 13–15.

3. "Darmede hy onder ende boven water varen kan als hy wilt." Beeckman, *Journal Tenu par Isaac Beeckman*, 25.

4. "Hem die gheen Const ontbreeckt / die sich sou onderwinden / Te swemmen g'licjk een Visch, te vlieghen op de winden, / Te stijghen tot de Maen, in d'Afgront van de Zeé / Te zeylen sonder Mast, Stuer, Riemen, Zeyl, oft' Reé." Mon heur est en gerbe [Balthasar Gerbier], *Clacht-Dicht*, 11.

5. Doorman, *Octrooien Voor Uitvindingen in de Nederlanden*, 114; my translation.

6. Martín, "Did Naval Artillery Really Exist," 73; Norton, *The Gunner*, 128.

7. "Unde non arduum est coniicere, quis audacis inventis usus in re bellicâ esset." C. Huygens, "Fragment Eener Autobiographie," 117.

8. "Si hoc pacto (quod asserentem Drebbelium non semel audivi) securas in statione naves hostium aggredi clam daretur applicatoque tormento pensili (cuius hodie effringendis portis pontibusque civitatum terribils usus est) improvisas perdere." C. Huygens, "Fragment Eener Autobiographie," 117.

9. "Nam et pulveris nitrati vires sic Daedalus iste regit, ut aquâ non magis quam aëre compescantur." C. Huygens, "Fragment Eener Autobiographie," 117.

10. "Un navire qui va entre deux eaux." Bibliothèque Municipale Inguimbertine, Carpentras, Ms. 1774, 407r. These notes by Peiresc were the result of his meeting with Jacob Küffler, the brother of the son-in-law of Drebbel, Abraham Küffler, in 1622.

11. Gerbier asked why the Dutch did "not blow up the ships on the coast, by such a like engine, as Cornelius Drebbell invented? which was to swim betwixt two waters; and the which is practised by too many wicked men in this age." Gerbier, *A Sea-Cabbin Dialogue*, 2.

12. The source of 1651 refers to the "recently deceased King in England," which would thus refer to Charles rather than James: "in welchem der jüngst-verstorbene König in Engeland auff der Teims selbst gefahren welcher auch eins an den Gross-Fürsten in Moscau als eine seltene und unglaubliche Sache verehrt." Harsdörffer, *Delitiae mathematicae et physicae*, 2:493.

13. As Drebbel's son-in-law, Johann Sibbert Küffler, reported to Samuel Hartlib, Drebbel found out "by experience" that at "6. fathom deepe there is no abiding vnder the Water with Boates or Men. For both will bee squeezed together." HP 29/5/73A.

14. "Et y a faict un navire qui va entre deux eaux, cappable de porter neuf persones; aprez lequel ledit Roy luy en faict faire cent plus petits et cappables de porter seulement chascun son homme." Bibliothèque Municipale Inguimbertine, Carpentras, Ms. 1774, 407r. Other accounts describe submarines with capacities varying between eight and twenty-four passengers: Emery, "A Further Note on Drebbel's Submarine."

15. For example, Paquette, *The European Seaborne Empires*.

16. Roper, *Advancing Empire*.

17. Kopperman, "Profile of Failure"; Lorimer, "The Failure of the English Guiana Ventures."

18. Subrahmanyam, *Europe's India*, 10–11.

19. Roper, *Advancing Empire*, 54–55.

20. Loe, *The Kings Shoe*, 25–26.

21. Schreiber, "The First Carlisle Sir James Hay; *CSPC*, East India, China and Japan, 1622–24, lxiii–lxv; English East India Company Court Book, BL IOR/B/7, 404–6, 409, 469–70.

22. Razzari, "Through the Backdoor," 487.

23. Peck, *Consuming Splendor*, 81–82; C. Wilson, *England's Apprenticeship*.

24. Published in Baker, *The History and Antiquities*, 242.

25. Baker, *The History and Antiquities*, 242.

26. BL Harleian 1581, 118.

27. BL Harleian 1581, 647.

28. For Jahangir's letter, Robinson, *Readings in European History*, 333–35.

29. East India Company Court Book, BL IOR/B/7, 404–5.

30. Terry, *A Voyage to East-India*, 149.

31. Ettinghausen "The Emperor's Choice."

32. Thackston, *The Jahangirnama*; Lefèvre, "Recovering a Missing Voice"; Screech, "Pictures," 55; M. Beach, "The Mughal Painter Abu'l Hasan," 12; Skelton, "Imperial Symbolism in Mughal Painting"; M. Beach, "The Gulshan Album." P. Beach, "European Painting and Mughal Miniatures."

33. Walker, *Flowers Underfoot*, 37.

34. Ettinghausen, "The Emperor's Choice."

35. Woude, *Kronijcke Van Alckmaer*, 116.

36. "Refrigeratoria Instrumenta pro aestate et imprimis in locis calidioribus vti India." HP 29/3/55B.

37. East India Company Court Book, BL IOR/B/7, 405.

38. East India Company Court Book, BL IOR/B/7, 406.

39. East India Company Court Book, BL IOR/B/7, 406–7.

40. Craven, "The Earl of Warwick."

41. East India Company Court Book, BL IOR/B/7, 409–10.

42. East India Company Court Book, BL IOR/B/7, 469–70.

43. On the appeal of automatons in Jahangir's court, see Keating, *Animating Empire.*

44. Kew, National Archives, Colonial Office 77, 34 (renumbered as 60). According to a modern note pencilled into the document, this was dated 1627, based on the fact that Drebbel was paid for water engines in 1627. However, these were for the siege at the Isle of Ré discussed here in a later chapter.

45. Reid, *An Indonesian Frontier*, 102.

46. Subrahmanyam, *Europe's India*, 8–9.

47. SP 14/141/362.

48. SP 14/141/362; printed in Sanderson, *Foedera*, 407–10.

49. Sanderson, *Foedera*, 57.

50. *CSPC*, East Indies, China and Japan, 1617–21, xxii, 288, 291.

51. Scott, *A Collection*, 425–426.

52. SP 14/141/362.

53. SP 14/141/362.

54. "La derniere et plus excelente invention que Derbbel aye trouuee, est de faire un soleil artificiel, cest a dire en feu perpetuel qui bruslera et esclairer tousiours. Lorsque le Prince de Galles alla en Espagne Derbbel [*sic*] luy proposa que comme l'on avoit rempli Londres de fontaines par le moyen d'une petite riviere qu'on y avoit conduit et divisee par petits tuyeaus a toutes les maisons qui en avoient voulu. Qu'il vouloit entreprende de faire un feu sur une petite montagne aupres de Londres, d'ou tous ceux de Londres pourroient aller prandre du feu et le conduire a leurs maisons, Et avec ce feu faire bouiller et roytir leur viande sans avoir besoin de bois. . . . Je crois qu'il ne demandoit que vingt mil livres sterlins." Carpentras, Bibliothèque Municipale Inguimbertine, MS 1776, 412v.

55. Ward, *London's New River*, 19, 69; Skempton, "Edmund Colthurst."

56. Nye, *Pounds Sterling*.

57. Neumaier, "Salomon de Caus," 37.

58. HP 29/3/55B.

59. "Le voyage qu'entreprint lors ce Prince l'empescha de fournir ce qui estoit necessaire pour faire faire ce miracle" and "Ce voyage fait grand tort au public." Carpentras, Bibliothèque Municipale Inguimbertine, MS 1776, 412v.

60. Birdwood, *The First Letter Book*, 457–60.

61. Terry, *A Voyage to East-India*, 73.

62. Asher, *Architecture of Mughal India*; Mubārak, *Ain I Akbari*, 55–56.

63. Foster, *Letters Received*, 6:168.

64. Razzari, "Through the Backdoor."

65. *CSPD*, 1611–18, 260.

66. Foster, *Letters Received*, 6:142.

67. Foster, *Letters Received*, 6:xxvi.

68. John Browne to the East India Company, Original Correspondence, BL IOR/E/3/5/2, no. 609. February 10, 1618/9.

69. Sir Thomas Roe to the East India Company, Original Correspondence, BL IOR/E/3/5/2, No. 610. February 14, 1618/9.

70. Sir Thomas Roe to the East India Company, Original Correspondence, BL IOR/E/3/5/2, No. 611. February 1618/9.

71. Natif, *Mughal Occidentalism*, 224.

72. Sir Thomas Roe to the East India Company, Original Correspondence, BL IOR/E/3/5/2, 611. February 1618/9.

73. William Biddulph and John Willoughby at the Mogul's Camp to the Company, Original Correspondence, BL IOR/E/3/7/1, 831, December 25, 1619.

74. Thomas Kerridge and Thomas Rastell at Surat to the Company, February 9 and 15, 1619, in Foster, ed. *The English Factories in India, 1618–1623*, 60.

75. East India Company Court Book, BL IOR/B/6, 408–12. September 17, 1619.

76. East India Company Court Book, IOR/B/8, 283–84. November 26, 1623.

77. For example: "they do not believe any such instrument can be made, and are therefore unwilling to trouble themselves any further about it," *CSPC*, East Indies, China and Japan, 1617–21, 311; "the Court observing it would be exceeding chargeable besides uncertain, would proceed no further with the project," *CSPC*, East Indies, China and Persia, 1625–29, 552.

78. *CSPC*, East Indies, China and Persia, 1625–29, 559.

79. Brenner, *Merchants and Revolution*, 106.

80. Bacon, *Essayes*, 248.

81. Bacon, *Essayes*, 249.

82. Andrews, *Trade, Plunder and Settlement*, 258–59.

83. Nerlich, *Ideology of Adventure*, 133.

84. Foster, *Letters Received*, 5:321.

85. Terry, *A Voyage to East-India*, 395.

86. Foster, *Letters Received*, 5:111.

87. Steinmann, "Shah Abbas and the Royal Silk Trade," 71.

88. Faroqui et al., *An Economic and Social History*, 455; Matthee, *The Politics of Trade*.

89. Foster, *Letters Received*, 6:xiii–xiv.

90. Foster, *Letters Received*, 6:318.

91. *CSC*, East Indies China and Japan, 1622–24, 243–44; Shirley, *The Sherley Brothers*, 109–10.

92. *CSPC*, East Indies, China and Japan, 1622–24, 370.

93. Foster, *Letters Received*, 6:xv.

94. Foster, *Letters Received*, 6:xvi.

95. Andrews, *Trade, Plunder and Settlement*, 275.

96. Sir Thomas Roe to the East India Company, East India Company Original Correspondence, BL IOR/E/3/5/2, No. 610, received January 4, 1618/9.

97. Stevens, "Robert Sherley," 120.

98. East India Company Court Book, BL IOR/B/9, 21.

99. *CSPC*, East Indies, China and Persia, 1622–24, 360–61.

100. *CSPD*, 1623–25, 381.

101. *CSPC*, East Indies, China and Persia, 1625–29, 17.

102. *CSPC*, East Indies, China and Persia, 1625–29, 18.

103. *CSPV*, 1623–24, 474.

104. BL IOR/B/9, 31–11.

105. East India Company Court Book, BL IOR/B/9, 295.

106. East India Company Court Book, BL IOR/B/9, 381–82.

107. *CSPC*, East Indies, China and Persia, 1625–29, 64.

108. East India Court Book (May 30, 1625), BL IOR/B/10, 57.

109. Mishra, *A Business of State*, 273–301.

110. *CSPD*, 1635, 96.

111. Brenner, *Merchants and Revolution*, 170.

112. Foster, *The English Factories in India, 1637–1641*, 24.

113. Sainsbury, *A Calendar of the Court Minutes*, 241.

114. Sainsbury, *A Calendar of the Court Minutes*, 123, 131; Roper, *Advancing Empire*, 97.

115. Chaudhuri, *The English East India Company*, 73; *Journal of a voyage begun with the ships the Dragon, the Sun, the Katherine, the Planter, the Ann, and the Discovery, for East India, set forth by Sir William Courteen and others, adventurers*, SP 16/351/59v.

116. "King Charles to Prince Rupert" and "Project of Prince Rupert for Colonizing Madagascar," in Sainsbury, *A Calendar of the Court Minutes*, 244.

117. Sainsbury, *A Calendar of the Court Minutes*, 245.

118. Sainsbury, *A Calendar of the Court Minutes*, 245.

119. Games, *The Web of Empire*, 181–218.

120. Foster, "An English Settlement in Madagascar"; E. Smith, "Canaanising Madagascar."

121. Roper, *Advancing Empire*, 98–99.

Chapter 5 · *Transplanters of Empire*

1. Kupperman, "The Puzzle of the American Climate," 1284; Kupperman, "How to Make a Successful Plantation."

2. Kupperman, *The Jamestown Project.*

3. Kupperman, *The Jamestown Project*, 9.

4. Botero, *A Treatise*, 71–79.

5. Botero, *A Treatise*, 102. Botero did not originally publish his account of Cuzco in this text. It was taken from another work by Botero and inserted by the translator.

6. Botero, *A Treatise*, 302.

7. Andrews, *Trade, Plunder, and Settlement*, 339.

8. Kupperman, *The Jamestown Project*, 240.

9. B. Marsh, *Unravelled Dreams.*

10. Drayton, *Nature's Government*, 92; D. Smith, "Useful Knowledge," 556.

11. Fitzmaurice, "The Commercial Ideology of Colonization"; Haskell, *For God, King and People*, 141–49; Paul, *Counsel and Command*.

12. Kupperman, "The Puzzle," 1283.

13. Fitzmaurice, "The Dutch Empire in Intellectual History."

14. R. Johnson, *Nova Britannia*, [Dr].

15. R. Johnson, *Nova Britannia*, [D3v].

16. R. Johnson, *Nova Britannia*, [B3r].

17. R. Johnson, *Nova Britannia*, [B3v]; cited in V. Keller and McCormick, "Towards a History of Projects," 433.

18. Quitslund, *The Virginia Company*, 66.

19. Gray, *A Good Speed to Virginia*, [Bv]. The author of this text is given as "R. G."

20. Gray, *A Good Speed to Virginia*, [B2r].

21. Historical Manuscripts Commission, *Twelfth Report*, 135; in a manuscript formerly in the collection of John Henry Gurney.

22. Symonds, *Virginia*, 25 and D. Price, *Sauls Prohibition Staide*, [F2v].

23. William Crashaw, *A Sermon*, [K2v–K3r].

24. D. Price, *Sauls Prohibition Staide*, [F2].

25. Virginia Company, *A True Declaration*, 58.

26. Virginia Company, *A True Declaration*, 59.

27. Bonoeil, *Obseruations*, 19.

28. Kupperman, *The Jamestown Project*, 303.

29. Pastorino, "Beyond Recipes"; V. Keller, "Into the Unknown"; V. Keller, "Everything Depends Upon the Trial."

30. Bonoeil, *Obseruations*, 19–20. On the continued circulation of this process into the eighteenth century, see B. Marsh, *Unravelled Dreams*, 5.

31. H. Cook, *Matters of Exchange* and Drayton, *Nature's Government*.

32. Harrison, *The Bible*, 239.

33. Cited in Samson, "Introduction," 13.

34. "Copy in Hartlib's Hand, Thomas Tempest's 'Of Gardens and Flowers,'" The James Marshal and Marie-Louise Osborn Collection, Beinecke Library, Document 41, accessed through *HP*.

35. Strong, *The Renaissance Garden in England*, 112; Fischer, Remmert, and Wolschke-Bulmahn, *Gardens, Knowledge and the Sciences*; Baldassarri and Matei, "Manipulating Flora"; Svensson, "And Eden from the Chaos Rose"; Jalobeanu and Matei, "Treating Plants as Laboratories"; V. Keller, "A 'wild swing to phantsy.'"

36. Bushnell, *Green Desire*; Willes, *The Making of the English Gardener*.

37. Child, "A Large Letter," 11.

38. Child, "A Large Letter," 82, 19.

39. Parkinson, *Theatrum Botanicum*, 11, 27, 75, 100, 119, 132, 244, 264, 752, 884, 1519.

40. See further, V. Keller, "A 'wild swing to phantsy.'"

41. Schiebinger, *Secret Cures of Slaves*.

42. Botero, *On the Causes*, 71; cited in McCormick, "Food, Population and Empire," 65.

43. R. Johnson, *Nova Britannia*, [Dv].

44. Davies, *Historical Tracts*, 290.

45. *A Publication of Guiana's Plantation*, 11.

46. *A Publication of Guiana's Plantation*, 2.

47. William Petty's "Phytological Letter," HP 8/22/1A–4B.

48. Kupperman, "The Puzzle," 1285; Zilberstein, *A Temperate Empire*; Mulry, *An Empire Transformed*.

49. *RVC*, 1:267.

50. *RVC*, 1:350–51.

51. Daston and Park, *Wonders and the Order of Nature*.

52. MacGregor, "The Cabinet of Curiosities," 149.

53. MacGregor, "The Tradescants," 6.

54. "Virginia" appears ninety-five times, compared to seventy-five for "India" or "Indian," despite the multiple meanings of the latter term: Tradescant, *Musaeum Tradescantium*.

55. MacGregor, *Naturalists in the Field*, 919.

56. Leith-Ross, *The John Tradescants*, 79–81.

57. Svalastog, *Mastering the Worst of Trades*, 75–6.

58. MacGregor, *Naturalists in the Field*, 920. This note has usually been dated 1625 on the basis of Tradescant's other list for Villiers.

59. Tradescant, *Musaeum Tradescantium*, 46, 51, 53.

60. Tradescant, *Musaeum Tradescantium*, 179–80. Tradescant does not note Humphrey Slaney, although he was mentioned in an early manuscript list of plants in Tradescant's collection. Leith-Ross, *The John Tradescants,* 199. On Claiborne and Cloberry, Brenner, *Merchants*, 185.

61. Sturdy, "The Tradescants at Lambeth."

62. Leith-Ross, *The John Tradescants*, 93.

63. Leith-Ross, *The John Tradescants*, 94.

64. Leith-Ross, *The John Tradescants*, 97.

65. Macgregor, "The Tradescants as Collectors of Rarities," *Tradescant's Rarities,* 17–23; 21.

66. Leith-Ross, *The John Tradescants*, 177 suggests that Pergins might be the royal goldsmith, Admondisham Perkins, but I argue that the context of Courten, Boeve and Trion suggests an identification with the merchant Jacob Pergens.

67. Brenner, *Merchants*, 171.

68. Mundy, *The Travels of Peter Mundy*, 1–2.

69. Mundy, *The Travels of Peter Mundy*, 2.

70. HP 42/6/6A.

71. HP 28/2/80A.

72. R. Davies, *Chelsea Old Church*, 14. On the forced movement of children around the world and the important roles they played, see Kupperman, *Pocahontas and the English Boys*.

73. Chester, *The Reiester Booke*, 212. For the possible identity of these individuals, see Vaughan, *Transatlantic Encounters*, 52.

74. Coryate, *Crudities*, [k4v]; Grigson, *Menagerie*, 18–22. For an image of the cassowary, see Wolfenbüttel, Herzog August Bibliothek Cod. Guelf. 235 Blankenburg, *Album amicorum* of Jakob Fetzer, vol. 1, 83r.

75. Edinburgh University Library, La. II.283, *Album amicorum* of Michael van Meer, 254v. Vaughan, *Transatlantic Encounters*, 53–54.

76. Vaughan, *Transatlantic Encounters*, 55.

77. Virga and Brinkley, *Eyes of the Nation*, 47. On another etching that Hollar made that year in Antwerp, of an Ottoman man, he wrote that he had first sketched it from the life in London in 1637: "ad vivum deli[neavit] Londoni 1637 et fecit Antuerpiae A[nn]o 1645."

78. Harkness, *The Jewel House*, 15–56.

79. Oestreich, *Antiker Geist*, 269–70.

80. "An Animadversion," 30.

81. Sturdy, "The Tradescants at Lambeth," 2.

82. Thomas, *Man and the Natural World*, 226.

83. SP 14/161 f. 104, "Statement of Losses sustained by Sir John Danvers." For brickmaking, tile-making, and lime, see *VCA*, Ferrar Papers 723. Hunt, *Garden and Grove*, 126–30; Strong, *The Renaissance Garden*, 11.

84. Pestana, *The English Atlantic*, 16.

85. Davies, *Chelsea Old Church*, 131; Kroyer, *The Story of Lindsey House*, 23. The Dutch resident envoy, Albert Joachimi, built a house at Chelsea in 1629, and Mayerne married his daughter in 1630.

86. Trevor-Roper, *Europe's Physician*, 347.

87. Mayerne called Danvers's niece, Dorothy Osborne, his "daughter" and was much admired by Danvers's stepson Edward Herbert, Lord Cherbury; D. Osborne, *Letters to Sir William Temple*, 254. HP 31/22/37B.

88. Aubrey, *Brief Lives*, 75.

89. J. Hunt, *Garden and Grove*, 127; Gomme and Maguire, *Design and Plan in the Country House*, 258–59.

90. J. Hunt, *Garden and Grove*, 126–30.

91. J. Hunt, *Garden and Grove*, 130.

92. HP 28/2/34A

93. Bodleian Library, Ms. Add. C. 287, 151.

94. Leith-Ross, *The Tradescants*, 101.

95. Leith-Ross, *The Florilegium of Alexander Marshal*, 324.

96. Parkinson, *Theatrum Botanicum*, 1117.

97. Parkinson, *Theatrum Botanicum*, 1118; Bekkers, *Correspondence of John Morris*, 14, 18; HP 18/2/9A.

98. L. Huygens, *The English Journal*, 71.

99. Evelyn, *Directions for the Gardiner*, 11.

100. Aubrey, *Brief Lives*, 1:75.

101. Webster, *Great Instauration*, 468. For a fuller discussion, see V. Keller, "A 'wild swing to phantsy.'"

102. Bennett, "John Aubrey and the Printed Book," 399. Dymock noted Bushell's paper on mines of 1649: HP 28/1/19A.

103. HP 28/1/30B; HP 28/2/43A.

104. *RVC*, 1:365, 1:370.

105. *RVC*, 1:404.

106. Wither, "To the Honorable Sir John Danvers Knight," 26.

107. Sir John Danvers to John Ferrar, Ferrar Papers 1221, VCA, May 7, 1653; see also VCA, Ferrar Papers 1218, 1224. M. Cook, "Governor Samuel Mathews, Junior."

108. V. Keller, "A 'wild swing to phantsy.'"

109. V. Keller, "Into the Unknown."

110. Bonoeil, *A Treatise*, 60; Ransome, "John Ferrar"; V. Keller, "Into the Unknown," 94–97.

111. Bonoeil, *A Treatise*, 81.

112. Bonoeil, *A Treatise*, 82–83.

113. Bonoeil, *A Treatise*, 86.

114. Donne, *A Sermon upon the VIII*, 18–19; Stanley Johnson, "John Donne and the Virginia Company."

115. Quinn, "A List of Books," 359; Virginia Company, *A Declaration*.

116. Quinn, "A List of Books," 360.

117. Ash, *The Draining of the Fens*.

118. Donne, *A Sermon*, 27. Discussed in Fitzmaurice, *Humanism and America*, 143; Shami, "Love and Power," 98.

119. Tomlins, *Freedom Bound*, 144.

120. Raymond Williams notes later questioning of improvement through enclosures in *The Country and the City*, 66–67.

121. Goodman, *The Fall of Man*, 164, 249.

122. Powell, *Depopulation*, 7. Ted McCormick discusses Powell's arguments in "Population: Modes of Seventeenth-Century Demographic Thought," 25–45.

123. Powell, *Depopulation*, 43.

124. Powell, *Depopulation*, 45.

125. Loe, *The Kings Shoe*, 36–37.

126. For similar critiques, see: Whately, *A Caveat For The Couetous*, 61–62; Huit, *The Anatomy of Conscience*, 389; Hall, *One of the Sermons Preacht at Westminster*, 50–51.

127. Norden, *The Surveyors Dialogue*, 3; Thirsk, "The Crown as Projector," 328.

128. Norden, *The Surveyors Dialogue*, 4.

129. Norden, *The Surveyors Dialogue*, 5.

130. Norden, *The Surveyors Dialogue*, 30.

131. Cf. Slack, *The Invention of Improvement*, 60.

132. Robson, "Improvement and Epistemologies of Landscape." More generally, Yamamoto similarly cautions against taking claims to public utility and social benefit in the projects at face value: Yamamoto, *Taming Capitalism*, 91–94.

133. Bigelow, "Gendered Language"; Bigelow, "Colonial Industry"; Skeehan, *The Fabric of Empire*.

134. Peck, *Consuming Splendor*, 91–92; Virginia Company, "A Note of the Shipping, Men and Provisions, sent to Virginia by the treasurer and company in the yeere 1619," in *A Declaration*, 4–5.

135. Bigelow, "Colonial Industry," 15.

136. Bigelow, "Colonial Industry," 15. On the cecropia moth as the species that colonial promoters called native silkworms, see Ransome and Lees, "The Virginian Silkworm."

137. Virginia Company, *A Declaration*, 4.

138. Bonoeil, *A Treatise*, [A3v].

139. Bonoeil, *A Treatise*, 73.

140. Bonoeil, *A Treatise*, 74.

141. Bonoeil, *A Treatise*, 75.

142. Bonoeil, *A Treatise*, 74.

143. Bonoeil, *A Treatise*, 77.

144. Botero, *Relations, Of The Most Famous Kingdoms and Commonweales* (1616), 40; B. Marsh, *Unravelled Dreams*, 128.

145. Bonoeil, *A Treatise*, 86.

146. The limited success that some Virginian planters, such as Edward Digges (1620–1674/5), realized in cultivating silkworms seems to have relied on the labor of enslaved Africans; V. Keller, "Into the Unknown." In 1688–89, Sir Nathaniel Johnson (1644–1712) sent twenty-two black slaves (likely from his own Antiguan sugar plantation) to Carolina to labor on the plantation he would later name "Silk Hope." Johnson's servants and slaves planted twenty-four thousand mulberry trees for silk production; Rugemer, *Slave Law*, 63–4.

147. Virginia Company, "A Note of the Shipping, Men and Provisions, sent to Virginia by the treasurer and company in the yeere 1619," in *A Declaration*, 2–3.

148. "A Proclamation concerning Soap and Soap-makers," in Larkin, *Stuart Royal Proclamations*, 405.

149. *RVC*, 1:253.

150. *RVC*, 1:259.

151. *RVC*, 1:271.

152. *RVC*, 1:288–89.

153. R. C. Johnson, "The Transportation."

154. *RVC*, 1:268, 1:271.

155. *RVC*, 1:387.

156. *RVC*, 1:389.

157. *RVC*, 1:391.

158. *RVC*, 1:391.

159. *RVC*, 1:413.

160. Virginia Company, "A declaration of the Supplies intended to be sent to Virginia in this yeare 1620," in *A Declaration*, 11 and 13.

161. Ransome, "Wives for Virginia," 6, 7, 8, 11.

162. House of Commons, *Journal of the House of Commons*, 1:596–97; *RVC*, 1:489.

163. VCA, Ferrar Papers 166, April 1620. Danvers advertised the favor he enjoyed with Villiers in Danvers to Secretary Conway, July 31, 1623, State Papers 14/149, fol. 151. On February 2, 1619/20, Danvers had previously sent "A letter to the Marquis of Buckingham, concerning a proposition for his Majesty's profit." Historical Manuscripts Commission, "Catalogue of Manuscripts," 57.

164. Land, "Henrico and Its College," 487; Stanwood, "Captives and Slaves."

165. Virginia Company, "A Note of the Shipping, Men and Provisions, sent to Virginia by the treasurer and company in the yeere 1619," in *A Declaration*, 2–3.

166. Land, "Henrico and Its College," 477; Gethyn-Jones, *George Thorpe and the Berkeley Company*, 36.

167. Land, "Henrico and Its College," 492.

168. Land, "Henrico and Its College," 495.

169. *RVC*, 2:91; J. Smith, *The Generall Historie of Virginia*, 149.

170. J. Smith, *Captain John Smith*, 198.

171. BL Additional MS 12496, 459–60, reprinted in *RVC*, 3:704–6.

172. Rountree, *The Powhatan Indians of Virginia*, 65; B. Marsh, *Unravelled Dreams*, 115.

173. BL Additional MS 12496, 459–60, reprinted in *RVC*, 3:704–6.

174. Williams, *Virginia's Discovery of Silke-Wormes*; Williams and Ferrar, *Virgo Triumphans*, [Bv].

175. Williams and Ferrar, *Virgo Triumphans*, 44.

176. Williams and Ferrar, *Virgo Triumphans*, 25.

177. Williams and Ferrar, *Virgo Triumphans*, 24.

178. Williams and Ferrar, *Virgo Triumphans*, 37–38.

179. Williams and Ferrar, *Virgo Triumphans*, 16.

180. Williams and Ferrar, *Virgo Triumphans*, 33.

181. Williams and Ferrar, *Virgo Triumphans*, 39.

182. Williams and Ferrar, *Virgo Triumphans*, 39.

183. Williams and Ferrar, *Virgo Triumphans*, 39.

184. Williams and Ferrar, *Virgo Triumphans*, 43.

185. Williams and Ferrar, *Virgo Triumphans*, 40.

186. Williams and Ferrar, *Virgo Triumphans*, 34.

187. Tradescant, *Musaeum Tradescantium*, 37.

188. Williams and Ferrar, *Virgo Triumphans*, 35.

189. Williams and Ferrar, *Virgo Triumphans*, 16.

190. For example, Fitzmaurice, *Humanism and America*, 55–56, 68. On Lipsian ideas of greatness, see Tuck, *Philosophy and Government*; Lipsius, *Admiranda*.

191. *CSPC*, East Indies, China and Japan, 1622–24, 226; K. Lindley, "Irish Adventurers and Godly Militants."

192. Colwall, "An Account."

193. HP 28/2/58A.

194. Evelyn, *Diary*, 2:79, 2:143.

195. R. Porter, "The Crispe Family," 59; Hunter, *The Royal Society and Its Fellows*, 160.

196. *A Proclamation Concerning the Trade of Ginney.*

197. Andrews, *Ships, Money, and Politics*, 18.

198. Svalastog, *Mastering the Worst of Trades*, 105.

199. Dantzig, *Forts and Castles of Ghana*, 35; Svalastog, *Mastering the Worst of Trades*, 92–94.

200. Roper, *Advancing Empire*, 65.

201. Roper, *Advancing Empire*, 73.

202. Roper, *Advancing Empire*, 75.

203. R. Porter, "The Crispe Family," 59.

204. *A Copie of The Kings Commission.*

205. Andrews, *Trade, Plunder and Settlement*,113.

206. V. Keller, "Scarlet Letters."

207. "Bois Rouge de Guinée rapport d'un tincturier de Londres faict a Sr. Nicholas Crisp - 1649." Royal Society Archives, Classified Papers 24/81/126, transcribed by Robert Hooke.

208. Crispe, *The Humble Petition.*

209. HP 9/4/16A.

210. Evelyn, *Diary*, 1:383.

211. Lloyd, *Memoires*, 629.

212. Lloyd, *Memoires*, 627.

213. HP 60/8/1A. In this text, Crispe refers to it being eighteen years since he established the trade in Africa, referring to the 1631 license. He also refers to twelve years remaining on the thirty-one-year license. HP 60/8/7B.

214. HP 60/8/8A.

215. HP 60/8/2A.

216. HP 60/8/2B.

217. HP 60/8/6A–B.

218. HP 60/8/4A.

219. HP 60/8/4A.

220. Karklins, Dussubieux, Hancock, "A 17th-Century Glass Bead Factory," 20, 22; Egan, "Evidence."

221. HP 60/8/3A.

222. HP 60/8/7B.

223. Brenner, *Merchants and Revolution*, 174.

224. HP 60/8/8B.

225. HP 60/8/2A.

226. HP 60/8/6B.

227. Kew, UK National Archives, PROB 11/319/561.

228. Zook, *Company of Royal Adventurers*, 16.

229. Oldys, *Biographia Britannica*, 1522.

230. For example, Parkinson, *Theatrum Botanicum*, 138, 638, 642.

231. Parkinson, *Paradisi in Sole Paradisus Terrestris*, 104.

232. Ferrar, *Virgo Triumphans*, 35–36.

233. T. Porter, "Quantification and the Accounting Ideal," 640.

234. Mukherjee, *Penury into Plenty*.

235. Mukherjee, *Penury into Plenty*, 124, 127.

236. Mukherjee, *Penury into Plenty*, 110–11.

237. Cf. Werrett, *Thrifty Science*.

238. R. Johnson, *Nova Britannia*, [Dv-Dr].

Chapter 6 · Active Knowledge

1. Peacham, *Minerva Britanna*, 158.

2. Cited in Salmon, "Stoicism and Roman Example," 223, who points out that James dropped the individual reference to Lipsius for the 1603 edition of *Basilikon Doron* while still criticizing the notion of Neostoic constancy.

3. Pollnitz, *Princely Education*, 288.

4. Pollnitz, *Princely Education*, 313.

5. James I, *Basilikon Doron*, 93.

6. James I, *Basilikon Doron*, 94.

7. James I, *Basilikon Doron*, 95.

8. Onslow, *The Life of Dr. George Abbot*, 16.

9. Heylyn, *Microcosmus*, 2r.

10. Bulstrode, *Memoirs and Reflections*, 184.

11. Warwick, *Memoires of the Reign of King Charles I*, 65; cited in Feingold, *The Mathematicians' Apprenticeship*, 203.

12. *CSPD*, 1611–18, 136.

13. Donagan. "Halcyon Days and the Literature of War"; Lawrence, *The Complete Soldier.*

14. Cohen, "The Nimble Gunner"; Cooke, *The Character of Warre*, [E2]; Cooke, *The Prospective Glasse of Warre*, [E2r–F2r].

15. Gouge, *Dignitie*, 14.

16. Manning, *An Apprenticeship in Arms*, 146.

17. Jonson, "A speach according to Horace," in *Workes*, 215.

18. Jonson, "A speach according to Horace," in *Workes.*

19. Walton, "The Tower Gunners," 61. For example, J. White, *Militant Protestantism*, 31.

20. Walton, "The Tower Gunners," 62.

21. *CSPD*, 1611–18, 153.

22. *CSPD*, 1611–18, 398.

23. Stow, *Survey of London*, 437.

24. *CSPD*, 1611–18, 397.

25. Inderwick, *A Calendar of the Inner Temple Records*, xxxv.

26. *CSPV*, 1621–23, 452.

27. *CSPD*, 1625–26, 408. For a critical account of the master gunner's dereliction of his duties, see BL Sloane 871, 150v, discussed in Bull, *"The Furie of the Ordnance,"* 16–17.

28. SP 16/30 f.86.

29. Walton, "The Tower Gunners."

30. Highmore, *The History of the Honourable Artillery Company*, 70.

31. Townshend, *Life and Letters of Mr. Endymion Porter*, 90.

32. Milward, *The Souldiers Triumph*, [B]; Highmore, *The History of the Honourable Artillery Company*, 65.

33. Highmore, *The History of the Honourable Artillery Company*, 61.

34. Feingold, *The Mathematicians' Apprenticeship*, 166–89.

35. Robson, "Improvement and Epistemologies of Landscape," 614.

36. Rathborne, *The Surveyor*, [A4r].

37. "Project for an Academy Royal in England," Bodleian Library, MS Tanner 94. According to the Bodleian catalog, a former owner of the manuscript, William Sancroft (1617–93), noticing the similarities between this plan and Gerbier's later academy, entitled it "Sir Balth. Gerbier's project," which is why it was published as John Gutch, "Sir Balthazar Gerbier's Project for an Academy Royal in England," in *Collectanea Curiosa*, 214. It is now usually ascribed to Bolton. For more correspondence between Bolton and Buckingham concerning the academy, see also: Bodleian Library, MS Tanner 89, 56; Bodleian Library, MS Tanner 74, 418, 419; Bodleian Library, MS Tanner 73, 420.

38. Portal, "The Academ Roial of King James I."

39. C. Kennedy, "Those who Stayed," 66; Sharpe, *Sir Robert Cotton.*

40. Grafton and Jardine, "Studied for Action"; Popper, *Walter Ralegh's History of the World.*

41. Cited in Osmond, "In Defense of Tiberius," 606.

42. Bolton, *Elements of Armories*, 8.

43. Edmund Bolton, "Projects for raising money, &c," BL MS Cotton Titus B. V., 201.

44. Pettegrew, *The Isthmus of Corinth*, 183–86.

45. Sandys, *A Relation of a Journey*, 281–82.

46. Bolton, *Nero Caesar* (1624), 269.

47. Bolton, *Nero Caesar* (1624), 270; Heylyn, *Mikrokosmus*, 794–95.

48. Bolton, *Nero Caesar* (1624), 272.

49. Bolton, *Nero Caesar* (1627), [A3v–A4].

50. Blackburn, "Edmund Bolton's The Cabanet Royal," 161.

51. Blackburn, "Edmund Bolton's The Cabanet Royal," 168.

52. Blackburn, "Edmund Bolton's The Cabanet Royal," 171.

53. *CSPD* 1629–31, 300.

54. L. Hotson, "The Projected Amphitheatre," 26

55. Rice, "Poussin's Elephant."

56. *CSPD*, 1623–25, 9, 24. Williams surrendered the office of serjeant-at-arms in 1624. *CSPD*, 1623–25, 282. Devon, *Issues of the Exchequer*, 274–75.

57. Adams, *The Dramatic Records of Sir Henry Herbert*, 46.

58. Bodleian MS Tanner 89, 51–54.

59. Bentley, "The Proposed Amphitheatre," 293–94.

60. Jonson, *The Characters of Two royall Masques*, [A4v].

61. Rosenfeld, *Short History of Scene Design*; Southern, *Changeable Scenery*.

62. Bentley, "The Proposed Amphitheatre," 294.

63. Bentley, "The Proposed Amphitheatre," 293.

64. Townshend, *Life and Letters of Mr. Endymion Porter*, 13.

65. J. Williams, "Olympism and Pastoralism," 42.

66. Bentley, "The Proposed Amphitheatre," 298.

67. Bentley, "The Proposed Amphitheatre," 300.

68. *RVC*, 2:42.

69. Bentley, "The Proposed Amphitheatre," 303. Murray and Porter were remembered together as his "friends" in the will of the global merchant Sir William Courten: *Hinc Illæ Lacrymæ*, 19.

70. Bentley, "The Proposed Amphitheatre," 302.

71. Cust, "Charles I's Noble Academy." On Kynaston more generally: Cuttica, "Sir Francis Kynaston"; Webster, *The Great Instauration*, 218; C. Williamson, "The Life and Works of Sir Francis Kinaston"; Ryan, "Chaucer's Criseyde in Neo-Latin Dress"; Beadle, "The Virtuoso's Troilus."

72. HP 30/4/17A.

73. HP 29/3/35A.

74. HP 47/9/32A–B. For Hartlib's interactions with Danvers, and evidence of Danvers' interest in education, see: HP 30/4/56A; HP 30/4/61B; HP 30/4/54A; HP 28/1/66B. Danvers also discussed education with Aubrey: Aubrey, *Aubrey on Education*, 162.

75. Jonston and Kynaston, *Musae Aulicae*. This is not explicitly related to the academy but seems to describe the range of knowledge offered in the curriculum.

76. Heylyn, *Microcosmus*, 2r.

77. Heylyn, *Microcosmus*, 2r.

78. Symonds, *Virginia*, 13.

79. Heylyn, *Microcosmus*, [B5r].

80. Gerbier, *A Brief Discourse*, 40. On the York House theater, see Britland, "Henry Killigrew and Dramatic Patronage."

81. Gerbier, *A Brief Discourse*, 42.

82. Cust, "Charles I's Noble Academy," 353.

83. Cust, "Charles I's Noble Academy," 350.

84. Motley, *Becoming a French Aristocrat*; Lockyer, *Buckingham*, 11.

85. "It is most certaine that the French nation is noe lesse sencible of the commoditie and riches which they gaine by the education of Englishmen in Paris and their academies then they are of the moneyes which they yearely receive from England for theire wines." Cust, "Charles I's Noble Academy," 347.

86. Kynaston, *The Constitutions Of The Musaeum Minervae*, unpaginated preface.

87. HP 29/3/61A.

88. Kynaston, *The Constitutions Of The Musaeum Minervae*, 4.

89. Cust, "Charles I's Noble Academy," 349.

90. Kynaston, *The Constitutions Of The Musaeum Minervae*, 19.

91. Kynaston, *The Constitutions Of The Musaeum Minervae*, 6.

92. Kynaston, *The Constitutions Of The Musaeum Minervae*, 7.

93. Kynaston, *The Constitutions Of The Musaeum Minervae*, 13.

94. Kynaston, *The Constitutions Of The Musaeum Minervae*, 17.

95. HP 29/3/47B.

96. Kynaston, *The Constitutions Of The Musaeum Minervae*, 17.

97. Kynaston, *The Constitutions Of The Musaeum Minervae*, 18.

98. Cuttica, "Sir Francis Kynaston," 142; Webster, *The Great Instauration*, 218.

99. HP 29/3/47B.

100. Kynaston, *Troilus & Creseid*, Bodleian Library MS. Add. C. 287. Kynaston dated his commentary on Chaucer as beginning on August 21, 1629, and concluding on May 4, 1640, with an imprimatur of June 2, 1640 (although the work never reached the press; Kynaston died in 1642).

101. Saintsbury, *Minor Poets of the Caroline Period*, 136, 167, 168.

102. Kynaston, *Troilus & Creseid*, Bodleian MS. Add. C. 287, 385, 404.

103. Kynaston, *Troilus & Creseid*, Bodleian MS. Add. C. 287, 122.

104. Kynaston, *Troilus & Creseid*, Bodleian MS. Add. C. 287, 122.

105. Kynaston, *Troilus & Creseid*, Bodleian MS. Add. C. 287, 110.

106. Kynaston, *Troilus & Creseid*, Bodleian MS. Add. C. 287, 151–52.

107. Kynaston, *Troilus & Creseid*, Bodleian MS. Add. C. 287, 151–52.

108. Kynaston, *Troilus & Creseid*, Bodleian MS. Add. C. 287, 152.

109. *CSPD*, 1636–37, 251. December 31, 1636.

110. Kynaston, *Troilus & Creseid*, Bodleian MS. Add. C. 287, 370.

111. Kynaston, *Troilus & Creseid*, Bodleian MS. Add. C. 287, 458–59.

112. *CSPD* 1637, 102.

113. HP 29/3/47B.

114. HP 8/63/3A–4B. Kynaston and Hartlib partnered for the production of a "folium fabrilium," a "statick instrument" that could be used in a number of crafts as well as for lifting whales into ships, and for which Kynaston was to receive a royal privilege. It was perhaps the device called the "Omnipotent" invented by Simon Stevin and allegedly improved by Berger: HP 29/2/32A.

115. HP /4/1A–2B.

116. HP 71/12/16A.

117. "Ex ore eius auditam sibi referre vocem: se reperturum aut neminem." Comenius, "De Arte Spontanei Motus," 315. Berger told Hartlib, "hee made cum Comenio Primum Mobile": HP 29/2/23B. See also HP 29/3/47B.

118. "Kuffler successor Drebbeli" and "Sir Christopher is very intimate with Cuffler." HP 29/3/44A–B.

119. Doorman, *Octrooien*, 176–77.

120. HP 71/12/11A–12B. See also HP 71/12/13A–14B.

121. HP 29/3/56B.

122. HP 31/23/1A–B; 11/1/66A–B; Nováková, "Der Brief des Johann Christoph Pergar."

123. HP 29/3/52A.

124. H. Hotson, *The Reformation of Common Learning*, 253–55; HP 29/2/63B.

125. HP 29/3/42B.

126. HP 29/3/62A; HP 29/3/48B.

127. HP 29/3/64B.

128. HP 29/3/43A.

129. Kynaston, *The Constitutions Of The Musaeum Minervae*, 12.

130. Winkler, *Music, Dance, and Drama*, 85.

131. HP 29/3/59B. See also HP 29/3/58B.

132. HP 8/60/4A.

133. HP 29/3/41B.

134. HP 71/12/9A.

135. HP 30/4/46B. See also HP 29/3/41B.

136. HP 2/10/15B.

137. Rossi, *Francis Bacon*; Long, *Openness, Secrecy, Authorship*; Pastorino, "The Philosopher and the Craftsman."

138. Shapin, "A Scholar and a Gentleman."

139. Kynaston, *Corona Minervae*, [A3v].

140. Kynaston, *Corona Minervae*, [C3v].

141. Kynaston, *Corona Minervae*, [C3v].

Chapter 7 · Unlimited Invention

1. Jonson, *The Staple of Newes*, 39–40.

2. Werrett, *Fireworks*.

3. Bull, *"The Furie of the Ordnance,"* 25.

4. Jonson, *Workes*, 39.

5. Jonson, *Workes*, 40.

6. Brown, "A Jewel of Great Value."

7. "Petition of Arnold Rotsypen to Prince Charles, to be sworn his servant, according to promise, and to have a patent for his invention of a piece of ordnance which His Highness has approved." SP 14/159/198. February 27, 1624.

8. Sec. Conway to Att. Gen. Coventry. SP 14/163/105. April 28, 1624.

9. Docquet Book, cited in Ruellet, *Les privilèges d'invention en France*, 45.

10. Rymer, *Foedera*, 8:75–76; W. Hardy, *Forty-Third Annual Report*, 29. This date is given incorrectly as July 13, 1628, in Woodcroft, *Subject-matter Index of Patents*, 879. It should be July 13, 1626.

11. SP 16/30/86, June 26, 1626.

12. SP 16/30/85.

13. SP 16/31/14.

14. SP 16/31/14.

15. Devon, *Issues of the Exchequer*, 350, dated July 16, 1626. Nye, *Pounds Sterling to Dollars*.

16. "Warrant to pay to Cornelius Drebble and Arnold Rotispen 100 lb. as a reward for forging divers water engines." *CSPD*, 1627–28, 206. June 5, 1627. Nye, *Pounds Sterling to Dollars*.

17. Kirsch, *Fireships*, 51.

18. Dooley, "Making it Present," 106–14.

19. BL Additional MS 64892, 38v.

20. "Wee have appointed Cornelius Drubble to receive the powder and *granados* specified in this our letter." Lyle, *Acts of the Privy Council of England*, 476, August 9, 1627.

21. BL Add. MS 64893, 48.

22. BL Add. MS 64893, 66.

23. R. Stewart, *The English Ordnance Office*, 18.

24. SP 16/101/144a. April 22, 1628.

25. Cited in C. White, "The Jenks Area and the Tailrace," 174; *CSPD*, 1671–72, 54.

26. SP 16/102/66. April 27, 1628.

27. Siddons, *The Heraldry of Foreigners*, 104.

28. SP 16/106/43. See also *CSPD*, 1628, 147.

29. Capt. John Heydon to Nicholas. SP 16/106/40. June 3, 1628.

30. Tomlinson, *A History of the Minories*, 136, 400.

31. Cited in K. Hill, "Juglers or Schollers?," 268.

32. SP 16/363/0105; SP 16/377/198.

33. Feingold, *The Mathematician's Apprenticeship*, 203. Delamain, *Grammalogia*, [A7r].

34. Hunt, "Book Trade Patents," 51–52. Doquet Book, cited in Ruellet, *Les privilèges d'invention*, 46. 3/10 March 1633 [1632].

35. S. Marsh, "The Construction and Arming of London's Defences," 290.

36. K. Hill, "Juglers or Schollers?," n2.

37. Delamain, *Gram[m]elogia*, [A2r].

38. Delamain, *The making, description, and use of a small portable Instrument*, [A3r].

39. Gunter, *The Description*.

40. Oughtred, *The Just Apologie*, [B4v].

41. Oughtred, *Circles*, [A4r].

42. Delamain *Grammelogia*, [A5v].

43. Delamain *Grammelogia*, [A3r].

44. Oughtred, *The Just Apologie*, [B4v].

45. Delamain, *Grammelogia*, [A7v].

46. Delamain, *Grammelogia*, [A8v]; K. Hill, "Juglers or Schollers?"

47. John Heydon to Thomas Browne, 23 September, 1652, Bodleian Rawlinson MS D 391, 26; V. Keller, "The Authority of Practice."

48. HP 29/2/31B.

49. HP 29/5/74A. 1656.

50. HP 29/8/13B.

51. HP 29/8/13B.

52. V. Keller, "The Authority of Practice."

53. HP 29/2/32B.

54. V. Keller, "The Authority of Practice."

55. HP 29/5/102B.

56. "Instruments for Cutting, Polishing, Filing, Turning, Boring, Grinding, Sawing, and Planing Metals, for Cutting Screws, for Making Files and Bullets, for Making and Rifling Barrels, for Pressing or Printing, and a Hammer moved by Water or Horse Power." Woodcroft, *Reference Index of Patents of Invention*, 15.

57. Samuel Hartlib to Robert Boyle, in Boyle, *The Works of the Honourable Robert Boyle*, 6:88.

58. *CSPD* 1666–67, 445.

59. Jenkins, "The Vauxhall Ordnance Factory," 30.

60. Jenkins, "The Vauxhall Ordnance Factory," 32.

61. Jenkins, "The Vauxhall Ordnance Factory," 28.

62. G. Wilson, *The Vauxhall Operatory*, 51.

63. G. Wilson, *The Vauxhall Operatory*, 32.

64. Webster associated Vauxhall entirely with Somerset: "Vauxhall had been the base for the activities of the Marquis of Worcester and his technicians before the Civil War." Webster, *Great Instauration*, 364.

65. Thorpe, "The Marquis of Worcester and Vauxhall."

66. MacGregor, "A Magazin of all Manner of Inventions," 209; *CSPD*, 1666–67, 445. *CSPD*, 1666–67, 445.

67. Shaw, *Letters of Denization*, 44; Bull, *"The Furie of the Ordnance,"* 10; G. Wilson, *The Vauxhall Operatory*, 57; Norman, "Arms, Armour and Militaria," 363.

68. *CSPD*, 1633–34, 474–75.

69. *CSPD*, 1631–33, 554. March 1, 1633.

70. G. Wilson, *The Vauxhall Operatory*, 9.

71. G. Wilson, *The Vauxhall Operatory*, 12; Vollgraff, "De rol van den Nederlander Caspar Calthoff."

72. Jenkins, "The Vauxhall Ordnance Factory of King Charles I," 30.

73. G. Wilson, *The Vauxhall Operatory*, 12.

74. G. Wilson, *The Vauxhall Operatory*, 12.

75. G. Wilson, *The Vauxhall Operatory*, 88; Malcolm and Stedall, *John Pell*, 65; HP 30/4/6B.

76. HP 30/4/35B; HP 30/4/50B.

77. HP 30/4/50B.

78. HP Royal Society MSs, Boyle Letters, 7.2 1A–2B.

79. HP 8/50/1A–2B. Discussed in V. Keller, "Communicated Only to Good Friends and Philosophers."

80. G. Wilson, *The Vauxhall Operatory*, 58–59.

81. Norman, "Arms, Armour and Militaria," 363.

82. HP 29/6/14B–29/6/15A, 29/6/16B.

83. *CSPD*, 1633–34, 18; *CSPD*, 1635–36, 73.

84. "A privilege for 14 yeares graunted unto John Lanyon gent for the sole use and benefitt within the kingdome of England and dominion of Wales of a new invention of

his owne for the covering of houses. And another invention for freyming of clothes after the indian manner." Docquet Book, April 3/11, 1635, cited in Ruellet, *Les privilèges d'invention*, 47. "Mr Lanion hath almost perfected the designe of making Pintadoes in England. Mr. Worsley." HP 28/2/60A.

85. Jenner, "Another Epocha?" 360.

86. HP 28/2/12A.

87. HP 28/2/14B; 28/2/18A.

88. HP 8/50/1B–2A.

89. Dijksterhuis, "Harnessing the Elements."

90. P. Smith, "Making the Edition of Ms. Fr. 640."

91. HP Royal Society MSs, Boyle Letters, 7.1 1A–3B; HP 8/64/4A.

92. Alberts, "Francesco I's Museum"; Kaufmann, *The Mastery of Nature*.

93. Ruellet, *Les fabricants d'instruments mathématiques*.

94. Cf. London, "Musaeum Tradescantium," 39, who associates Calthof and Lambert only with Worcester, and does not identify Lanyon.

95. Rosenthal, "Plus Ultra, Non Plus Ultra."

96. Bull, *"The Furie of the Ordnance,"* 50.

97. Heywood, *A True description of his Majesties royall ship*.

98. Fulton, *The Sovereignty of the Sea*.

99. "Augustissimo & invictissimo Carolo Maximo magnae Britanniae Franciae & Hiberniae, principi & Oceani Monarchae unico." Kynaston, Bodleian Library, Oxford. MS. Add. C. 287, 321.

100. Sobecki, "Introduction," 26; Cf. Selden, *Titles of Honor* (1614), 35 and Selden, *Titles of Honor* (1631), 18. Dee, *General and rare memorials*, 55–58; Parry, "John Dee and the Elizabethan British Empire."

101. SP 16/407/184.

102. *CSPD*, 1637–38, 121–22.

103. SP 16/383/64. On Delamain, see Cressy, "A Mathematical Petitioner."

104. Herbert, *Memoirs of the Two Last Years*, 187; cited by Feingold, *Mathematician's Apprenticeship*, 203.

105. HP 29/2/11A. HP 29/3/17B. "Appendix de Physica Watts."

106. Watts, "To the venerable Artists."

107. Roper, *Advancing Empire*, 97.

108. James, *The Strange and Dangerous Voyage*, 2.

109. Maclaren, "'Zealous Sayles' and Zealous Sales," 257.

110. James, *The Strange And Dangerous Voyage*, 4.

111. Schotte, *Sailing School*, 23.

112. Schotte, *Sailing School*, 271.

113. James, *The Strange And Dangerous Voyage*, [Qr-v].

114. Iliffe, "Capitalizing Expertise."

115. Maclaren, "'Zealous Sayles' and Zealous Sales," 262.

116. McElligot, "Watts, William (c. 1590–1649)," *ODNB*.

117. Maclaren, "'Zealous Sayles' and Zealous Sales," 258.

118. Kelley, "The Problem of Knowledge," 15.

119. Watts, "To the venerable Artists," [S2v].

120. Watts, "To the venerable Artists," [Sr].

121. Watts, "To the venerable Artists," [S2v].

122. Watts, "To the venerable Artists," [S2v]. Discussed in Feingold, "And Knowledge Shall be Increased."

123. *Instructions for the increasing of Mulberie trees.*

Conclusion

1. Donne, *Works*, 539, 536.

2. Donne, *Works*, 536.

3. Virginia Company, *A True Declaration*, 59.

4. P. Smith, *The Body of the Artisan*, 238.

5. Rusu and Lüthy, "Extracts from a Paper Laboratory"; Jalobeanu, "Bacon's Apples."

6. Harkness, *The Jewel House*, 241–53.

7. For example, Henry, *Knowledge is Power.*

8. On Kynaston and Baconian influence, see G. Turnbull, "Samuel Hartlib's Connection; Cust, "Charles I's Noble Academy"; Beadle, "The Virtuoso's Troilus." On Norton as the "Baconian artilleryman," see Cressy, *Saltpeter*, 14. On Norden as applying a "Baconian framework," see Robson, "Improvement and Epistemologies of Landscape," 615.

9. G. Wilson, *The Vauxhall Operatory*, 41.

10. Cust, "Charles I's Noble Academy."

11. HP 29/3/49A.

12. These were Cornelius Agrippa, Pierre d'Ailly, Albertus Magnus, Roger Bacon, Guillaume du Bartas, Caspar Bartholinus, Caspar Bauhinus, Jean Jacques Boissard, Anselm de Boodt, Johann Ernst Burggrav, Robert Burton, Tomasso Campanella, Girolamo Cardano, Oswald Croll, John Dee, John Donne, Domenico Fontana, George Fortescue, Jacques Gaffarel, Galileo Galilei, William Gilbert, Melchior Guilandinus, Ben Jonson, Andreas Kentzollius, Johannes Kepler, André du Laurens, Fortunio Liceti, Guido Pancirolli, Paracelsus, Picatrix, Giambattista della Porta, "the later Hebrew Rabbins" (probably via Gaffarel), Francois Rabelais, Heinrich Rantzow, Hercules a Saxonia, Christoph Scheiner, John Selden, and Ludovicus Vives.

13. Serjeantson, "The Philosophy of Sir Francis Bacon," 1100n75.

14. HP 29/3/13A; HP 29/3/20A–20B.

15. P. Johnson, "Proof of the Heavenly Iris," 58.

16. Colie, *"Some thankfulnesse to Constantine,"* 94. There is no evidence that Drebbel had a laboratory at all. Cf. Preston, *The Poetics of Scientific Investigation*, 95. Most stories and the only depiction (in *On the Nature of the Elements*) of Drebbel's working setup show him out of doors, in gardens, in the river, in wells, under trees, and in the street. The furnace he built was designed for portability. He may, like many contemporaries, have used a kitchen for activities that required a more enclosed environment, rather than a specialized laboratory.

17. Colie, "Cornelis Drebbel and Salomon de Caus."

18. Webster, *The Great Instauration*, 147, 347, 391.

19. C. Huygens, *Mijn Jeugd*, 133.

20. H. Hotson, *Commonplace Learning*, 231.

21. Farrington, *Francis Bacon, Philosopher of Industrial Science*, and Farrington, *Francis Bacon: Pioneer of Planned Science.*

22. For example, Rossi, "Truth and Utility."

23. Hill, *The Intellectual Origins*, 88.

24. For example: Slack, *The Invention of Improvement*, 46, 74, 99, 114; Mokyr, *A Culture of Growth*, 71, 246, 248, 275; Irving, *Natural Science*; Drayton, *Nature's Government*.

25. HP 53/14/24A–B.

26. Bacon, *The Instauratio magna*, 171.

27. V. Keller, "Deprogramming Baconianism," 2.

28. Bacon, *Works*, 66.

29. Bacon, *The Instauratio magna*, 175.

30. Comenius, "An advertisement," 171.

31. HP 30/4/49A–B

32. "Jamais nous n'arriverons à ce point que de rendre notre intellect pareil à la nature des choses, c'est pourquoy je croy que le dessein de Verulamius est impossible." Mersenne, *La Verité Des Sciences*, 212–13.

33. L. Jardine, *Francis Bacon*. Thurs, "Myth 26"; Woodcock, "'The Scientific Method' as Myth and Ideal."

34. Bacon, *The Instauratio magna*, 165.

35. Bacon, *The Instauratio magna*, 169.

36. Bacon, *The Instauratio magna*, 173.

37. Universitätsbibliothek Kassel, Francis Segar, *A Practice of Corn*, 1r–v. Cited in V. Keller, "A Wild Swing."

38. Universitätsbibliothek Kassel, Francis Segar, *A Practice of Corn*, 1v–2r.

39. Brugis, *The Discovery of a Projector*, 3.

40. Ginzburg, "Clues."

41. Ginzburg, "Morelli, Freud and Sherlock Holmes," 12.

42. Eamon, *Science and the Secrets of Nature*, 281.

43. Carey, "Inquiries, Heads, and Directions," 26; Schleck, "Forming Knowledge," 58.

44. The questionnaires sent to the periphery of Spain's empire in the sixteenth century are a well-known example, but questionnaires were also in use at the time in England; Sellers-García, *Distance and Documents*. For the use of interrogatories, see, e.g. *Actes made in the parliament*.

45. *OED* Online, s.v. "clue, n.," accessed August 18, 2020, https://www.oed.com; *OED* Online, s.v. "clew, n.," accessed August 18, 2020, https://www.oed.com.

46. *OED* Online, s.v. "hint, n.," accessed August 18, 2020, https://www.oed.com.

47. Hobbes, *De cive*, [A7v-A8r].

48. Hobbes, *De cive*, [A7v].

49. HP 71/7/1/2A; HP 8/22/3B; HP 67/22/18B.

50. In Early English Books Online, a text search yields 1,118 results for all variants of "clue" and "clew" combined between 1600 and 1700, compared to 6,885 for "hint" over the same period.

51. Watts, "To the venerable Artists," [S2v].

52. Williams and Ferrar, *Virgo Triumphans*, 42.

53. Sir Isaac Newton to Henry Oldenburg, February 10, 1671/2, in Newton, *The Correspondence of Isaac Newton*, 109; quoted in Schilt, "To Improve upon Hints of Things," 77.

54. Hartlib, *The Reformed Virginian Silk-Worm*, 29.

55. Boothby, *A Briefe Discovery*, 56–57.

56. Shapiro, *Probability and Certainty*; Dolby, *Uncertain Knowledge*.

57. Bacon, *The Instauratio magna*, 444–45.

58. O'Connell, "Bacon's Hints."

59. HP 53/14/24A.

60. Petty, *The Advice of W. P. to Mr. Samuel Hartlib*, 21.

61. Bennett, "John Aubrey, Hint-Keeper."

62. Hooke, *Micrographia*, 1.

63. Hooke, *Lectiones Cutlerianae*; Evelyn, *Sylva*; Robert Boyle, *Some Considerations*; Boyle, *Experiments and Considerations*; Boyle, *New Experiments*.

64. Boyle, *Some Considerations*, 113.

65. Newman and Principe, *Alchemy Tried in the Fire*, 204.

66. Charleton, *The Immortality Of The Human Soul*, 34.

67. Charleton, *The Immortality Of The Human Soul*, 35.

68. Charleton, *The Immortality Of The Human Soul*, 37.

69. Charleton, *The Immortality Of The Human Soul*, 41.

70. Sharp, "Walter Charleton's Early Life." V. Keller, "Scarlet Letters."

71. Charleton, *The Immortality Of The Human Soul*, 43.

Manuscript Sources

CARPENTRAS, BIBLIOTHÈQUE MUNICIPALE INGUIMBERTINE

Ms. 1774. Nicolas-Claude Fabrì de Peiresc, *Auctores antiqui*, 1622.

Ms. 1776. Nicolas-Claude Fabrì de Peiresc. *Relation de ce que j'ai appris de la vie et des inventions de Cornelius Derbbel, de la ville d'Alcmar en Hollande, par Abraham Kuffler son gendre et Gilles Kuffler son frere, à Paris, au commencement de septembre 1624.*

EDINBURGH, NATIONAL LIBRARY OF SCOTLAND

Adv. MS. 33.1.1. Cornelis Drebbel to James I.

THE HAGUE, KONINKLIJKE BIBLIOTHEEK

KA 47, Constantijn Huygens, Verzameling musica, medica, etc.

KEW, NATIONAL ARCHIVES

Colonial Office 77, 34.

LONDON, BRITISH LIBRARY

Additional 10038.
Additional 12496.
Additional 16371 h.
Additional 19402.
Additional 64893.
Cotton Titus B. V.
East India Company Court Book, IOR/B/6; IOR/B/7; IOR/B/8; IOR/B/9; IOR/B/10.
East India Company Original Correspondence, E/3/5/2; E/3/7/1.
Egerton 1140.
Harleian 1581.
Harleian 7011.
King's Topographical CXVII.
Sloane 203.
Sloane 871.
Stowe 325.

LONDON, ROYAL SOCIETY

Classified Papers 24/81/126.

LONDON, WELLCOME COLLECTION

MS 484.
MS 716.

OXFORD, BODLEIAN LIBRARY

Add. C. 287.
Ashmole 446.
Rawlinson MS D 391.
Tanner 89.
Tanner 94.

PRAGUE, CZECH NATIONAL ARCHIVE I (M. HORÁKOVÉ 133, PRAGUE 6)

Ms. A 215. Jindřich Michael Hýzrle z Chodů, *Život*.

UNIVERSITÄTSBIBLIOTHEK KASSEL, LANDESBIBLIOTHEK UND MURHARDSCHE
BIBLIOTHEK DER STADT KASSEL

Francis Segar. *A Practice of Corn*, ca. 1600. 4° Ms. philos. 13. Digitized at https://orka
.bibliothek.uni-kassel.de/viewer/image/1369122431709/1/LOG_0000/.

WOLFENBÜTTEL, HERZOG AUGUST BIBLIOTHEK

Cod. Guelf. 235 Blankenburg, Jakob Fetzer, *Album amicorum*, vol. 1.

Digital Resources

Early English Books Online. https://eebo.chadwyck.com/search.
East India Company: India Office Records from the British Library, 1599–1947. https://
www.eastindiacompany.amdigital.co.uk.
Greengrass, M., Leslie, M. and Hannon, M. *The Hartlib Papers*. The Digital Humanities
Institute, University of Sheffield. https://www.dhi.ac.uk/hartlib.
Nye, Eric. *Pounds Sterling to Dollars: Historical Conversion of Currency*. https://www.uwyo
.edu/numimage/currency.htm.
OED Online. Oxford: Oxford University Press. https://www-oed-com.
State Papers Online. https://www.gale.com/primary-sources/state-papers-online.
"Tagebuch Christian II. von Anhalt-Bernburg, Jahr 1623," *Digitale Edition und
Kommentierung der Tagebücher des Fürsten Christian II. von Anhalt-Bernburg
(1599–1656)*, Wolfenbüttel, Herzog August Bibliothek, 2013 (Editiones Electronicae
Guelferbytanae). LAZA, Z 18 A 9b Nr. 14 III, 8r–8v. http://diglib.hab.de/edoc
/ed000228/start.htm.
Virginia Company Archives. https://www.amdigital.co.uk/primary-sources/virginia
-company-archives.

Printed Primary Sources

Actes Made in the Parliament. London: Berthelet, 1545.

Adams, Joseph Quincy. *The Dramatic Records of Sir Henry Herbert, Master of the Revels, 1623–1673.* New Haven, CT: Yale University Press, 1917.

"An Animadversion upon the Letter from Dublin." In Samuel Hartlib, *The Reformed Virginian Silk-Worm.* London: Streater, 1655.

Atkinson, E. G., ed. *Acts of the Privy Council of England*, Vol. 34, *1615–1616.* London: HMSO, 1925.

Aubrey, John. *Brief Lives*, Vol. 1. Edited by Andrew Clark. Oxford: Clarendon, 1898.

———. *Aubrey on Education: A Hitherto Unpublished Manuscript by the author of Brief Lives.* Edited by J. E. Stephens. London: Routledge, 2011.

Babington, John. *Pyrotechnia.* London: Mab, 1635.

Bacon, Francis. *Novum Organum.* London: Bill, 1620.

———. *Essayes.* London: Haviland, 1625.

———. "New Atlantis: A Work Unfinished." In *Sylva Sylvarum*, 42–44. London: Lee, 1628.

———. *Works*, Vol. XI. Edited by James Spedding. London: Longmans et al., 1868.

———. *The Instauratio magna Part II: Novum organum and Associated Texts.* Edited and translated by Graham Rees and Maria Wakely. Oxford: Oxford University Press, 2004.

Baker, George. *The History and Antiquities of the County of Northampton*, Vol. 2. London: Bowyer Nichols, 1836.

Beeckman, Isaac. *Journal Tenu Par Isaac Beeckman de 1604 à 1634*, Vol. 2. Edited by Cornelis de Waard. The Hague: Nijhoff, 1942.

———. *Journal Tenu Par Isaac Beeckman de 1604 à 1634*, Vol. 3. Edited by Cornelis de Waard. The Hague: Nijhoff, 1945.

Bekkers, J. A. F., ed. *Correspondence of John Morris with Johannes de Laet (1634–1649).* Assen: Van Gorcum, 1970.

Bickley, Francis L., ed. "The Manuscripts of the Hon. Frederick Lindley Wood." In *Report on Manuscripts in Various Collections*, Vol. 8. London: HMSO, 1913.

Birch, Thomas. *The Court and Times of Charles the First*, Vol. 2. London: Colburn, 1849.

Birdwood, George. *The First Letter Book of the East India Company: 1600–1619.* London: Quaritch, 1893.

———. *The Register of Letters of the Governour and Company of Merchants of London Trading into the East Indies, 1600–1619.* London: Quaritch, 1893.

Blount, Thomas. *Glossographia.* London: Moseley, 1656.

Bolton, Edmund. *Elements of Armories.* London: Eld, 1610.

———. *Nero Caesar, or Monarchie Depraved.* London: Walkley, 1624.

———. *Nero Caesar, or Monarchie Depraved.* London: Thomas Snodham and Bernard Alsop, 1627.

[Bolton, Edmund]. *The Cities Advocate in This Case or Question of Honor and Armes; Whether Apprentiship extinguisheth Gentry?* London: Lee, 1629.

Bonoeil, Jean. *Obseruations To Be Followed, For The making of fit roomes, to keepe Silk-wormes in as also, For The Best Manner Of planting of Mulbery trees, to feed them.* London: Kyngston, 1620.

———. *A Treatise of the Art of making Silke.* Translated by John Ferrar. London: Kyngston, 1622.

Boothby, Richard. *A Briefe Discovery or Description Of the most Famous Island of Madagascar.* London: Hardesty, 1647.

Botero, Giovanni. *Della Ragion Di Stato.* Venice: Gioliti, 1589.

———. *A Treatise Concerning the causes of the Magnificencie and greatness of Cities.* Translated by Robert Peterson. London: Purfoot, 1606.

———. *Relations, Of The Most Famous Kingdoms and Common-Weales Thorough the World, Discoursing of Their Scituations, Manners, Customes, Strengthes and Pollicies.* Translated by Robert Johnson. London: Jaggard, 1611.

———. *Relations, of the Most Famous Kingdoms and Commonweales Thorough the World.* Translated by Robert Johnson. London: Jaggard, 1616.

———. *On the Causes of the Greatness and Magnificence of Cities.* Translated by Gregory Symcox. Toronto: University of Toronto Press, 2012.

Bourne, William. *Inuentions or Deuises Very necessary for all Generalles and Captaines, or Leaders of men, as wel by Sea as by Land.* London: Woodcock, 1590.

Boyle, Robert. *Some Considerations Touching the Usefulnesse of Experimental Naturall Philosophy.* Oxford: Davis, 1663.

———. *Experiments and Considerations Touching Colours.* London: Herringman, 1664.

———. *New Experiments and Observations Touching Cold.* London: Crook, 1665.

———. *Some Considerations Touching the Usefulness of Experimental Natural Philosophy.* Oxford: Hall, 1671.

———. *Of the Reconcileableness of Specifick Medicines to the Corpuscular Philosophy.* London: Smith, 1685.

———. *The Works of the Honourable Robert Boyle*, Vol. 6. Edited by Thomas Birch. London, 1772.

Brock, R. A. *Abstract of the Proceedings of the Virginia Company of London, 1619–1624*, Vol. 2. Richmond: Virginia Historical Society, 1889.

Brugis, Thomas. *The Discovery of a Projector.* London: R. H., 1641.

Bullokar, John. *An English Expositor.* London: Legatt, 1616.

Bulstrode, Richard. *Memoirs and Reflections upon the Reign & Government of King Charles the Ist and K. Charles the II.* London: Mist, 1721.

Calendar of State Papers, Colonial, America and West Indies, Vol. 9, *1675–1676 and Addenda 1574–1674.* Edited by W. Noel Sainsbury. London: HMSO, 1893.

Calendar of State Papers, Colonial, East Indies, China and Japan, Vol. 3, *1617–1621.* Edited by W. Noel Sainsbury. London: HMSO, 1870.

Calendar of State Papers, Colonial, East India, China and Japan, Vol. 4, *1622–1624.* Edited by W. Noel Sainsbury. London: HMSO, 1878.

Calendar of State Papers, Colonial, East Indies, China and Persia, Vol. 6, *1625–1629.* Edited by W. Noel Sainsbury. London: HMSO, 1884.

Calendar of State Papers Domestic: Charles I, 1625–1626. Edited by John Bruce. London: HMSO, 1858.

Calendar of State Papers Domestic: Charles I, 1627–1628. Edited by John Bruce. London: HMSO, 1858.

Calendar of State Papers Domestic: Charles I, 1628–1629. Edited by John Bruce. London: HMSO, 1859.

Calendar of State Papers Domestic: Charles I, 1629–1631. Edited by John Bruce. London: HMSO, 1860.

Calendar of State Papers Domestic: Charles I, 1631–1633. Edited by John Bruce. London: HMSO, 1862.

Calendar of State Papers Domestic: Charles I, 1633–1634. Edited by John Bruce. London: HMSO, 1863.

Calendar of State Papers Domestic: Charles I, 1634–1635. Edited by John Bruce. London: HMSO, 1864.

Calendar of State Papers Domestic: Charles I, 1636–1637. Edited by John Bruce. London: HMSO, 1868.

Calendar of State Papers Domestic: Charles I, 1637. Edited by John Bruce. London: HMSO, 1868.

Calendar of State Papers Domestic: Charles I, 1637–1638. Edited by John Bruce. London: HMSO, 1869.

Calendar of State Papers Domestic: Charles I, 1638–1639. Edited by John Bruce and William Douglas Hamilton. London: HMSO, 1871.

Calendar of State Papers Domestic: Charles I, 1639. Edited by William Douglas Hamilton. London: HMSO, 1873.

Calendar of State Papers Domestic: Charles I, 1639–1640. Edited by William Douglas Hamilton. London: HMSO, 1877.

Calendar of State Papers Domestic: Charles I, Addenda 1625–1649. Edited by William Douglas Hamilton and Sophie Crawford Lomas. London: HMSO, 1897.

Calendar of State Papers Domestic: Charles II, 1666–1667. Edited by Mary Anne Everett Green. London: HMSO, 1864.

Calendar of State Papers Domestic: Charles II, 1671–1672. Edited by F. H. Blackburne Daniell. London: HMSO, 1897.

Calendar of State Papers Domestic: James I, 1603–1610. Edited by Mary Anne Everett Green. London: HMSO, 1857.

Calendar of State Papers Domestic: James I, 1611–1618. Edited by Mary Anne Everett Green. London: HMSO, 1858.

Calendar of State Papers Domestic: James I, 1619–1623. Edited by Mary Anne Everett Green. London: HMSO, 1858.

Calendar of State Papers Domestic: James I, 1623–1625. Edited by Mary Anne Everett Green. London: HMSO, 1859.

Calendar of State Papers, Venice, Vol. 17, *1621–1623*. Edited by Allen B. Hinds. London: HMSO, 1911.

Calendar of State Papers, Venice, Vol. 18, *1623–1624*. Edited by Allen B. Hinds. London: HMSO, 1912.

Camden, William. *Britain*. London: Bishop, 1610.

The Case of the heir of Sir Anthony Thomas Kt., deceased. N.p.: n.p., 1661.

Chamberlain, John. *The Letters of John Chamberlain, Part 1*. Edited by Norman Egbert McClure. Philadelphia: American Philosophical Society, 1939.

Charleton, Walter. *The Immortality Of The Human Soul*. London: Herringman, 1657.

Chester, Joseph Lemuel. *The Reiester Booke of Saynte De'nis Backchurch Parishe*. London: Harleian Society, 1878.

Child, Robert. "A Large Letter Concerning the Defects and Remedies of English Husbandry." In *Samuel Hartlib His Legacie*, edited by Samuel Hartlib, 1–108. London: Wodnothe, 1651.

Cochran-Patrick, Robert William. *Early Records Relating to Mining in Scotland*. Edinburgh: Douglas, 1878.

Coldham, Peter Wilson. *English Adventurers and Emigrants, 1609–1660: Abstracts of Examinations in the High Court of Admiralty with Reference to Colonial America*. Baltimore: Clearfield, 2002.

Colwall, Daniel. "An Account of the Way of Making English Green Copperas, Communicated by the Same." *Philosophical Transactions* 12, no. 142 (1677): 1056–59.

Comenius, Jan Amos. "An Advertisement Touching the Scenography, or Shaddowed Description of the Work of Pansophy." In *A Patterne of Universall Knowledge, in a Plaine and True Draught*, translated by Jeremy Colliers, 170–80. Collins: London, 1651.

———. "De Arte Spontanei Motus." In *Opera Omnia*, Vol. 12. Prague: Czech National Academy of Sciences, 1978.

Cooke, Edward. *The Character of Warre, or The Image of Martiall Discipline*. London: Purfoot, 1626.

———. *The Prospectiue Glasse of Warre. Shewing a glimps of Warres Mystery, in her admirable Strategems, Policies, Wayes*. London: Sparke, 1628.

A Copie of The Kings Commission, Granted To Sir Nicolas Crispe, Making him Admirall of the Sea-Pirats. London: Austin, 1645.

Coryate, Thomas. *Crudities*. London: William Stansby, 1611.

Cotgrave, Randle. *A Dictionarie of the French and English Tongues*. London: Islip, 1611.

County of Middlesex. *Calendar to the Sessions Records: New Series*, Vol. 1, *1612–1614*. London: Clerk of the Peace, 1635.

Courten, Sir William. *Hinc Illæ Lacrymæ, Or, An Epitome Of The Life and Death Of Sir Wlliam Courten And Sir Paul Pyndar*. London: n.p., 1671.

Crashaw, William. *A Sermon Preached In London*. London: Welby, 1610.

Crispe, Nicholas. *The Humble Petition of Sir Nicholas Crisp Knight*. London: n.p., 1660.

Davies, John. *Historical Tracts*. London: Stockdale, 1786.

Dee, John. *General and Rare Memorials Pertayning to the Perfect Arte of Nauigation*. London: Daye, 1577.

Defoe, Daniel. *An Essay upon Projects*. London: Cockerill, 1697.

Delamain, Richard. *Gram[m]elogia, or, The Mathematicall Ring*. London: Haviland, 1630.

———. *Grammelogia. or, The Mathematicall Ring*. N.p: n.p., n.d.

———. *The making, description, and use of a small portable Instrument . . . called a Horizontall Quadrant*. London: Hawkins, 1632.

Devon, Frederick. *Issues of the Exchequer . . . During the Reign of King James I*. London: Rodwell, 1836.

Digby, Kenelm. *Two Treatises in the One of Which the Nature of Bodies, in the Other, the Nature of Mans Soule*. Paris: Gilles Blaizot, 1644.

Donne, John. *Pseudo-martyr*. London: Stansby, 1610.

———. *A Sermon upon the VIII. Verse of the I Chapter of the Acts of the Apostles Preach'd to the Honourable Company of the Virginian Plantation*. London: Jones, 1622.

———. *Poems*. London: Marriot, 1633.

———. *Poems*. London: Marriot, 1639.

———. *Works, 1621–1631*, Vol. 1. London: Parker, 1839.

Drebbel, Cornelis. *Een Kort Tractaet van de Natuere der Elementen*. Haarlem: Casteleyn, 1621.

———. *Wonder-vondt*. Alkmaar: de Meester, 1607.

Evelyn, John. *Sylva, or A Discourse of Forest-Trees*. London: Martyn and Allestry, 1670.

———. *Diary*, Vol. 2. Edited by William Bray. London: Bickers, 1879.

———. *Diary*, Vol. 1. Edited by William Bray. London: Bohn, 1885.

———. *Directions for the Gardiner at Says-Court*. Edited by Geoffrey Keynes. London: Nonesuch Press, 1932.

Ferrar, Nicholas. *Memoirs of the Life of Mr. Nicholas Ferrar*. Edited by Peter Peckard. Cambridge: Cambridge University Press, 1790.

Forster, John. *The Debates on the Grand Remonstrance, November and December 1641*. London: Murray, 1860.

Foster, William, ed. *Letters received by the East India Company . . . 1617 (January to June)*. East India Company's Records, Vol. 5. London: Sampson, Low, Marston & Co, 1901.

———. *Letters received by the East India Company . . . 1617 (July to December)*. East India Company's Records, Vol. 6. London: Sampson, Low, Marston & Co., 1902.

———. *The English Factories in India, 1618–1623*. Oxford: Clarendon, 1906.

———. *The English Factories in India, 1637–1641*. Oxford: Clarendon, 1912.

Gassendi, Pierre. *De Vita Peireskii*. Paris: Cramoisy, 1641.

Gerbier, Balthasar. *Clacht-Dicht*. The Hague: Meurs, 1620.

———. *Baltazar Gerbier Knight to all Men that Loves Truth*. N.p: n.p., 1646.

———. *The Second Part, of the Holland-Sea-Cabbin-Dialogue*. London: T.M., 1652.

———. *A Sea-Cabbin Dialogue, Between Two Travellers Lately Come from Holland*. London: T.M., 1653.

———. *A Sommary Description Manifesting That Greater Profits Are to Bee Done in the Hott Then in the Could Parts off the Coast of America and How Much the Public Good Is Concerned Therein*. Rotterdam: For the author, 1660.

———. *A Brief Discourse Concerning the Three chief Principles of Magnificent Building*. London: n.p., 1662.

———. *Subsidium peregrinantibus*. Oxford: Gascoigne, 1665.

Goodman, Godfrey. *The Fall of Man, or the Corruption of Nature*. London: Kyngston, 1616.

Gorges, Arthur. *A True Transcript and Publication of His Maiesties Letters Patent for an Office to Be Erected, and Called the Publicke Register for Generall Commerce*. London: Budge, 1611.

———. *A True Transcript and Publication of his Maiesties Letters Patent for an Office to be Erected, and called the Publike Register for generall Commerce*. London: Budge, 1612.

Gray, Robert. *A Good Speed to Virginia*. London: Kyngston, 1609.

Gouge, William. *The Dignitie of Chiualrie set forth a sermon preached before the Artillery Company*. London: Mab, 1626.

Gunter, Edmund. *The Description And Use Of His Maiesties Dials In White-Hall Garden*. London: Norton and Bill, 1624.

Gutch, John. "Sir Balthazar Gerbier's Project for an Academy Royal in England." In *Collectanea Curiosa*, 209–15. Oxford: Clarendon, 1781.

Hall, Joseph. *Quo vadis? A Iust Censure of Travell as it is commonly vndertaken by the Gentlemen of Our Nation*. London: Butter, 1617.

———*One of the Sermons Preacht at Westminster . . . to the Lords of the High Court of Parliament*. London: Butter, 1628.

Halse, Nicholas. *Patent of Sir Nicholas Halse, Kilns for Drying Malt and Hops*. London: Spottiswoode, 1857.

Hardy, Thomas Duffus. *Annual Report of the Deputy Keeper of the Public Records*, Vol. 38. London: Spottiswoode, 1877.

Hardy, William. *Forty-Third Annual Report of the Deputy Keeper of the Public Records*. London: Eyre and Spottiswoode, 1882.

Harsdörffer, Georg Philipp. *Delitiae mathematicae et physicae*. Nuremberg: Dümler, 1651.

Hartlib, Samuel. *A Rare and New Discovery of a Speedy Way and Easie Means, Found out by a Young Lady in England, She Having Made Full Proofe Thereof in May, Anno 1652. For the Feeding of Silk-Worms in the Woods*. London: Wodenothe, 1652.

———. *The Reformed Virginian Silk-Worm*. London: Streater, 1655.

Herbert, Thomas. *Newes out of Islington*. London: Lambert, 1641.

———. *Memoirs of the Two Last Years of the Reign of King Charles I*. London: Nicol, 1815.

Heylyn, Peter. *Microcosmus: or, A little description of the great world*. Oxford: Lichfield, 1621.

———. *Mikrokosmos A little description of the great world. Augmented and reuised*. Oxford: Lichfield, 1625.

Heywood, Thomas. *A True Description of His Majesties Royall Ship, Built This Year 1637*. London: Okes, 1637.

Highmore, Anthony. *The History of the Honourable Artillery Company of London*. London: Printed privately, 1804.

Historical Manuscripts Commission. "Catalogue of Manuscripts in the Possession of the Honble. G. M. Fortescue, Dropmore, Maidenhead." *Second Report, Appendix*, 49–63. London: Eyre and Spottiswoode, 1874.

Historical Manuscripts Commission, *Twelfth Report, Appendix, Part IX: The Manuscripts of the Duke of Beaufort, K. G., The Earl of Donoughmore, and Others*. London: HMSO, 1891.

Hobbes, Thomas. *De cive*. London: Royston, 1651.

Hooke, Robert. *Lectiones Cutlerianae, or, A Collection of Lectures*. London: Martyn, 1679.

———. *Micrographia*. London: Martyn and Allestry, 1665.

House of Commons. *Journal of the House of Commons*, Vol. 1, *1547–1629*. London: HMSO, 1802.

———. *Journals of the House of Commons*, Vol. 2, *1640–1642*. London: House of Commons, 1803.

House of Lords. *Journal of the House of Lords*, Vol. 3, *1620–1628*. London: HMSO, 1767.

Hubert, Robert. *A Catalogue of Many Natural Rarities*. London: Ratcliffe, 1664.

Huit, Ephraim. *The Anatomy of Conscience*. London: Sheffard, 1626.

Huygens, Constantijn. "Fragment Eener Autobiographie Van Constantijn Huygens." Edited by J. A. Worp, *Bijdragen en Mededeelingen van het Historisch Genootschap* 18 (1897); 1–122.

———. *Gedichten*, Vol. 8, *1671–1687*. Edited by J. A. Worp. Groningen: Wolters, 1898.

———. *Mijn Jeugd*. Translated by C. L. Heesakkers. Amsterdam: Querido, 1987.

Huygens, Lodewijck. *The English Journal, 1651–1652*. Edited by A. G. H Bacharach and R. G. Collmer. Leiden: Leiden University Press, 1982.

Inderwick, Frederick Andrew, ed. *A Calendar of the Inner Temple Records*, Vol. 2. London: Chiswick, 1898.

Instructions for the increasing of Mulberie trees, and the breeding of Silke-Wormes, for the making of silke in This Kingdome Whereunto is Annexed his Maiesties Letters to the Lords Liefetenants of the seuerall Shiers of England, tending to that purpose. London: Edgar, 1609.

James I. *Basilikon Doron.* London: White, 1603.

———. *A Declaration of His Maiesties Royall pleasure, in what sort He thinketh fit to enlarge, or reserue Himselfe in matter of Bountie.* London: Barker, 1611.

James, Thomas. *The Strange And Dangerous Voyage Of Captaine Thomas Iames, in his intended Discouery of the Northwest Passage into the South Sea.* London: Legatt, 1633.

Johnson, Robert. *Nova Britannia Offering Most Excellent fruites by Planting in Virginia.* London: Macham, 1609.

Johnson, Samuel. *The Works of Samuel Johnson*, Vol. 3. Edited by Arthur Murphey. London, 1823.

Jonson, Ben. *The Characters of Two royall Masques. The one of Blacknesse, the other of Beautie.* London: Thorpe, 1608.

———. *The Masque of Augures.* London: n.p., 1621.

———. *The Staple of Newes.* London: Allot, 1631.

———. *Workes.* London: Andrew Crooke, 1640 [1641].

———. *Poems.* London: Griffen, 1870.

Jonston, Arthur, and Francis Kynaston. *Musae Aulicae.* London: Harper, 1635.

Keeler, Mary Frear, Maija Jansson Cole, and William Bidwell, eds. *Commons Debates 1628*, Vol. 6. New Haven, CT: Yale University Press, 1978.

Kingsbury, Susan Myra, ed. *Records of the Virginia Company of London*, 4 vols. Washington, DC: Government Printing Office, 1906–1935.

Knowler, William. *The Earl of Strafforde's Letters and Dispatches*, Vol. 2. London: Bowyer, 1739.

Knowles, James. "Jonson's Entertainment at Britain's Burse." In *Re-presenting Ben Jonson: Text, History, Performance*, edited by Martin Butler, 114–151. London: Palgrave Macmillan, 1999.

Kynaston, Francis. *Corona Minervae.* London: Sheares, 1635.

———. *The Constitutions Of The Musaeum Minervae.* London: Spencer, 1636.

Larkin, James F. *Stuart Royal Proclamations*, Vol. 2: *Royal Proclamations of King Charles I 1625–1646.* Oxford: Oxford University Press, 1983.

Le Surveillant de Charenton au Duc de Bouckinghan. N.p: n.p., 1627.

Lipsius, Justus. *Admiranda, sive de magnitudine Romana.* Antwerp: Moretus, 1598.

Lloyd, David. *Memoires.* London: Speed, 1668.

Loe, William. *The Kings Shoe Made, and Ordained To Trample On And To Treade Downe Edomites.* London: Legat, 1623.

Lyle, J. V. *Acts of the Privy Council of England, 1627–1628*, Vol. 43. London: HMSO, 1940.

———. *Acts of the Privy Council of England, 1627, Jan.–Aug.* Nendeln: Kraus Reprint, 1974.

Major, Johann Daniel. *Dissertatio epistolica de cancris.* Jena: Fellgiebel, 1664.

Malynes, Gerard. *Consuetudo, vel Lex Mercatoria.* London: Islip, 1622.

———. *The Center of the Circle of Commerce.* London: Bourne, 1623.

Matthews, Harold Evan. *Proceedings, Minutes and Enrollments of the Company of Soapmakers, 1562–1642.* Bristol: Printed for the Bristol Record Society, 1940.

McGowan, Alan Patrick, ed. *The Jacobean Commissions of Enquiry, 1608 and 1618.* London: Navy Records Society, 1971.

Melton, John. *A sixe-folde politician.* London: Busby, 1609.

Mersenne, Marin. *La Verité Des Sciences.* Paris: du Bray, 1625.

Milward, Matthias. *The Souldiers Triumph.* London: Clark, 1641.

Monson, William. *Naval Tracts.* London: Churchill, 1703.

Montaigne, Michel de. *Essais.* Paris: L'angelier, 1588.

Morton, Thomas. *New English Canaan.* Amsterdam: Jacob Fredrick Stam, 1637.

Mubārak, Abū al-Fazl ibn. *Ain I Akbari.* Translated by H. Blochmann. Calcutta: Baptist Mission Press, 1873.

Mundy, Peter. *The Travels of Peter Mundy, In Europe and Asia, 1608–1667,* Vol. 3, Part 1. Edited by Richard Carnac Temple. London: Hakluyt Society, 1919.

Newton, Sir Issac. *The Correspondence of Isaac Newton,* Vol. 2, *1676–1687.* Edited by Herbert W. Turnbull. Cambridge: Cambridge University Press, 1960.

Norden, John. *The Surveyors Dialogue.* London: Astley, 1607.

North, Roger. *The Life of the Honourable Sir Dudley North.* London: Whiston, 1744.

Norton, Robert. *The Gunner.* London: Robinson, 1628.

Notestein, Wallace, ed. *The Journal of Sir Simonds D'Ewes.* New Haven, CT: Yale University Press, 1923.

Oldys, William, ed. *Biographia Britannica, or the Lives of the Most Eminent Persons Who Have Flourished in Great Britain and Ireland,* Vol. 3. London: Innys, et al, 1750.

Onslow, Arthur. *The Life of Dr. George Abbot.* Guildford: Russell, 1777.

Oughtred, William. *The Circles of Proportion and the Horizontall Instrument.* Translated by William Forster. London: Mathewes, 1632.

———. *The Just Apologie . . . against the Slaunderous Insimulations of Richard Delamain.* London: Mathewes, 1634.

Overall, W. H. and H. C. Overall, eds. *Analytical Index to the Series of Records known as the Remembrancia 1579–1664.* London: Francis & Co., 1878.

Owen, G. Dyfnallt, ed. *Calendar of the Manuscripts of the Most Honourable the Marquess of Salisbury,* Part 22 (1609–12). London: HMSO, 1970.

Palmer, Thomas. *An Essay of the Means How to Make Our Travels into Foreign Countries the More Profitable and Honourable.* London: Lownes, 1606.

Parkinson, John. *Paradisi in Sole Paradisus Terrestris.* London: Lownes and Young, 1629.

———. *Theatrum Botanicum.* London: Cotes, 1640.

Patterson, Samuel. *A Catalogue of . . . the Collection of Manuscripts of . . . Sir Julius Caesar.* London: Brindley, 1757.

Peacham, Henry. *Minerva Britanna.* London: Dight, 1612.

"Petitions to the Westminster Quarter Sessions: 1620s." In *Petitions to the Westminster Quarter Sessions, 1620–1799,* edited by Brodie Waddell. British History Online. http://www.british-history.ac.uk/petitions/westminster/1620s.

Petty, William. *The Advice of W. P. to Mr. Samuel Hartlib for the Advancement of Some Particular Parts of Learning.* London: n.p., 1647.

Powell, Robert. *Depopulation Arraigned, Convicted and Condemned, By The Lawes of God and Man.* London: R.B. 1636.

Price, Daniel. *Sauls Prohibition Staide.* London: Law, 1609.

A Proclamation Concerning the Trade of Ginney, and Binney, in the Parts of Africa. London: Barker, 1631.

A Proclamation concerning certain Kilnes for the sweet and speedy drying of Mault and Hops at a small charge. London: Barker, 1637.

The Proiectors Down-Fall, or, Times Changeling. Wherein the Monopolists and Patentees Are Unmasked to the View of the World. The Chiefe of Which Are These; Vix. Custemers, Vintners, Refiners of Salt, Soap Boylers. London: Paine, 1642.

A Publication of Guiana's Plantation Newly undertaken by the Right Honble. the Earl of Barkshire . . . and Company for that most famous River of the Amazones in America. London: Paine, 1632.

Rathborne, Aaron. *The Surveyor in Foure Bookes*. London: Stansby, 1616.

Reisch, Gregor. *Natural Philosophy Epitomised: Books 8–11 of Gregor Reisch's Philosophical Pearl (1503)*. Translated by Andrew Cunningham and Sachiko Kusukawa. Burlington, VT: Ashgate, 2010.

Ripa, Cesare. *Iconologia*. London: Motte, 1709.

Robinson, James Harvey. *Readings in European History*, Vol. 2. Boston: Ginn and Co., 1904–1906.

Rooke, W. "Report of the Clerk of the Chapel of the Rolls." In *A Report from the Committee appointed to View the Cottonian Library*, 170–73. London: William, 1732.

Ruellet, Aurélien. *Les Privilèges d'invention En France et En Angleterre (ca 1600–1660): Base de Données Provisoire*. 2014. https://halshs.archives-ouvertes.fr/halshs-01116703v2.

Rushworth, John. "A Decree Concerning Soap-Boilers, 1633." In *Historical Collections of Private Passages of State*, Vol. 3, *1639–40*, 109–15. London, 1721.

Russell, Thomas. *To the Kings most Excellent Maiestie, the Lords Spirituall and Temporall, and the Commons in this present Parliament assembled, The humble Petition of Thomas Russell, Esquire*. London, 1626.

Rymer, Thomas. *Foedera*, Vol. 7. London: Tonson, 1727.

———. *Foedera*, Vol. 8. The Hague: Neaulme, 1742.

Saledinus, Valerius [Daniel Mögling]. *De perpetuo mobile*. Frankfurt: Jennis, 1625.

Sanderson, Robert. *Foedera, conventiones, literae, et cujuscunque generis acta publica*, Vol. 12. London: Churchill, 1717.

Sandys, George. *A Relation of a Journey begun An. Dom. 1610*. London: Barren, 1615.

Scot, Thomas. *The Projector*. London: n.p., 1623.

Scott, Walter, ed. *A Collection of Scarce and Valuable Tracts*, Vol. 2. London: Cadell and Davies, 1809.

Selden, John. *Titles of Honor*. London: Stansby, 1614.

———. *Titles of Honor*. London: Stansby, 1631.

Shirley, James. *Triumph of Peace*. London: Norton, 1634.

A Short and True Relation Concerning the Soap-Busines. Containing the Severall Patents, Proclamations, Orders, Whereby the Soape-Makers of London, and Other His Majesties Subjects, Were Damnified, by the Gentlemen That Were the Patentees for Soape at Westminster, with the Particular Proceedings Concerning the Same. London: Bourne, 1641.

Smith, John. *Captain John Smith: A Select Edition of His Writings*. Edited by Karen Ordahl Kupperman. Chapel Hill: University of North Carolina Press, 2012.

———. *The Generall Historie of Virginia*. London: Haviland, 1626.

St. John, John, ed. "Norden's Project for the Improving Some of His Majesty's Forests, Parks, Chases and Wastes, Presented to Sir Julius Caesar." In *Observations on the Land Revenue of the Crown*, 289–94. London: Debrett, 1792.

Stow, John. *Survey of London*. London: Bourne, 1633.

Sturtevant, Simon. *Metallica*. London: Eld, 1612.

Symonds, William. *Virginia*. London: Windet, 1609.

Terry, Edward. *A Voyage to East-India*. London: Martin and Allestrye, 1655.

Thackston, W. M., ed. *The Jahangirnama: Memoirs of Jahangir, Emperor of India*. New York: Oxford University Press, 1999.

Thomas, Anthony. *The Propositions of Sir Anthony Thomas, Knight, and Iohn Worsop, Esquire for making of the bargaine with the Country, and Henry Briggs, Professor of the Mathematicks in the Vniuersitie of Oxford, Heldebrand Pruson, Citizen and Salter of London, and Cornelius Drible, Engeneere with the rest of the Undertakers for the drayning of the Levell within the six Counties of Norfolke, Suffolke, Cambridge, Isle of Elie, Huntington, North-hampton and Lincolne-shire*. London: N.A., 1629.

Tradescant, John. *Musaeum Tradescantium or, A collection of rarities preserved at South-Lambeth neer London*. London: Printed by John Grismond, 1656.

Trésor Politique, Divisé En Trois Livres; Contentant Les Relations, Instructions, Traictez, et Divers Discours Appartenans à La Parfaicte Intelligence de La RAISON D'ESTAT, & de Tres-Grande Importance à l'entiere Cognoissance Des Interests, Pretentions, Desseins, & Reuenus, Des plus Grands Princes & Seigneurs Du Monde. Paris: Nicolas du Fossé, 1608.

Virginia Company of London. *A True Declaration of the estate of the Colonie in Virginia*. London: Barret, 1610.

———. *A Declaration of the State of the Colonie and Affaires of Virginia: With the Names of the Adventurors, and Summes Adventured in That Action*. London: T. S. 1620.

Warwick, Philip. *Memoires of the Reign of King Charles I*. London: Chiswell, 1701.

Watts, William. "To the venerable Artists and younger Students in Divinity." In *The Strange And Dangerous Voyage Of Captaine Thomas Iames, in his intended Discouery of the Northwest Passage into the South Sea . . . And an Aduise concerning the Philosophy of these late Discouereyes*. London: Legatt, 1633.

Whately, William. *A Caveat For The Couetous*. London: Man, 1609.

Whitelocke, Bulstrode. *Memorials of the English Affairs*. London: Nathaniel Ponder, 1682.

———. *The Whitelocke Papers at Longleat House*, Vol. 12. Microform Academic Publishers.

Williams, Edward. *Virginia's Discovery of Silke-Wormes*. London: Stephenson, 1650.

Williams, Edward, and John Ferrar. *Virgo Triumphans, or Virginia in Generall*. London: Harper, 1650.

Wither, George. "To the Honorable Sir John Danvers Knight, Governor of the Sommer Islands Company." In *Josias Foster, Copy of a Petition from the Governor and Company of the Sommer Islands*, 26–30. London: Husband, 1651.

W.M. *The Queens closet opened incomparable secrets in physick, chyrurgery, preserving, and candying &c*. London: Brooke, 1659.

Woodcroft, Bennet. *Titles of Patents of Invention, Chronologically Arranged*, Part I. London: Eyre and Spottiswoode, 1854.

———. *Appendix to Reference Index of Patents of Invention*. London: Patent Office, 1855.

———. *Subject-Matter Index of Patents, Part II*. London: Eyre and Spottiswoode, 1857.

————. "A Commission Directed to Sir Richard Wynne, Sir Thomas Hatton, Sir Henry Spiller, and Lawrence Whitaker Esq. to Enquire upon Oath Whether Nicholas Page, Clerk, or Sir Nicholas Halse Was the First Inventor of Certaine Kilns for the Drying of Malt, Dated 2nd June 1637." In *Supplement to the Series of Letters Patent*, 53–54. London: Spottiswoode, 1858.

————. *Reference Index of Patents of Invention*. London: Patent Office, 1862.

Worm, Ole. *Epistolae*, Vol. 2. Copenhagen: n.p., 1751.

Woude, Cornelis van der. *Kronijcke Van Alckmaer*. Alkmaar: Breken-Geest, 1645.

Zeiller, Martin. *Handbuch von Allerley Nützlichen Erinneringen*. Ulm: Balthasar Kühn, 1655.

Secondary Sources

Aaron, Melissa. *Global Economics: A History of the Theater Business, the Chamberlain's/ King's Men, and Their Plays, 1599–1642*. Newark: University of Delaware Press, 2005.

Alberts, Lindsay. "Francesco I's Museum: Cultural Politics at the Galleria Degli Uffizi." *Journal of the History of Collections* 30, no. 2 (2018): 203–16.

Amundsen, Karin. "Thinking Metallurgically: Metals and Empire in the Projects of Edward Hayes." *Huntington Library Quarterly* 79, no. 4 (2016): 561–90.

Andrews, Kenneth R. *The Spanish Caribbean: Trade and Plunder, 1530–1630*. New Haven, CT: Yale University Press, 1978.

————. *Trade, Plunder and Settlement: Maritime Enterprise and the Genesis of the British Empire, 1480–1630*. Cambridge: Cambridge University Press, 1984.

————. *Ships, Money and Politics: Seafaring and Naval Enterprise in the Reign of Charles I*. Cambridge: Cambridge University Press, 1991.

Ash, Eric. *Power, Knowledge, and Expertise in Elizabethan England*. Baltimore: Johns Hopkins University Press, 2004.

————. *The Draining of the Fens: Projectors, Popular Politics, and State Building in Early Modern England*. Baltimore: Johns Hopkins University Press, 2017.

————. "By Any Other Name: Early Modern Expertise and the Problem of Anachronism." *History and Technology* 35, no. 1 (2019): 3–30.

Asher, Catherine Ella Blanchard. *Architecture of Mughal India*, Part 1. Vol. 4. Cambridge: Cambridge University Press, 1992.

Aylmer, G. E. *The King's Servants: The Civil Service of Charles I, 1625–1642*. London: Routledge & Kegan Paul, 1961.

Bäcklund, Jan. "In the Footsteps of Edward Kelley." In *John Dee: Interdisciplinary Studies in English Renaissance Thought*, edited by Stephen Clucas, 295–330. Dordrecht: Springer, 2006.

Baldassarri, Fabrizio, and Oana Matei. "Manipulating Flora: Seventeenth-Century Botanical Practices and Natural Philosophy; Introduction." *Early Science and Medicine* 23, no. 5–6 (2018): 413–19.

Bauer, Leonhard. "Zur entwicklung des 'homo oeconomicus.'" In *Von der Glückseligkeit des Staates: Staat, Wirtschaft, und Gesellschaft in Österreich im Zeitalter des aufgeklärten Absolutismus*, edited by Herbert Matis and Leonhard Bauer, 39–76. Berlin: Duncker & Humblot, 1981.

Bauer, Ralph, and Marcy Norton. "Introduction: Entangled Trajectories: Indigenous and European Histories." *Colonial Latin American Review* 26, no.1 (2017): 1–17.

Beach, Milo Cleveland. "The Gulshan Album and Its European Sources." *Bulletin of the Museum of Fine Arts* 63, no. 332 (1965): 63–91.

———. "The Mughal Painter Abu'l Hasan and Some English Sources for His Style." *The Journal of the Walters Art Gallery* 38 (1980): 6–33.

Beach, P. "European Painting and Mughal Miniatures." In *Intercultural Encounters in Mughal Miniatures*, edited by Khalid Anis Ahmed, 29–32. Lahore: National College of Arts, 1995.

Beadle, Richard. "The Virtuoso's Troilus," In *Chaucer Traditions: Studies in Honour of Derek Brewer*, edited by Ruth Morse and Barry Windeatt, 213–33. Cambridge: Cambridge University Press, 1990.

Bellany, Alastair. "Singing Libel in Early Stuart England: The Case of the Staines Fiddlers, 1627." *The Huntington Library Quarterly* 69, no. 1 (2006): 177–94.

———. "'Naught But Illusion'? Buckingham's Painted Selves." In *Writing Lives: Biography and Textuality, Identity and Representation in Early Modern England*, edited by Kevin Sharpe and Steven Zwicker, 127–60. Oxford: Oxford University Press, 2008.

Bennett, Kate. "John Aubrey, Hint-Keeper: Life-Writing and the Encouragement of Natural Philosophy in the Pre-Newtonian Seventeenth Century." *The Seventeenth Century* 22, no. 2 (2007): 358–80.

———. "John Aubrey and the Printed Book." *The Huntington Library Quarterly* 76, no. 3 (2013): 393–411.

Bentley, Gerald Eades. "The Proposed Amphitheatre." In *The Jacobean and Caroline Stage*, Vol. 6, 291–304. London: Oxford University Press, 1968.

Biagioli, Mario. *Galileo, Courtier: The Practice of Science in the Culture of Absolutism*. Chicago: University of Chicago Press, 1993.

———. "Etiquette, Interdependence, and Sociability in Seventeenth-Century Science." *Critical Inquiry* 22, no. 2 (1996): 193–238.

———. "Patent Republic: Representing Inventions, Constructing Rights and Authors," *Social Research* 73, no. 4 (2006): 1129–72.

Bigelow, Allison Margaret. "Gendered Language and the Science of Colonial Silk." *Early American Literature* 49, no. 2 (2014): 271–325.

———. "Colonial Industry and the Language of Empire: Silkworms in the Virginia Colony, 1607–1655." In *European Empires in the American South: Colonial and Environmental Encounters*, edited by Joseph P. Ward, 8–36. Jackson: University Press of Mississippi, 2017.

Blackburn, Thomas H. "Edmund Bolton's The Cabanet Royal: A Belated Reply to Sidney's Apology for Poetry." *Studies in the Renaissance* 14 (1967): 159–71.

Britland, Karen. "Henry Killigrew and Dramatic Patronage at the Stuart Courts." In *Thomas Killigrew and the Seventeenth-Century English Stage: New Perspectives*, edited by Philip Major, 91–112. New York: Routledge, 2016.

Bod, Rens, Bart Karstens, Jeroen van Dongen, Emma Mojet, and Sjang L. ten Hagen. "The Flow of Cognitive Goods: A Historiographical Framework for the Study of Epistemic Transfer." *Isis* 110, no. 3 (2019): 483–96.

Borkenau, Franz. "Zur Soziologie des mechanistischen Weltbildes." *Zeitschrift Für Sozialforschung* 1, no. 3 (1932): 311–35.

———. *Der Übergang vom feudalen zum bürgerlichen Weltbild: Studien zur Geschichte der Philosophie der Manufakturperiode*. Paris: Alcan, 1934.

————. "The Sociology of the Mechanistic World-Picture." Translated by Richard Hadden. *Science in Context* 1, no. 1 (1987): 109–27.

Borrelli, Arianna. "The Weatherglass and its Observers in the Early Seventeenth Century." In *Philosophies of Technology: Francis Bacon and His Contemporaries*, edited by Claus Zittel, Romano Nanni, Gisela Engel, and Nicole Karafyllis, 67–130. Leiden: Brill, 2008.

Braddick, Michael and John Walter. *Negotiating Power in Early Modern Society: Order, Hierarchy, and Subordination in Britain and Ireland*. Cambridge: Cambridge University Press, 2001.

Brendecke, Arndt. *The Empirical Empire: Spanish Colonial Rule and the Politics of Knowledge*. Berlin: De Gruyter, 2016.

Brenner, Robert. *Merchants and Revolution: Commercial Change, Political Conflict, and London's Overseas Traders, 1550–1653*. Princeton, NJ: Princeton University Press, 1993.

Brooke, Christopher. *Philosophic Pride: Stoicism and Political Thought from Lipsius to Rousseau*. Princeton, NJ: Princeton University Press, 2012.

Brown, Ruth. "'A Jewel of Great Value': English Iron Gunfounding and Its Rivals, 1550–1650." In *Ships and Guns: The Sea Ordnance in Venice and in Europe between the 15th and the 17th Centuries*, edited by Carlo Beltrame and Renato Gianni Ridella, 98–105. Oxford: Oxbow Books, 2011.

Brunsman, Denver Alexander. *The Evil Necessity: British Naval Impressment in the Eighteenth-Century Atlantic World*. Charlottesville: University of Virginia Press, 2013.

Bull, Stephen. "Pearls from the Dungheap: English Saltpetre Production 1590–1640." *Journal of the Ordnance Society* 2 (1990): 5–10.

————. *"The Furie of the Ordnance": Artillery in the English Civil Wars*. Woodbridge: Boydell, 2008.

Buning, Marius. "Inventing Scientific Method: The Privilege System as a Model for Scientific Knowledge-Production." *Intellectual History Review* 24, no. 1 (2014): 59–70.

Burckhardt, Jacob. *Die Cultur der Renaissance in Italien*. Basel: Schweighauser, 1860.

Burke, Peter. "The Renaissance Translator as Go-Between." In *Renaissance Go-Betweens: Cultural Exchange in Early Modern Europe*, edited by Andreas Höfele and Werner von Koppenfels, 17–31. Munich: De Gruyter, 2011.

Burke, Peter, and Ronnie Po-chia Hsia. *Cultural Translation in Early Modern Europe*. Cambridge: Cambridge University Press, 2007.

Bushnell, Rebecca. *Green Desire: Imagining Early Modern English Gardens*. Ithaca, NY: Cornell University Press, 2003.

Carey, Daniel. "Inquiries, Heads, and Directions: Orienting Early Modern Travel." In *Travel Narratives, the New Science, and Literary Discourse, 1569–1750*, edited by Judy Hayden, 25–52. New York: Routledge, 2012.

Cavert, William M. *The Smoke of London: Energy and Environment in the Early Modern City*. Cambridge: Cambridge University Press, 2016.

Chaudhuri, Kirti N. *The English East India Company: The Study of an Early Joint-Stock Company, 1600–1640*. New York: Augustus M. Kelley, 1965.

Clark, Alice. *Working Life of Women in the Seventeenth Century*. London: Routledge, 1919.

Clegg, Cyndia Susan. *Press Censorship in Jacobean England*. Cambridge: Cambridge University Press, 2001.

Clow, Archibald, and Nan L. Clow. "The Natural and Economic History of Kelp." *Annals of Science* 5, no. 4 (1947): 297–316.

Clucas, Stephen. "Poetic Atomism in Seventeenth-Century England: Henry More, Thomas Traherne and Scientific Imagination." *Renaissance Studies* 5, no. 3 (1991): 327–40.

Coffey, Donna. "'As in a Theatre': Scientific Spectacle in Bacon's New Atlantis." *Science as Culture* 13, no. 2 (2004): 259–90.

Cogswell, Thomas. "'The Symptomes and Vapors of a Diseased Time': The Earl of Clare and Early Stuart Manuscript Culture." *The Review of English Studies* 57 (2006): 310–36.

Cohen, Adam Max. "The Nimble Gunner and the Versatile Prince: Agility and the Early Modern Military Revolution." In *Technology and the Early Modern Self,* 115–34. New York: Palgrave Macmillan, 2009.

Colie, Rosalie L. "Cornelis Drebbel and Salomon de Caus: Two Jacobean Models for Salomon's House." *The Huntington Library Quarterly* 18, no. 3 (1955): 245–60.

———. *"Some Thankfulnesse to Constantine": A Study of English Influence upon the Early Works of Constantijn Huygens.* The Hague: Nijhoff, 1956.

Connor, Steven. *The Madness of Knowledge: On Wisdom, Ignorance and Fantasies of Knowing.* London: Reaktion Books, 2019.

Cook, Harold. *Matters of Exchange: Commerce, Medicine, and Science in the Dutch Golden Age.* New Haven, CT: Yale University Press, 2007.

Cook, Minnie. "Governor Samuel Mathews, Junior." *The William and Mary Quarterly* 14 (1934): 105–13.

Cormack, Lesley B. "Twisting the Lion's Tail: Practice and Theory at the Court of Henry Prince of Wales." In *Patronage and Institutions,* edited by Bruce Moran, 67–83. Rochester: Boydell, 1991.

Cowen, David L. "Boom and Bust: Sassafras." *Apothecary's Cabinet* 8 (2004): 9.

Cramsie, John. *Kingship and Crown Finance under James VI and I, 1603–1625.* Woodbridge: Boydell Press, 2002.

Craven, W. Frank. "The Earl of Warwick, A Speculator in Piracy." *The Hispanic American Historical Review* 10, no. 4 (1930): 457–79.

Cressy, David. *Dangerous Talk: Scandalous, Seditious and Treasonable Speech in Pre-Modern England.* Oxford: Oxford University Press, 2010.

———. "Saltpetre, State Security and Vexation in early modern England." *Past & Present* 212 (2011): 73–111.

———. *Saltpeter: The Mother of Gunpowder.* Oxford: Oxford University Press, 2013.

———. "A Mathematical Petitioner." In *Charles I and the People of England,* 186–90. Oxford: Oxford University Press, 2015.

Cruyningen, Piet van. "Dealing with Drainage: State Regulation of Drainage Projects in the Dutch Republic, France, and England during the Sixteenth and Seventeenth Centuries." *Economic History Review* 68, no. 2 (2015): 420–40.

Cust, Richard. *Charles I and the Aristocracy, 1625–1642.* Cambridge: Cambridge University Press, 2013.

———. "Charles I's Noble Academy." *The Seventeenth Century* 29, no. 4 (2014): 337–57.

Cuttica, Cesare. "Sir Francis Kynaston: The Importance of the 'Nation' for a 17th-Century English Royalist." *History of European Ideas* 32, no. 2 (2006), 139–61.

Dacos, Nicole. *La Découverte de La Domus Aurea et La Formations Des Grotesques à La Renaissance*. London: Warburg Insititute, 1969.

Davidson, Alan and Rosemary Sgroi, "Brooke, Sir John (1575–1669)." In *The History of Parliament*, Vol. 3. Edited by Andrew Thrush and John P. Ferris, 333–34. Cambridge: Cambridge University Press, 2010.

Dantzig, Albert van. *Forts and Castles of Ghana*. Accra: Sedco, 1999.

Darley, Gillian. *John Evelyn: Living for Ingenuity*. New Haven, CT: Yale University Press, 2006.

Daston, Lorraine. "The Nature of Nature in Early Modern Europe." *Configurations* 6, no. 2 (1998): 149–72.

Daston, Lorraine and Katharine Park. *Wonders and the Order of Nature, 1150–1750*. New York: Zone Books, 1998.

Davids, Karel. "Amsterdam as a Centre of Learning in the Dutch Golden Age, c. 1580–1700." In *Urban Achievement in Early Modern Europe: Golden Ages in Antwerp, Amsterdam and London*, edited by Patrick O'Brien, Derek Keene, Marjolein 't Hart, and Herman van der Wee, 305–25. Cambridge: Cambridge University Press, 2001.

Davies, Randall. *Chelsea Old Church*. London: Duckworth, 1904.

Dear, Peter. *Discipline & Experience: The Mathematical Way in the Scientific Revolution*. Science and its Conceptual Foundations. Chicago: University of Chicago Press, 1995.

———. "A Mechanical Microcosm: Bodily Passions, Good Manners and Cartesian Mechanism." In *Science Incarnate: Historical Embodiments of Natural Knowledge*, edited by Christopher Lawrence and Steven Shapin, 51–82. Chicago: University of Chicago Press, 1998.

———. "Mysteries of State, Mysteries of Nature: Authority, Knowledge and Expertise in the Seventeenth century." In *States of Knowledge: The Co-production of Science and Social Order*, edited by Sheila Jasanoff, 206–24. London: Routledge, 2004.

———. "The Meaning of Experience." In *The Cambridge History of Science*, edited by K. Park and L. Daston, 106–31. Cambridge: Cambridge University Press, 2006.

Deringer, William. *Calculated Values: Finance, Politics, and the Quantitative Age*. Cambridge, MA: Harvard University Press, 2018.

Descendre, Romain. *L'État Du Monde: Giovanni Botero Entre Raison d'État et Géopolitique*. Geneva: Droz, 2009.

Dijksterhuis, Fokko Jan. "Magi from the North. Instruments of Fire and Light in the Early Seventeenth Century." In *The Optics of Giambattista Della Porta (ca. 1535–1615): A Reassessment*, edited by Arianna Borelli, Giora Hon, and Yaakov Zik, 125–43. Cham: Springer, 2017.

———. "Harnessing the Elements: Beeckman and Atmosphere Instruments." In *Knowledge and Culture in the Early Dutch Republic: Isaac Beeckman in Context*, edited by Klaas van Berkel, Albert Clement and Arjan van Dixhoorn, 369–92. Amsterdam: Amsterdam University Press, 2022.

DiMeo, Michelle. *Lady Ranelagh: The Incomparable Life of Robert Boyle's Sister*. Chicago: University of Chicago Press, 2021.

Doering, Oscar. *Des Augsburger Patriciers Philip Hainhofer Reisen Nach Innsbruck und Dresden*. Vienna: Graeser, 1901.

Dolby, R. G. A. *Uncertain Knowledge: An Image of Science for a Changing World*. New York: Cambridge University Press, 1996.

Donagan, Barbara. "Halcyon Days and the Literature of War: England's Military Education before 1642." *Past & Present* 147, no. 1 (1995): 65–100.

Dooley, Brendan. "Making It Present." In *The Dissemination of News and the Emergence of Contemporaneity in Early Modern Europe*, edited by Brendan Dooley, 95–114. Farnham: Ashgate, 2010.

Doorman, Gerard. *Octrooien Voor Uitvindingen in de Nederlanden uit de 16e-18e Eeuw.* The Hague: Nijhoff, 1940.

Drayton, Richard Harry. *Nature's Government: Science, Imperial Britain, and the "Improvement" of the World.* New Haven, CT: Yale University Press, 2000.

Eamon, William. *Science and the Secrets of Nature: Books of Secrets in Medieval and Early Modern Culture.* Princeton, NJ: Princeton University Press, 1996.

Edmunds, Lewis. *Law and Practice of Letters Patent for Inventions.* London: Stevens, 1897.

Egan, Geoff. "Evidence for Early Seventeenth-Century Glass Bead-Making from Hammersmith, West London." *Post-Medieval Archaeology* 42, no. 2 (2008): 332–33.

Eggert, Katherine. *Disknowledge: Literature, Alchemy, and the End of Humanism in Renaissance England.* Philadelphia: University of Pennsylvania Press, 2015.

Elias, Norbert. *The Civilizing Process*, Vol. 1. Oxford: Blackwell, 1969.

———. *The Civilizing Process*, Vol. 2. Oxford: Blackwell, 1982.

Emery, Clark. "A Further Note on Drebbel's Submarine." *Modern Language Notes* 57, no. 6 (1942): 421–25.

Ettinghausen, Richard. "The Emperor's Choice." In *De Artibus Opuscula-Essays in Honor of Erwin Panofsky*, edited by M. Meiss, 98–120. New York: New York University Press, 1961.

Faroqui, Suraiya, Donald Quataert, Bruce McGowan, and Sevket Pamu. *An Economic and Social History of the Ottoman Empire, 1600–1914*, Vol. 2. Cambridge: Cambridge University Press, 1994.

Farrington, Benjamin. *Francis Bacon: Philosopher of Industrial Science.* New York: Schuman, 1949.

———. *Francis Bacon: Pioneer of Planned Science.* New York: Praeger, 1963.

Feingold, Mordechai. *The Mathematicians' Apprenticeship: Science, Universities and Society in England, 1560–1640.* Cambridge: Cambridge University Press, 1984.

———. "'And Knowledge Shall Be Increased': Millenarianism and the Advancement of Learning Revisited." *The Seventeenth Century* 28, no. 4 (2013): 363–93.

———. "'Experimental Philosophy': Invention and Rebirth of a Seventeenth-Century Concept." *Early Science and Medicine* 21, no. 1 (2016): 1–28.

Finkelstein, Andrea. *Harmony and the Balance: An Intellectual History of Seventeenth-Century English Economic Thought.* Ann Arbor: University of Michigan Press, 2000.

Fischer, Hubertus, Volker R. Remmert, and Joachim Wolschke-Bulmahn. *Gardens, Knowledge and the Sciences in the Early Modern Period.* Cham: Springer, 2016.

Fitzmaurice, Andrew. *Humanism and America: An Intellectual History of English Colonisation, 1500–1625.* Cambridge: Cambridge University Press, 2003.

———. "The Commercial Ideology of Colonization in Jacobean England: Robert Johnson, Giovanni Botero, and the Pursuit of Greatness." *The William and Mary Quarterly* 64, no. 4 (2007): 791–820.

———. "The Dutch Empire in Intellectual History." *BMGN - Low Countries Historical Review* 132, no. 2 (2017): 97–109.

Forman, Paul. "On the Historical Forms of Knowledge Production and Curation: Modernity Entailed Disciplinarity, Postmodernity Entails Antidisciplinarity." *Osiris* 27 (2012): 56–97.

Foster, William. "An English Settlement in Madagascar in 1645–6." *The English Historical Review* 27, no. 106 (1912): 239–50.

Foucault, Michel. *Surveiller et punir: naissance de la prison.* Bibliothèque des histoires. Paris: Gallimard, 1975.

French, H. R. *The Middle Sort of People in Provincial England, 1600–1750.* Oxford: Oxford University Press, 2007.

Freudenthal, Gideon and Peter McLaughlin ed. and trans. *The Social and Economic Roots of the Scientific Revolution: Texts by Boris Hessen and Henryk Grossman.* Boston: Springer, 2009.

Friedeburg, Robert von, and John Morrill, eds. *Monarchy Transformed: Princes and their Elites in Early Modern Western Europe.* Cambridge: Cambridge University Press, 2017.

Fulton, Thomas Wemyss. *The Sovereignty of the Sea.* London: Blackwood, 1911.

Gabbey, Alan. "The Mechanical Philosophy and Its Problems: Mechanical Explanations, Impenetrability, and Perpetual Motion." In *Change and Progress in Modern Science,* edited by Joseph C. Pitt, 9–84. Dordrecht: D. Reidel, 1985.

Games, Alison. *The Web of Empire: English Cosmopolitans in an Age of Expansion, 1560–1660.* Oxford: Oxford University Press, 2008.

Gaukroger, Stephen. *Francis Bacon and the Transformation of Early-Modern Philosophy.* Cambridge: Cambridge University Press, 2001.

Gethyn-Jones, Eric. *George Thorpe and the Berkeley Company: A Gloucestershire Enterprise in Virginia.* Gloucester: Sutton, 1982.

Ginzburg, Carlo. "Clues: Roots of a Scientific Paradigm." *Theory and Society* 7, no. 3 (1979): 273–88.

———. "Morelli, Freud and Sherlock Holmes: Clues and Scientific Method." *History Workshop Journal* 9, no. 1 (1980): 5–36.

Goldgar, Anne. *Tulipmania: Money, Honor, and Knowledge in the Dutch Golden Age.* Chicago: University of Chicago Press, 2007.

Gomme, Andor and Alison Maguire. *Design and Plan in the Country House: From Castle Donjons to Palladian Boxes.* New Haven, CT: Yale University Press, 2008.

Gough, John Wiedhofft. *Sir Hugh Myddelton, Entrepreneur and Engineer.* New York: Oxford University Press, 1964.

———. *The Rise of the Entrepreneur.* New York: Schocken Books, 1970.

Gover, J. E. B. *The Place-Names of Middlesex.* Cambridge: Cambridge University Press, 1942.

Govier, Mark. "The Royal Society, Slavery and the Island of Jamaica: 1660–1700." *Notes and Records of the Royal Society of London* 53, no. 2 (1999): 203–17.

Grant, Edward. *Planets, Stars, and Orbs: The Medieval Cosmos, 1200–1687.* Cambridge: Cambridge University Press, 1994.

Gregg, Pauline. *King Charles I.* Berkeley: University of California Press, 1984.

Grigson, Caroline. *Menagerie: The History of Exotic Animals in England, 1100–1837.* Oxford: Oxford University Press, 2016.

Guillory, John. "The Bachelor State: Philosophy and Sovereignty in Bacon's New Atlantis." In *Politics and the Passions, 1500–1850,* edited by Victoria Kahn, Neil

Saccamano, and Daniela Coli, 49–74. Princeton, NJ: Princeton University Press, 2006.

Hadden, Richard. "Social Relations and the Content of Early Modern Science." *The British Journal of Sociology* 39, no. 2 (1988): 255–80.

———. *On the Shoulders of Merchants: Exchange and the Mathematical Conception of Nature in Early Modern Europe.* Albany: State University of New York Press, 1994.

Hammersley, George. *Daniel Hechstetter the Younger, Memorabilia and Letters, 1600–1639.* Wiesbaden: Steiner, 1988.

Hanna, Mark G. *Pirate Nests and the Rise of the British Empire, 1570–1740.* Chapel Hill: The University of North Carolina Press, 2015.

Harkness, Deborah E. *The Jewel House: Elizabethan London and the Scientific Revolution.* New Haven, CT: Yale University Press, 2007.

———. "Accounting for Science: How a Merchant Kept His Books in Elizabethan London." In *The Self-Perception of Early Modern Capitalists*, edited by Margaret C. Jacob and Catherine Secretan, 205–28. New York: Palgrave Macmillan, 2008.

Harrison, Peter. *The Bible, Protestantism, and the Rise of Natural Science.* Cambridge: Cambridge University Press, 1998.

Haskell, Alexander B. *For God, King & People: Forging Commonwealth Bonds in Renaissance Virginia.* Chapel Hill: University of North Carolina Press, 2017.

Haycock, David Boyd. "Living Forever in Early Modern Europe: Sir Francis Bacon and the Project for Immortality." In *The Age of Projects*, edited by Maximillian E. Novak, 166–84. Toronto: University of Toronto Press, 2008.

Heal, Felicity and Clive Holmes. *The Gentry in England and Wales, 1500–1700.* Stanford, CA: Stanford University Press, 1994.

Healy, Simon. "CHALONER, Sir Thomas (?1564–1615), of Richmond Palace, Surr., Steeple Claydon, Bucks. and Clerkenwell, Mdx." In *The History of Parliament: The House of Commons 1604–1629*, Vol. 3, edited by Andrew Thrush and John P. Ferris, 485–88. Cambridge: Cambridge University Press, 2010.

Hebb, David. *Piracy and the English Government, 1616–1642.* New York: Routledge, 1994.

Heesen, Anke te. "Accounting for the Natural World: Double-Entry Bookkeeping in the Field." In *Colonial Botany: Science, Commerce, and Politics in the Early Modern World*, edited by Londa Schiebinger and Claudia Swan, 237–51. Philadelphia: University of Pennsylvania Press, 2005.

Henry, John. *Knowledge Is Power: How Magic, the Government and an Apocalyptic Vision Inspired Francis Bacon to Create Modern Science.* Cambridge: Icon Books, 2003.

Heringman, Noah. *Sciences of Antiquity: Romantic Antiquarianism, Natural History, and Knowledge Work.* Oxford: Oxford University Press, 2013.

Heuvel, Danielle van den. *Women and Entrepreneurship: Female Traders in the Northern Netherlands, c. 1560–1815.* Amsterdam: Aksant, 2007.

Heyd, Michael. *Be Sober and Reasonable: The Critique of Enthusiasm in the Seventeenth and Early Eighteenth Centuries.* Leiden: Brill, 1995.

Higton, Hester. "Portrait of an Instrument-Maker: Wenceslaus Hollar's Engraving of Elias Allen." *The British Journal for the History of Science* 37, no. 2 (2004): 147–66.

Hilaire-Pérez, Liliane. "Patents et Privilèges (Angleterre et France, XVIIe et XVIIIe Siècles)." In *Les Projets: Une Histoire Politique (XVIe-XXIe Siècles)*, edited by Frédéric Graber and Martin Giraudeau, 137–48. Paris: Presses des Mines, 2018.

Hill, Christopher. *The Intellectual Origins of the English Revolution*. Oxford: Clarendon Press, 1965.

Hill, Katherine. "'Juglers or Schollers?': Negotiating the Role of a Mathematical Practitioner." *The British Journal for the History of Science* 31, no. 3 (1998): 253–74.

Hille, Christiane. *Visions of the Courtly Body: The Patronage of George Villiers, First Duke of Buckingham, and the Triumph of Painting at the Stuart Court*. Berlin: Akademie Verlag, 2012.

Hindle, Steve. *The State and Social Change in Early Modern England, c. 1550–1640*. Houndmills: Palgrave, 2000.

Hirschman, Albert O. *The Passions and the Interests: Political Arguments for Capitalism before Its Triumph*. Princeton, NJ: Princeton University Press, 1977.

Hodgetts, E. A. Brayley. *The Rise and Progress of the British Explosives Industry*. London: Whittaker, 1909.

Hotson, Howard. *Commonplace Learning: Ramism and Its German Ramifications, 1543–1630*. Oxford: Oxford University Press, 2007.

———. *The Reformation of Common Learning: Post-Ramist Method and the Reception of the New Philosophy, 1618–c.1670*. Oxford: Oxford University Press, 2021.

Hotson, Leslie. *I, William Shakespeare Do Appoint Thomas Russell, Esquire*. New York: Oxford University Press, 1938.

———. "The Projected Amphitheatre." *Shakespeare Survey* 2 (1949): 24–35.

House, Anthony Paul. "The City of London and the Problem of the Liberties, c1540–c1640." DPhil diss., University of Oxford, 2006.

Hoyle, Richard. "Disafforestation and Drainage: The Crown as Entrepreneur?" In *The Estates of the English Crown, 1558–1640*, edited by R. W. Hoyle, 353–89. Cambridge: Cambridge University Press, 1992.

Hunneyball, Paul. "Russell, Thomas (fl. 1613–1623) of Truro, Cornw. and London." In *The History of Parliament: The House of Commons, 1604–1629*, Vol. 6, edited by Andrew Thrush and John P. Ferris, 115–16. Cambridge: Cambridge University Press, 2010.

Hunt, John Dixon. *Garden and Grove: The Italian Renaissance Garden and the English Imagination 1600–1750*. London: Dent, 1986.

Hunt, Arnold. "Book Trade Patents, 1603–1640." In *The Book Trade and its Customers, 1450–1900: Historical Essays for Robin Myers*, edited by Arnold Hunt, Giles Mandelbrote, and Alison Shell, 27–54. New Castle: Oak Knoll Press, 1997.

Hunter, Michael. *The Royal Society and Its Fellows, 1660–1700: The Morphology of an Early Scientific Institution*. Oxford: Alden, 1994.

Iliffe, Rob. "Capitalizing Expertise: Philosophical and Artisan Expertise in Early Moderne London, 1650–1750." In *Fields of Expertise: A Comparative History of Expert Procedures in Paris and London, 1600 to Present*, edited by Christelle Rabier, 55–84. Newcastle: Cambridge Scholars Publishing, 2007.

Irving, Sarah. *Natural Science and the Origins of the British Empire*. London: Pickering & Chatto, 2008.

Jalobeanu, Dana. *The Art of Experimental Natural History: Francis Bacon in Context* (Bucharest: Zeta, 2015).

———. "Bacon's Apples: A Case Study in Baconian Experimentation." In *Francis Bacon on Motion and Power*, edited by Guido Giglioni, James Lancaster, Sorana Corneanu, and Dana Jalobeanu, 83–114. Cham: Springer, 2016.

———. "Disciplining Experience: Francis Bacon's Experimental Series and the Art of Experimenting." *Perspectives on Science* 24, no. 3 (2016): 324–42.

Jalobeanu, Dana, and Oana Matei. "Treating Plants as Laboratories: A Chemical Natural History of Vegetation in 17th-century England." *Centaurus* 62, no. 3 (2020): 542–61.

Jardine, Boris. "Instruments of Statecraft: Humphrey Cole, Elizabethan Economic Policy and the Rise of Practical Mathematics." *Annals of Science* 75, no. 4 (2018): 304–29.

Jardine, Lisa. *Francis Bacon: Discovery and the Art of Discourse.* Cambridge: Cambridge University Press, 1974.

Jardine, Lisa, and Anthony Grafton. "'Studied for Action': How Gabriel Harvey Read His Livy." *Past & Present* 129 (1990): 30–78.

Jasanoff, Sheila. "Citizens At Risk: Cultures of Modernity in the US and EU." *Science as Culture* 11, no. 3 (2002): 363–80.

Jenkins, Rhys. "Paper-Making in England, 1588–1860." In *Library Association Record*, 577–88. London: Marshall, 1900.

———. "The Vauxhall Ordnance Factory of King Charles I." In *Links in the History of Engineering and Technology from Tudor Times: The Collected Papers of Rhys Jenkins*, 28–33. Cambridge: Newcomen Society, 1936.

Jenner, Mark. "'Another Epocha'? Samuel Hartlib, John Lanyon and the Cleansing of London." In *Samuel Hartlib and Universal Reformation*, edited by Mark Greengrass, Michael Leslie, and Timothy Raylor, 343–56. Cambridge: Cambridge University Press, 1994.

Jitschin, Adrian. "Family Background of Norbert Elias." *Figurations—Newsletter of the Norbert Elias Foundation* 40 (July 2013): 5.

Johns, Adrian. *Piracy: The Intellectual Property Wars from Gutenberg to Gates.* Chicago: University of Chicago Press, 2009.

Johnson, Paige. "Proof of the Heavenly Iris: The Fountain of Three Rainbows at Wilton House, Wiltshire." *Garden History* 35, no. 1 (2007): 51–67.

Johnson, Robert C. "The Transportation of Vagrant Children from London to Virginia, 1618–1622." In *Early Stuart Studies: Essays in Honor of David Harris Wilson*, edited by Howard S. Reinmuth, Jr., 137–151. Minneapolis: University of Minnesota Press, 1970.

Johnson, Stanley. "John Donne and the Virginia Company." *English Literary History* 14, no. 2 (1947): 127–38.

Jowitt, Claire. *The Culture of Piracy, 1580–1630: English Literature and Seaborne Crime.* Burlington: Ashgate, 2010.

———. "'To Sleep, Perchance to Dream': The Politics of Travel in the 1630s." *The Yearbook of English Studies* 44 (2014): 249–64.

Kamen, Henry. *Early Modern European Society.* London: Routledge, 2000.

Karklins, Karlis, Laure Dussubieux, and Ron G. V. Hancock. "A 17th Century Glass Bead Factory at Hammersmith Embankment, London, England." *Beads: Journal of the Society of Bead Researchers* 27 (2015): 16–24.

Kaufmann, Thomas DaCosta. *The Mastery of Nature: Aspects of Art, Science, and Humanism in the Renaissance.* Princeton, NJ: Princeton University Press, 1993.

Keating, Jessica. *Animating Empire: Automata, the Holy Roman Empire, and the Early Modern World.* University Park: The Pennsylvania State University Press, 2018.

Keblusek, Marika, and Badeloch Noldus. *Double Agents: Cultural and Political Brokerage in Early Modern Europe.* Leiden: Brill, 2011.

Keller, Alex. "The Age of Projectors." *History Today* 16, no. 7 (1966): 467–74.

Keller, Vera. "Cornelis Drebbel (1572–1633): Fame and the Making of Modernity." PhD diss., Princeton University, 2008.

———. "Drebbel's Living Instruments, Hartmann's Microcosm, and Libavius's Thelesmos: Epistemic Machines before Descartes." *History of Science* 48, no. 1 (2010): 39–74.

———. "Accounting for Invention: Guido Pancirolli's Lost and Found Things and the Development of Desiderata." *Journal of the History of Ideas* 73, no. 2 (2012): 223–45.

———. "The Authority of Practice in the Alchemy of Sir John Heydon (1588–1653)." *Ambix* 59, no. 3 (2012): 197–217.

———. "The 'Framing of a New World': Sir Balthazar Gerbier's 'Project for Establishing a New State in America,' ca. 1649." *The William and Mary Quarterly* 70, no. 1 (2013): 147–76.

———. *Knowledge and the Public Interest, 1575–1725.* New York: Cambridge University Press, 2015.

———. "Deprogramming Baconianism: The Meaning of *Desiderata* in the Eighteenth Century." *Notes and Records of the Royal Society of London* 72, no. 2 (2018): 119–37.

———. "Scarlet Letters: Sir Theodore de Mayerne and the Early Stuart Color World in the Royal Society." In *Archival Afterlives: Life, Death, and Knowledge-Making in Early Modern British Scientific and Medical Archives*, edited by Vera Keller, Anna Marie Roos, and Elizabeth Yale, 72–119. Leiden: Brill, 2018.

———. "Into the Unknown: Clues, Hints, and Projects in the History of Knowledge." *History and Theory* 59, no. 4 (2020): 86–110.

———. "Pennetrek: Sir Balthazar Gerbier (1592–1663) and the Calligraphic Aesthetics of Commercial Empire." In *Early Modern Knowledge Societies as Affective Economies*, edited by Inger Leemans and Anne Goldgar, 58–86. New York: Routledge, 2020.

———. "'Everything Depends Upon the Trial (Le tout gist à l'essay)': Four Manuscripts Between the Recipe and the Experimental Essay." In *Secrets of Craft and Nature in Renaissance France: A Digital Critical Edition and English Translation of BnF Ms. Fr. 640*, edited by Pamela H. Smith, Naomi Rosenkranz, Tianna Helena Uchacz, Tillmann Taape, Clément Godbarge, Sophie Pitman, Jenny Boulboullé, et al. New York: Making and Knowing Project, 2020. https://edition640.makingandknowing.org/#/essays/ann_320_ie_19.

———. "Happiness and Projects between London and Vienna: Wilhelm von Schröder on the London Weavers' Riot of 1675, Workhouses, and Technological Unemployment," *History of Political Economy* 53, no. 3 (2021): 407–23.

———. "A 'Wild Swing to Phantsy': The Philosophical Gardener and Emergent Experimental Philosophy in the Seventeenth-Century Atlantic World," *Isis* 112, no. 3 (2021): 507–30.

Keller, Vera, and Ted McCormick. "Towards a History of Projects." *Early Science and Medicine* 21, no. 5 (2016): 423–44.

Kelley, Donald. "The Problem of Knowledge and the Concept of Discipline." In *History and the Disciplines: The Reclassification of Knowledge in Early Modern Europe*, edited by Donald R. Kelley, 14–28. Rochester, NY: University of Rochester Press, 1997.

Kelly, Duncan. "The Politics of Intellectual History in Twentieth-Century Europe." In *Palgrave Advances in Intellectual History*, edited by R. Whatmore and B. Young, 210–30. Basingstoke: Palgrave, 2006.

Kennedy, Claire. "Those Who Stayed: English Chorography and the Elizabethan Society of Antiquaries." In *Motion and Knowledge in the Changing Early Modern World: Orbits, Routes and Vessels*, edited by Ofer Gal and Yi Zheng, 47–72. Dordrecht: Springer, 2014.

Kennedy, Mark E. "Charles I and Local Government: The Draining of the East and West Fens." *Albion* 15, no. 1 (1983): 19–31.

Kenny, Neil. *The Uses of Curiosity in Early Modern France and Germany*. Oxford: Oxford University Press, 2004.

Kirchner, Ernst Daniel Martin. *Das Schloss Boytzenburg und seine Besitzer*. Berlin: Duncker, 1860.

Kirk, R. E. G., and Ernest F. Kirk, eds. *Returns of Aliens Dwelling in the City and Suburbs of London, Part III, 1598–1625*. Aberdeen: University of Aberdeen Press, 1907.

Kirsch, Peter. *Fireships: The Terror Weapon of the Age of Sail*. Annapolis: Naval Institute Press, 2009.

Knowles, James. "Jonson's Entertainment at Britain's Burse." In *Re-Presenting Ben Jonson: Text, History, Performance*, edited by Martin Butler, 114–51. London: Palgrave Macmillan, 1999.

Koenigsberger, Helmut Georg. *Politicians and Virtuosi: Essays in Early Modern History*. London: Hambledon Press, 1986.

Kopperman, Paul. "Profile of Failure: The Carolana Project, 1629–1640." *The North Carolina Historical Review* 59, no. 1 (1982): 1–23.

Kostylo, Joanna. "From Gunpowder to Print: The Common Origins of Copyright and Patent." In *Privilege and Property: Essays on the History of Copyright*, edited by Ronan Deazley, Martin Kretschmer, and Lionel Bently, 21–50. Cambridge: Open Book Publishers, 2010.

Koyré, Alexandre. *From the Closed World to the Infinite Universe*. Baltimore: Johns Hopkins Press, 1957.

Kroyer, Peter. *The Story of Lindsey House, Chelsea*. London: Country Life, 1956.

Kupperman, Karen Ordahl. "How to Make a Successful Plantation: Colonial Experiment in America." In *Ireland: 1641, Contexts and Reactions*, edited by Micheál ó Siochrú and Jane Ohlmeyer, 219–235. Manchester: Manchester University Press, 2013.

———. *The Jamestown Project*. Cambridge, MA: Harvard University Press, 2007.

———. "The Love-Hate Relationship with Experts in the Early Modern Atlantic." *Early American Studies* 9, no. 2 (2011): 248–67.

———. *Pocahontas and the English Boys: Caught between Cultures in Early Virginia*. New York: University Press, 2019.

———. "The Puzzle of the American Climate in the early Colonial Period." *The American Historical Review* 87, no. 15 (1982): 1262–89.

Lancaster, Henry, and Ben Coates. "Howard, Sir Thomas (1587–1669), of Charlton Park, Charlton, Wilts.; Later of St. Martin-in-the-Fields, Westminster; Berkshire House, Mdx.; Newark Castle, Notts. and Ewelme Park, Oxon." In *The History of Parliament: The House of Commons 1604–1629*, Vol. 4, edited by Andrew Thrush and John P. Ferris, 817–20. Cambridge: Cambridge University Press, 2010.

Land, Robert Hunt. "Henrico and Its College." *The William and Mary Quarterly* 18, no. 4 (1938): 453–98.

Laslett, Peter. "Size and Structure of the Household in England over Three Centuries." *Population Studies* 23, no. 2 (1969): 199–223.

Lawrence, David R. *The Complete Soldier: Military Books and Military Culture in Early Stuart England, 1603–1645.* Leiden: Brill, 2009.

Leary, John E., Jr. *Francis Bacon and the Politics of Science.* Ames: Iowa State University Press, 1994.

Leemans, Inger, and Anne Goldgar, eds. *Early Modern Knowledge Societies as Affective Economies.* New York: Routledge, 2020.

Lefèvre, Corinne. "Recovering a Missing Voice from Mughal India: The Imperial Discourse of Jahāngīr (R. 1605–1627) in His Memoirs." *Journal of the Economic and Social History of the Orient* 50, no. 4 (2007): 452–89.

Lefroy, J. H. *Memorials of the Discovery and Early Settlement of the Bermudas or Somers Islands, 1515–1685.* London: Longmans, Green, and Co., 1877.

Leith-Ross, Prudence. *The John Tradescants: Gardeners to the Rose and Lily Queen.* London: Peter Owen, 1984.

———. *The Florilegium of Alexander Marshal in the Collection of Her Majesty the Queen at Windsor Castle.* London: Royal Collection, 2000.

Leng, Thomas. "Interlopers and Disorderly Brethren at the Stade Mart: Commercial Regulations and Practices amongst the Merchant Adventurers of England in the Late Elizabethan Period." *The Economic History Review* 69, no. 3 (2016): 823–43.

Leong, Elaine. *Recipes and Everyday Knowledge: Medicine, Science, and the Household in Early Modern England.* Chicago: University of Chicago Press, 2018.

Lindley, David, ed. *Court Masques: Jacobean and Caroline Entertainments, 1605–1640.* Oxford: Oxford University Press, 1995.

Lindley, Keith. "Irish Adventurers and Godly Militants in the 1640s." *Irish Historical Studies* 29, no. 13 (1994): 1–12.

Lindsay, David. *The Three Estates: A Pleasant Satire in Commendation of Virtue and in Vituperation of Vice.* Aldershot: Ashgate, 1998.

Linklater, Andrew, and Stephen Mennell. "Norbert Elias, The Civilizing Process: Sociogenetic and Psychogenetic Investigations-an Overview and Assessment." *History and Theory* 49, no. 3 (2010): 384–411.

Lockyer, Roger. *Buckingham, the Life and Political Career of George Villiers, First Duke of Buckingham, 1592–1628.* New York: Longman, 1981.

London, April. "Musaeum Tradescantium and the Benefactors to the Tradescants' Museum." In *Tradescant's Rarities: Essays on the Foundation of the Ashmolean Museum, 1683, with a Catalogue of the Surviving Early Collections,* edited by Arthur Macgregor, 24–39. Oxford: Clarendon, 1983.

Long, Pamela. *Openness, Secrecy, Authorship: Technical Arts and the Culture of Knowledge from Antiquity to the Renaissance.* Baltimore: Johns Hopkins University Press, 2001.

———. "Trading Zones in Early Modern Europe." *Isis* 106, no. 4 (2015): 840–47.

———. "Multi-Tasking 'Pre-Professional' Architect/Engineers and Other Bricolagic Practitioners as Key Figures in the Elision of Boundaries between Practice and Learning in Sixteenth-Century Europe: Some Roman Examples." In *The Structures of Practical Knowledge,* edited by Matteo Valleriani, 223–46. Cham: Springer International Publishing, 2017.

Lorimer, Joyce. "The Failure of the English Guiana Ventures 1595–1667 and James I's Foreign Policy." *Journal of Imperial and Commonwealth History* 21, no. 1 (1993): 1–30.

MacGregor, Arthur. "The Tradescants as Collectors of Rarities." In *Tradescant's Rarities: Essays on the Foundation of the Ashmolean Museum*, edited by Arthur MacGregor, 17–23. Oxford: Clarendon, 1983.

———. "The Cabinet of Curiosities in Seventeenth-Century Britain." In *The Origins of Museums: The Cabinet of Curiosities in Sixteenth- and Seventeenth-century Europe*, edited by Oliver Impey and Arthur MacGregor, 147–58. Oxford: Clarendon, 1985.

———. "'A Magazin of all Manner of Inventions': Museums in the Quest for 'Salomon's House' in Seventeenth-Century England." *Journal of the History of Collections* 1, no. 2 (1989): 207–12.

MacGregor, Arthur, ed. *Naturalists in the Field: Collecting, Recording and Preserving the Natural World from the Fifteenth to the Twenty-First Century*. Leiden: Brill, 2018.

MacIntyre, Jean. "Buckingham the Masquer." *Renaissance and Reformation* 34, no. 3 (1998): 59–81.

MacLaren, I. S. "'Zealous Sayles' and Zealous Sales: Bookings on the Northwest Passage." *The Princeton University Library Chronicle* 64, no. 2 (2003): 253–87.

MacLeod, Christine. *Inventing the Industrial Revolution: The English Patent System, 1660–1800*. Cambridge: Cambridge University Press, 1988.

Mahoney, Edward P. "Metaphysical Foundations of the Hierarchy of Being According to Some Late-Medieval and Renaissance Philosophers." In *Philosophies of Existence, Ancient and Medieval*, edited by Morewedge Parviz, 165–257. New York: Fordham University Press, 1982.

———. "Lovejoy and the Hierarchy of Being." *Journal of the History of Ideas* 48, no. 2 (1987): 211–30.

Malcolm, Noel, and Jacqueline Stedall. *John Pell (1611–1685) and His Correspondence with Sir Charles Cavendish: The Mental World of an Early Modern Mathematician*. Oxford: Oxford University Press, 2005.

Malcolmson, Cristina. *Studies of Skin Color in the Early Royal Society: Boyle, Cavendish, Swift*. London: Routledge, 2016.

Manning, Roger B. *Swordsmen: The Martial Ethos in the Three Kingdoms*. Oxford: Oxford University Press, 2003.

———. *An Apprenticeship in Arms: The Origins of the British Army 1585–1702*. Oxford: Oxford University Press, 2006.

Marchitello, Howard and Evelyn B. Tribble, eds. *The Palgrave Handbook of Early Modern Literature and Science*. London: Palgrave, 2017.

Markley, Robert. "Riches, Power, Trade and Religion: The Far East and the English Imagination, 1600–1720." *Renaissance Studies* 17, no. 3 (2003): 494–516.

Marr, Alexander, Raphaële Garrod, José Ramón Marcaida and Richard J. Oosterhoff. *Logodaedalus: Word Histories of Ingenuity in Early Modern Europe*. Pittsburgh: University of Pittsburgh Press, 2019.

Marsh, Ben. *Unravelled Dreams: Silk and the Atlantic World, 1500–1840*. Cambridge: Cambridge University Press, 2020.

Marsh, Simon. "The Construction and Arming of London's Defences 1642–1645." *Journal of the Society for Army Historical Research* 91, no. 368 (2013): 275–98.

Marshall, Peter. "Forgery and Miracles in the Reign of Henry VIII." *Past & Present* 178 (2003): 39–73.

Martín, Javier López. "Did Naval Artillery Really Exist during the Modern Period? A Brief Note on Cannon Design." In *Ships & Guns: The Sea Ordnance in Venice and Europe between the 15th and 17th Centuries,* edited by Carlo Beltrame and Renato Gianni Ridella, 73–84. Oxford: Oxbow Press, 2011.

Martin, Julian. *Francis Bacon, the State and the Reform of Natural Philosophy.* Cambridge: Cambridge University Press, 1992.

Mason, Thomas. *Serving God and Mammon: William Juxon, 1582–1663, Bishop of London, Lord High Treasurer of England, and Archbishop of Canterbury.* Newark: University of Delaware Press, 1985.

Matar, Nabil. *British Captives from the Mediterranean to the Atlantic, 1563–1760.* Leiden: Brill, 2014.

Matthee, Rudolph. *The Politics of Trade in Safavid Iran: Silk for Silver, 1600–1730.* Cambridge: Cambridge University Press, 1999.

Maydom, Katrina. "New World Drugs in England's Early Empire." PhD diss., University of Cambridge, 2018.

McCormick, Ted. "Food, Population, and Empire in the Hartlib Circle, 1639–1660." *Osiris* 35 (2020): 60–83.

———. "Population: Modes of Seventeenth-Century Demographic Thought," in *Mercantilism Reimagined: Political Economy in Early Modern Britain and its Empire,* edited by Philip J. Stern and Carl Wennerlind, 25–45. Oxford: Oxford University Press, 2014.

———. *William Petty and the Ambitions of Political Arithmetic.* Oxford: Oxford University Press, 2009.

McDonald, Kevin P. *Pirates, Merchants, Settlers, and Slaves: Colonial America and the Indo-Atlantic World.* Oakland: University of California Press, 2015.

McElligot, Jason. "Watts, William (c.1590–1649)." *Oxford Dictionary of National Biography,* edited by H. C. G. Matthew and B. Harrison. Oxford: Oxford University Press, 2004.

McEvansoneya, Philip. "A Note on Cornelius Drebbel." *Journal of Garden History* 6, no. 1 (1986): 19–20.

McGowen, Alan Patrick. "The Royal Navy under the First Duke of Buckingham, Lord High Admiral 1618–1628." PhD diss., Royal Holloway University of London, 1967.

Merchant, Carolyn. *The Death of Nature: Women, Ecology, and the Scientific Revolution.* San Francisco: Harper & Row, 1980.

Miller, Peter. "Nazis and Neo-Stoics: Otto Brunner and Gerhard Oestreich before and after the Second World War." *Past & Present* 176 (2002): 144–86.

Miller, Timothy Earl. "Pleasure, Honor, And Profit: Samuel Hartlib in His Papers 1620–1662." PhD diss., Georgia State University, 2015.

Millstone, Noah. "Seeing like a Statesman in early Stuart England." *Past and Present* 223 (2014): 77–127.

Mishra, Rupali Raj. *A Business of State: Commerce, Politics, and the Birth of the East India Company.* Cambridge, Massachusetts: Harvard University Press, 2018.

Mokyr, Joel. *A Culture of Growth: The Origins of the Modern Economy.* Graz Schumpeter Lectures. Princeton, NJ: Princeton University Press, 2017.

Monagle, Clare. *The Scholastic Project*. Kalamazoo: ARC Humanities Press, 2017.

Mossoff, Adam. "Rethinking the Development of Patents: An Intellectual History, 1550–1800." *The Hastings Law Journal* 52, no. 6 (2001): 1255–1322.

Motley, Mark Edward. *Becoming a French Aristocrat: The Education of the Court Nobility, 1580–1715*. Princeton, NJ: Princeton University Press, 1990.

Muchmore, Lynn. "Gerrard de Malynes and Mercantile Economics." *History of Political Economy* 1, no. 2 (1969): 336–58.

Mukherjee, Ayesha. *Penury into Plenty: Dearth and the Making of Knowledge in Early Modern England*. London: Routledge, 2015.

Mulry, Kate Luce. *An Empire Transformed: Remolding Bodies and Landscapes in the Restoration Atlantic*. New York: New York University Press, 2021.

Mulsow, Martin and Lorraine Daston. "History of Knowledge." In *Debating New Approaches to History*, edited by Marek Tamm and Peter Burke, 159–87. London: Bloomsbury Academic, 2019.

Müller-Mahn, Detlef, Jonathan Everts, and Christiane Stephan. "Riskscapes Revisited: Exploring the Relationship between Risk, Space and Practice." *Erdkunde* 72, no. 3 (2018): 197–214.

Muraca, Barbara. "Décroissance: A Project for a Radical Transformation of Society." *Environmental Values* 22, no. 2 (2013): 147–69.

Murphy, Rhoads. "Merchants, Nations and Free-Agency: An Attempt at a Qualitative Characterization of Trade in the Eastern Mediterranean 1620–1640." In *Friends and Rivals in the East: Studies in Anglo-Dutch Relations in the Levant from the Seventeenth to the Early Nineteenth Century*, edited by Alastair Hamilton, Alexander H. De Groot, and Maurits H. Van Den Boogert, 25–58. Leiden: Brill, 2000.

Natif, Mika. *Mughal Occidentalism: Artistic Encounters between Europe and Asia at the Courts of India, 1580–1630*. Leiden: Brill, 2018.

Needham, Joseph. *Science and Civilisation in China*, Vol. 3. Cambridge: Cambridge University Press, 1976.

Nerlich, Michael. *Ideology of Adventure: Studies in Modern Consciousness, 1100–1750*. Minneapolis: University of Minnesota Press, 1987.

Neumaier, Marco. "Salomon de Caus (1576–1626): Stationen Seines Lebens." In *Magische Maschinen: Salomon de Caus' Erfindungen für Den Heidelberger Schlossgarten 1614–1619*, edited by Frieder Hepp, Richard Leiner, Rüdiger Mach, and Marcus Popplow, 28–37. Neustadt: Pollichia, 2008.

Newell, Margaret Ellen. *From Dependency to Independence: Economic Revolution in Colonial New England*. Ithaca: Cornell University Press, 1998.

Newman, William R., and Lawrence M. Principe. *Alchemy Tried in the Fire: Starkey, Boyle, and the Fate of Helmontian Chymistry*. Chicago: University of Chicago Press, 2002.

Norman, A. V. B. "Arms, Armour and Militaria." In *The Late King's Goods: Collections, Possessions and Patronage of Charles I in the Light of the Commonwealth Sale Inventories*, edited by Arthur MacGregor, 351–66. London: Alistair McAlpine, 1989.

Novak, Maximillian E., ed. *The Age of Projects*. Toronto: University of Toronto Press, 2008.

Nováková, Julie, ed. "Der Brief des Johann Christoph Pergar an Samuel Hartlib Vom 24. 12. 1638." *Acta Comeniana* 6 (1985): 179–83.

Oestreich, Gerhard. *Geist und Gestalt des Frühmodernen Staates: Ausgewählte Aufsätze*. Berlin: Duncker & Humblot, 1969.

———. *Neostoicism and the Early Modern State*. Translated by David McClintock. Cambridge: Cambridge University Press, 1982.

———. *Antiker Geist und moderner Staat bei Justus Lipsius (1547–1606): Der Neostoizismus als Politische Bewegung*. Göttingen: Vandenhoeck & Ruprecht, 1989.

O'Connell, Caryn. "Bacon's Hints: The 'Sylva Sylvarum's' Intimate Science." *Studies in Philology* 113, no. 3 (2016): 634–67.

Okamura, Keisuke. "Interdisciplinarity Revisited: Evidence for Research Impact and Dynamism." *Palgrave Communications* 5, no. 141 (2019). https://doi.org/10.1057/s41599-019-0352-4.

Osborne, Dorothy. *Dorothy Osborne: Letters to Sir William Temple, 1652–54: Observations on Love, Literature, Politics, and Religion*. Burlington: Ashgate, 2002.

Osborne, Toby. *Dynasty and Diplomacy in the Court of Savoy: Political Culture and the Thirty Years' War*. Cambridge: Cambridge University Press, 2002.

Osmond, Patricia. "In Defense of Tiberius: Edmund Bolton, Tacitean Scholarship, and Early Stuart Politics." *Huntington Library Quarterly* 83, no. 3 (2020): 591–613.

Östling, Johan, Erling Sandmo, David Larsson Heidenblad, Anna Nilsson Hammar, and Kari Hernæs Nordberg. *Circulation of Knowledge: Explorations in the History of Knowledge*. Lund: Nordic Academic Press, 2018.

Otremba, Eric. "Inventing Ingenios: Experimental Philosophy and the Secret Sugar-Makers of the Seventeenth-Century Atlantic." *History and Technology* 28, no. 2 (2012): 119–47.

Ovitt, George, Jr.. "The Status of the Mechanical Arts in Medieval Classifications of Learning." *Viator* 14 (1983): 89–105.

Page, William, ed. *The Victoria History of the County of Oxford*, Vol. 2. London: Archibald Constable, 1907.

Pal, Carol. *Republic of Women: Rethinking the Republic of Letters in the Seventeenth Century*. Cambridge: Cambridge University Press, 2012.

Palmer, Alfred Neobald. "A History of the Old Parish of Gresford: Additions and Corrections." *Archaeologia Cambrensis* 5, no. 4 (1905): 265–94.

Paquette, Gabriel B. *The European Seaborne Empires: From the Thirty Years' War to the Age of Revolutions*. New Haven, CT: Yale University Press, 2019.

Park, Katharine, and Lorraine Daston. "Introduction: The Age of the New." In *Early Modern Science, Cambridge History of Science*, Vol. 3, 1–17. Cambridge: Cambridge University Press, 2008.

Parry, Glyn. "John Dee and the Elizabethan British Empire in its European Context." *The Historical Journal* 49, no. 3 (2006): 643–75.

Pastorino, Cesare. "The Mine and the Furnace: Francis Bacon, Thomas Russell, and Early Stuart Mining Culture." *Early Science and Medicine*, 14, no. 5 (2009): 630–60.

———. "Weighing Experience: Francis Bacon, the Inventions of the Mechanical Arts, and the Emergence of Modern Experiment." PhD diss., Indiana University, 2011.

———. "The Philosopher and the Craftsman: Francis Bacon's Notion of Experiment and its Debt to Early Stuart Inventors." *Isis* 108, no. 4 (2017): 749–68.

———. "Beyond Recipes: The Baconian Natural and Experimental Histories as an Epistemic Genre." *Centaurus* 62, no. 3 (2020): 447–64.

Paul, Joanne. *Counsel and Command in Early Modern English Thought.* Cambridge: Cambridge University Press, 2020.

Peck, Linda Levy. *Consuming Splendor: Society and Culture in Seventeenth-Century England.* Cambridge: Cambridge University Press, 2005.

Pestana, Carla Gardina. *The English Atlantic in an Age of Revolution, 1640–1661.* Cambridge, MA: Harvard University Press, 2004.

Pettegrew, David. *The Isthmus of Corinth: Crossroads of the Mediterranean World.* Ann Arbor: University of Michigan Press, 2016.

Pollnitz, Aysha. *Princely Education in Early Modern Britain.* Cambridge: Cambridge University Press, 2015.

Popper, Nicholas. *Walter Ralegh's History of the World and the Historical Culture of the Late Renaissance.* Chicago: University of Chicago Press, 2012.

Portal, Ethel M. "The Academ Roial of King James I." *Proceedings of the British Academy* (1915): 189–208.

Porter, Harry Culverwell. "Alexander Whitaker: Cambridge Apostle to Virginia." *The William and Mary Quarterly* 14, no. 3 (1957): 317–43.

Porter, Roy. "The Crispe Family and the African Trade in the Seventeenth Century." *Journal of African History* 9, no. 1 (1968): 57–77.

Porter, Theodore M. "Quantification and the Accounting Ideal in Science." *Social Studies of Science* 22, no. 4 (1992): 633–51.

———. "Thin Description: Surface and Depth in Science and Science Studies." *Osiris* 27 (2012): 209–26.

Portuondo, María. *Spanish Disquiet: The Biblical Natural Philosophy of Benito Arias Montano.* Chicago: University of Chicago Press, 2019.

Preston, Claire. *The Poetics of Scientific Investigation in Seventeenth-Century England.* Oxford: Oxford University Press, 2015.

Price, Michael W. "Recovering Donne's Critique of the *Arcana Imperii* in the *Problems.*" *Studies in Philology* 101, no. 3 (2004): 332–55.

Price, William Hyde. *English Patents of Monopoly.* Cambridge, MA: Harvard University Press, 1906.

Proctor, Robert, and Londa Schiebinger. *Agnotology: The Making and Unmaking of Ignorance.* Stanford, CA: Stanford University Press, 2008.

Pumfrey, Stephen, and Frances Dawbarn. "Science and Patronage in England, 1570–1625: A Preliminary Study." *History of Science* 42, no. 2 (2004): 138–88.

Quinn, David B. "A List of Books Purchased for the Virginia Company." *The Virginia Magazine of History and Biography* 77, no. 3 (1969): 347–60.

Quitslund, Beth. "The Virginia Company, 1606–1624." In *Anglo-American Millennialism, from Milton to the Millerites,* edited by Richard Connors and Andrew Colin Gow, 43–114. Leiden: Brill, 2004.

Raeff, Marc. *The Well-Ordered Police State: Social and Institutional Change through Law in the Germanies and Russia, 1600–1800.* New Haven, CT: Yale University Press, 1983.

Ransome, David R. "John Ferrar: A Half-Hidden Propagandist for Virginia." *The Seventeenth Century* 35, no. 5 (2020): 611–24.

———. "Wives for Virginia, 1621." *The William and Mary Quarterly* 48, no. 1 (1991): 3–18.

Ransome, David R., and David C. Lees. "The Virginian Silkworm: From Myth to Moth. Or: How a Businessman Turned into a Naturalist." *Antenna* 41, no. 3 (2017): 120–27.

Ratcliff, Jessica. "Art to Cheat the Common-Weale: Inventors, Projectors, and Patentees in English Satire, ca. 1630–70." *Technology and Culture* 53, no. 2 (2012): 337–65.

Razzari, Daniel. "Through the Backdoor: An Overview of the English East India Company's Rise and Fall in Safavid Iran, 1616–1640." *Iranian Studies* 52 (2019): 485–511.

Reid, Anthony. *An Indonesian Frontier: Acehnese and Other Histories of Sumatra*. Singapore: Singapore University Press, 2005.

Renn, Jürgen. *The Evolution of Knowledge: Rethinking Science for the Anthropocene*. Princeton, NJ: Princeton University Press, 2020.

Rice, Louise. "Poussin's Elephant." *Renaissance Quarterly* 70, no. 2 (2017): 548–93.

Rieppel, Lukas, Eugenia Lean and William Deringer, "Introduction: The Entangled Histories of Science and Capitalism," *Osiris* 33 (2018): 1–24.

Robertson, Haileigh. "Reworking Seventeenth-Century Saltpetre." *Ambix* 63, no. 2 (2016): 145–61.

Robson, Elly. "Improvement and Epistemologies of Landscape in Seventeenth-Century English Forest Enclosure." *The Historical Journal* 60, no. 3 (2017): 597–632.

Roodenburg, Herman, Robert Muchembled, and William Monter, eds. *Forging European Identities, 1400–1700*. Cambridge: Cambridge University Press, 2007.

Roper, Louis H. *Advancing Empire: English Interests and Overseas Expansion, 1613–1688*. Cambridge: Cambridge University Press, 2017.

Rosenfeld, Sybil. *A Short History of Scene Design in Great Britain*. Totowa, NJ: Rowman and Littlefield, 1973.

Rosenthal, Earl. "Plus Ultra, Non plus Ultra, and the Columnar Device of Emperor Charles V." *Journal of the Warburg and Courtauld Institutes* 34 (1971): 204–28.

Rossi, Paolo. *Francis Bacon: From Magic to Science*. Translated by Sacha Rabinovitch. London: Routledge & Kegan Paul, 1968.

———. "Truth and Utility in the Science of Francis Bacon." In *Philosophy, Technology, and the Arts in the Early Modern Era*, translated by Salvator Attanasio, 146–73. New York: Harper and Row, 1970.

Rountree, Helen C. *The Powhatan Indians of Virginia: Their Traditional Culture*. Norman: University of Oklahoma Press, 1989.

Ruellet, Aurélien. *La Maison de Salomon: Histoire Du Patronage Scientifique et Technique en France et en Angleterre Au XVII Siècle*. Rennes: PU Rennes, 2016.

———. *Les fabricants d'instruments mathématiques logés dans la Grande Galerie Du Louvre (ca. 1600—ca. 1660): quelques notices biographiques*. 2014. halshs-01117458

Rugemer, Edward Bartlett. *Slave Law and the Politics of Resistance in the Early Atlantic World*. Cambridge, MA: Harvard University Press, 2018.

Rusu, Doina-Cristina, and Christoph Lüthy. "Extracts from a Paper Laboratory: The Nature of Francis Bacon's Sylva Sylvarum." *Intellectual History Review* 27, no. 2 (2017): 171–202.

Ryan, Lawrence V. "Chaucer's Criseyde in Neo-Latin Dress." *English Literary Renaissance* 17, no. 3 (1987): 288–302.

Saco, Louis D. "Sir Christopher Gardyner." *Transactions of the Colonial Society of Massachusetts* 38 (1959): 3–15.

Sainsbury, Ethel Bruce. *A Calendar of the Court Minutes of the East India Company, 1635–1679*. Oxford: Clarendon Press, 1907.

Saintsbury, George, ed. *Minor Poets of the Caroline Period*, Vol. 2. Oxford: Clarendon, 1906.

Salmon, J. H. M. "Stoicism and Roman Example: Seneca and Tacitus in Jacobean England." *Journal of the History of Ideas* 50, no. 2 (1989): 199–225.

Samson, Alexander. "Introduction." In *Locus Amoenus: Gardens and Horticulture in the Renaissance*, edited by Alexander Samson, 1–23. Chichester: Wiley, 2012.

———. *The Spanish Match: Prince Charles's Journey to Madrid, 1623*. Burlington: Ashgate, 2006.

Sandl, Marcus. "Development as Possibility: Risk and Chance in the Cameralist Discourse." In *Economic Growth and the Origins of Modern Political Economy: Economic Reasons of State, 1500–2000*, edited by Philipp Robinson Rössner, 139–55. New York: Routledge, 2016.

Schaffer, Simon, Lissa Roberts, Kapil Raj, and James Delbourgo, eds. *The Brokered World: Go-betweens and Global Intelligence, 1770–1820*. Sagamore Beach, MA: Science History Publications, 2009.

Schiebinger, Londa. *Secret Cures of Slaves: Peoples, Plants and Medicine in the Eighteenth-century Atlantic World*. Stanford, CA: Stanford University Press, 2017.

Schilt, Cornelis J. "'To Improve upon Hints of Things': Illustrating Isaac Newton." *Nuncius* 31, no. 1 (2016): 50–77.

Schleck, Julia. "Forming Knowledge: Natural Philosophy and English Travel Writing." In *Travel Narratives, the New Science, and Literary Discourse, 1569–1750*, edited by Judy Hayden, 53–70. New York: Routledge, 2012.

Schotte, Margaret. *Sailing School: Navigating Science and Skill, 1550–1800*. Baltimore: Johns Hopkins University Press, 2019.

Schreiber, Roy E. "The First Carlisle Sir James Hay, First Earl of Carlisle as Courtier, Diplomat and Entrepreneur, 1580–1636." *Transactions of the American Philosophical Society* 74, no. 7 (1984): 1–202.

Screech, Timon. "'Pictures (the Most Part Bawdy)': The Anglo-Japanese Painting Trade in the Early 1600s." *The Art Bulletin*, 87, no. 1 (2005): 50–72.

Secord, James A. "Knowledge in Transit." *Isis* 95, no. 4 (2004): 654–72.

Sellers-García, Sylvia. *Distance and Documents at the Spanish Empire's Periphery*. Stanford, CA: Stanford University Press, 2014.

Sellin, Paul R. "Michel Le Blon and England III: Gustav II Adolf, Sir Walter Raleigh's Gold Mine, and the Perfidy of George Villiers, Duke of Buckingham." *Dutch Crossing* 23, no. 1 (1999): 102–32.

Serjeantson, Richard. "The Philosophy of Sir Francis Bacon in Early Jacobean Oxford, with an Edition of an Unknown Manuscript of the 'Valerius Terminus.'" *The Historical Journal* 56, no. 4 (2013): 1087–106.

Shachtman, Tom. *Absolute Zero and the Conquest of Cold*. Boston: Houghton Mifflin Co, 1999.

Shami, Jeanne. "Love and Power: The Rhetorical Motives of John Donne's 1622 Sermon to the Virginia Company." In *Renaissance Papers 2004*, edited by Christopher Cobb and Thomas Hester, 85–106. Columbia, SC: Camden House, 2005.

Shapin, Steven. "The Invisible Technician." *American Scientist* 77, no. 6 (1989): 554–63.

———. "'A Scholar and a Gentleman': The Problematic Identity of the Scientific Practitioner in Early Modern England." *History of Science* 29, no. 3 (1991): 279–327.

———. *A Social History of Truth: Civility and Science in Seventeenth-Century England.* Chicago: University of Chicago Press, 1994.

———. *Never Pure: Historical Studies of Science as If It Was Produced by People with Bodies, Situated in Time, Space, Culture, and Society, and Struggling for Credibility and Authority.* Baltimore: Johns Hopkins University Press, 2010.

———. *The Scientific Revolution.* Chicago: University of Chicago Press, 2018.

Shapin, Steven, and Simon Schaffer. *Leviathan and the Air Pump: Hobbes, Boyle, and the Experimental Life.* Princeton: Princeton University Press, 1985.

Shapiro, Barbara J. *Probability and Certainty in Seventeenth-Century England: A Study of the Relationships between Natural Science, Religion, History, Law, and Literature.* Princeton, NJ: Princeton University Press, 1983.

———. "The Concept 'Fact': Legal Origins and Cultural Diffusion." *Albion: A Quarterly Journal Concerned with British Studies* 26, no. 2 (1994): 227–52.

Sharp, Lindsay. "Walter Charleton's Early Life 1620–1659, and Relationship to Natural Philosophy in Mid-Seventeenth Century England." *Annals of Science* 30, no. 3 (1973): 311–40.

Sharpe, Kevin. "The Earl of Arundel, His Circle and the Opposition to the Duke of Buckingham, 1618–1628." In *Faction and Parliament: Essays on Early Stuart History,* edited by Kevin Sharpe, 209–44. Oxford: Clarendon, 1978.

———. *Sir Robert Cotton 1586–1631: History and Politics in Early Modern England* (Oxford: Oxford University Press, 1979).

———. "Virtues, Passions, and Politics in Early Modern England." *History of Political Thought* 32, no. 5 (2011): 773–98.

Shaw, William. *Letters of Denization and Acts of Naturalization for Aliens in England and Ireland, 1603–1800.* London: Huguenot Society, 1911.

Sherman, William. "Patents and Prisons: Simon Sturtevant and the Death of the Renaissance Inventor." *Huntington Library Quarterly* 72 (2009): 239–56.

Shirley, Evelyn Philip. *The Sherley Brothers.* Chiswick: Whittingham, 1848.

Siddons, Michael P. *The Heraldry of Foreigners in England 1400–1700.* London: Harleian Society, 2010.

Skeehan, Danielle C. *The Fabric of Empire: Material and Literary Cultures of the Global Atlantic, 1650–1850.* Baltimore: Johns Hopkins University Press, 2020.

Skelton, Robert. "Imperial Symbolism in Mughal Painting." In *Content and Context of Visual Arts in the Islamic World,* edited by P. Soucek, 177–87. London: Pennsylvania State University Press, 1988.

Skempton, A. W. "Edmund Colthurst." In *Biographical Dictionary of Civil Engineers in Great Britain and Ireland, 1500–1830,* Vol. 1, 147–8. London: Telford, 2002.

Slack, Paul. *The Invention of Improvement: Information and Material Progress in Seventeenth-Century England.* Oxford: Oxford University Press, 2015.

Smith, David Chan. "Useful Knowledge, Improvement, and the Logic of Capital in Richard Ligon's True and Exact History of Barbados." *Journal of the History of Ideas* 78, no. 4 (2017): 549–70.

Smith, Edmond J. "'Canaanising Madagascar': Africa in English Imperial Imagination, 1635–1650." *Itinerario* 39, no. 2 (2015): 277–98.

Smith, Pamela H. *The Business of Alchemy: Science and Culture in the Holy Roman Empire.* Princeton, NJ: Princeton University Press, 1994.

———. *The Body of the Artisan: Art and Experience in the Scientific Revolution*. Chicago: University of Chicago Press, 2004.

———. "Science on the Move: Recent Trends in the History of Early Modern Science." *Renaissance Quarterly* 62, no. 2 (2009): 345–75.

———. "Itineraries of Materials and Knowledge in the Early Modern World." In *The Global Lives of Things: The Material Culture of Connections in the Early Modern World*, edited by Anne Gerritsen and Giorgio Riello, 31–61. London: Routledge, 2016.

———. "Making the Edition of Ms. Fr. 640." In *Secrets of Craft and Nature in Renaissance France: A Digital Critical Edition and English Translation of BnF Ms. Fr. 640*, edited by Pamela H. Smith, Naomi Rosenkranz, Tianna Helena Uchacz, Tillmann Taape, Clément Godbarge, Sophie Pitman, Jenny Boulboullé, et al. New York: Making and Knowing Project, 2020. https://edition640.makingandknowing.org/# /essays/ann_320_ie_19.

Smith, Pamela H., and Benjamin Schmidt. *Making Knowledge in Early Modern Europe: Practices, Objects, and Texts, 1400–1800*. Chicago: University of Chicago Press, 2007.

Smuts, Robert Malcolm. *Court Culture and the Origins of a Royalist Tradition in Early Stuart England*. Philadelphia: University of Pennsylvania Press, 1987.

Sobecki, Sebastian I. "Introduction: Edgar's Archipelago." In *The Sea and Englishness in the Middle Ages: Maritime Narratives, Identity and Culture*, edited by Sebastian I. Sobecki, 1–30. Cambridge: D.S. Brewer, 2011.

Solomon, Julie Robin. *Objectivity in the Making: Francis Bacon and the Politics of Inquiry*. Baltimore: Johns Hopkins University Press, 1998.

Sorell, Tom, G. A. J. Rogers, and Jill Kraye. Introduction. In *Scientia in Early Modern Philosophy: Seventeenth-Century Thinkers on Demonstrative Knowledge from First Principles*, edited by Tom Sorell, G. A. J. Rogers and Jill Kraye, vii–xiii. Dordrecht: Springer, 2010.

Southern, Richard. *Changeable Scenery, Its Origin and Development in the British Theatre*. London: Faber and Faber, 1952.

Speake, Jennifer. "The Wrong Kind of Wonder: Ben Jonson and Cornelis Drebbel." *The Review of English Studies* 66, no. 273 (2015): 60–70.

Speer, Andreas. "Schüler und Meister." In *Schüler und Meister*, edited by Andreas Speer and Thomas Jeschke, xi–xvii. Berlin: De Gruyter, 2016.

Srbik, Heinrich Ritter von. *Wilhelm von Schröder: Ein Beitrag zur Geschichte der Staatswissenschaften*. Vienna: Kommission bei Hölder, 1910.

Stanwood, Owen. "Captives and Slaves: Indian Labor, Cultural Conversion, and the Plantation Revolution in Virginia." *The Virginia Magazine of History and Biography* 114, no. 4 (2006): 434–63.

Stark, Laura. "Emergence." *Isis* 110, no. 2 (2019): 332–36.

Steen, Sara Jayne, ed. *The Letters of Lady Arabella Stuart*. New York: Oxford University Press, 1994.

Steinmann, Linda. "Shah 'Abbas and the Royal Silk Trade 1599–1629." *Bulletin for the British Society for Middle Eastern Studies* 14, no. 1 (1987): 68–74.

Stevens, Roger. "Robert Sherley: The Unanswered Questions." *Iran: Journal of the British Institute of Persian Studies* 17 (1979): 115–25.

Stevenson, Christine. "Occasional Architecture in Seventeenth-Century London." *Architectural History* 49 (2006): 35–74.

Stewart, Larry. "Global Pillage: Science, Commerce, and Empire." In *Cambridge History of Science*, Vol. 4, *Eighteenth-Century Science*, edited by Roy Porter, 825–844. Cambridge: Cambridge University Press, 2003.

Stewart, Richard Winship. *The English Ordnance Office, 1585–1625: A Case Study in Bureaucracy*. Woodbridge: Boydell Press, 1996.

Stollberg-Rilinger, Barbara. *Der Staat als Maschine: Zur Politischen Metaphorik des Absoluten Fürstenstaats*. Berlin: Duncker & Humblot, 1986.

Stone, Lawrence. *The Crisis of the Aristocracy, 1558–1641*. Oxford: Clarendon Press, 1965.

Strong, Roy C. *The Renaissance Garden in England*. London: Thames and Hudson, 1979.

Sturdy, David. "The Tradescants at Lambeth." *Journal of Garden History* 2, no. 1 (1982): 1–16.

Subrahmanyam, Sanjay. *Europe's India: Words, People, Empires, 1500–1800*. Cambridge, MA: Harvard University Press, 2017.

Sutto, Antoinette. *Loyal Protestants and Dangerous Papists: Maryland and the Politics of Religion in the English Atlantic, 1630–1690*. Charlottesville: University of Virginia Press, 2015.

Svalastog, Julie. *Mastering the Worst of Trades: England's Early Africa Companies and Their Traders, 1618–1672*. Leiden: Brill, 2021.

Svensson, Anna. "'And Eden from the Chaos Rose': Utopian Order and Rebellion in the Oxford Physick Garden." *Annals of Science* 76, no. 2 (2019): 157–83.

Tait, Hugh. "Southwark (Alias Lambeth) Delftware and the Potter, Christian Wilhelm." *The Connoisseur* 146 (1960): 36–42.

Testa, Simone. "From the 'Bibliographical Nightmare' to a Critical Bibliography: Tesori Politici in the British Library, and Elsewhere in Britain." *British Library Journal* (2008): 1–33.

Third Report of the Royal Commission on Historical Manuscripts. London: Eyre and Spottiswoode, 1872.

Thirsk, Joan. *Economic Policy and Projects: The Development of a Consumer Society in Early Modern England*. Oxford: Clarendon Press, 1978.

———. "The Crown as Projector on its own Estates, from Elizabeth I to Charles I." In *The Estates of the English Crown, 1558–1640*, edited by R. W. Hoyle, 297–352. Cambridge: Cambridge University Press, 1992.

Thomas, Keith. *Man and the Natural World: Changing Attitudes in England, 1500–1800*. New York: Oxford University Press, 1996.

Thorpe, W. H. "The Marquis of Worcester and Vauxhall." *Transactions of the Newcomen Society* 13 (1932): 75–88.

Thrush, Andrew Derek. "The Navy Under Charles I." PhD diss., University College London, 1990.

Thurs, Daniel P. "Myth 26: That the Scientific Method Accurately Reflects What Scientists Actually Do." In *Newton's Apple and Other Myths about Science*, edited by Ronald L. Numbers and Kostas Kampourakis, 210–18. Cambridge, MA: Harvard University Press, 2015.

Tillyard, Eustace M. *The Elizabethan World Picture: A Study of the Idea of Order in the Age of Shakespeare, Donne and Milton*. New York: Vintage, 1959.

Todt, Kim. "A Venture of Her Own: Early American Women in Business." *Early Modern Women: An Interdisciplinary Journal* 10, no. 1 (2015): 152–163.

Tomlins, Christopher. *Freedom Bound: Law, Labor, and Civic Identity in Colonizing English America, 1580–1865*. New York: Cambridge University Press, 2010.

Tomlinson, Edward Murray. *A History of the Minories, London*. London: John Murray, 1922.

Townshend, Dorothea. *Life and Letters of Mr. Endymion Porter*. London: Unwin, 1897.

Trevor-Roper, H. R. *Europe's Physician: The Various Life of Sir Theodore de Mayerne*. New Haven, CT: Yale University Press, 2006.

Trim, D. J. B. "English Military Émigrés and the Protestant Cause in Europe, 1603–c. 1640." In *British and Irish Emigrants and Exiles in Europe: 1603–1688*, edited by David Worthington, 237–260. Leiden: Brill, 2010.

Truitt, Elly Rachel. *Medieval Robots: Mechanism, Magic, Nature, and Art*. Philadelphia: University of Pennsylvania Press, 2015.

Tuck, Richard. *Philosophy and Government 1572–1651*. Cambridge: Cambridge University Press, 1993.

Tullock, Tom Gregorie. *The Rise and Progess of the British Explosives Industry*. London: Whittaker, 1909.

Turnbull, David. *Masons, Tricksters, and Cartographers: Comparative Studies in the Sociology of Scientific and Indigenous Knowledge*. New York: Routledge, 2000.

Turnbull, George. "Samuel Hartlib's Connection with Sir Francis Kynaston's 'Musaeum Minervae.'" *Notes and Queries* 197, no. 2 (1952): 33–37.

Turton, R. B. *The Alum Farm: Together with a History of the Origin, Development and Eventual Decline of the Alum Trade in North-East Yorkshire*. Whitby: Home & Son, 1938.

Upton, Anthony F. *Sir Arthur Ingram, c.1565–1642*. London: Oxford University Press, 1961.

Vaughan, Alden. *Transatlantic Encounters: American Indians in Britain, 1500–1776*. Cambridge: Cambridge University Press, 2006.

Virga, Vincent, with Alan Brinkley. *Eyes of the Nation: A Visual History of the United States*. Charlestown, MA: Bunker Hill, 2004.

Vollgraff, J.A. "De Rol van Den Nederlander Caspar Calthoff bij de Uitvinding van het Moderne Stoomwerktuig." *Physica* 12 (1932): 257–68.

Walker, Daniel. *Flowers Underfoot: Indian Carpets of the Mughal Era*. New York: Metropolitan Museum of Art, 1997.

Wakefield, Andre. *The Disordered Police State: German Cameralism as Science and Practice*. Chicago: University of Chicago Press, 2009.

Walton, Steven. "The Tower Gunners and the Artillery Company in the Artillery Garden before 1630." *Journal of the Ordance Society* 18 (2006): 53–66.

Ward, Robert. *London's New River*. London: Historical Publications, 2003.

Webster, Charles. *The Great Instauration: Science, Medicine, and Reform, 1626–1660*. London: Duckworth, 1974.

———. "William Harvey and the Crisis of Medicine in Jacobean England." In *William Harvey and His Age: The Professional and Social Context of the Discovery of the Circulation,* edited by J. Bylebyl, 1–27. Baltimore: Johns Hopkins University Press, 1979.

Weisheipl, James, A. "Classifications of the Sciences in Medieval Thought." *Mediaeval Studies* 27 (1965): 54–90.

Wennerlind, Carl. *Casualties of Credit: The English Financial Revolution, 1620–1720.* Cambridge: Harvard University Press, 2011.

Werrett, Simon. *Fireworks: Pyrotechnic Arts and Sciences in European History.* Chicago: University of Chicago Press, 2010.

———. *Thrifty Science: Making the Most of Materials in the History of Experiment.* Chicago: University of Chicago Press, 2019.

White, Curtis. "The Jenks Area and the Tailrace." In *Saugus Iron Works: The Roland W. Robbins Excavations, 1948–1953,* edited by William A. Griswold and Donald W. Linebaugh, 173–98. Washington: National Park Service, 2010.

White, Jason. *Militant Protestantism and British Identity, 1603–1642.* New York: Routledge, 2016.

Whitney, Elspeth. "Paradise Restored: The Mechanical Arts from Antiquity through the Thirteenth Century." *Transactions of the American Philosophical Society* 80, no. 1 (1990): 1–169.

Wijnman, H. F. "Vernatti, Sir Philibert (1)." In *Nieuw Nederlandsch Biografisch Woordenboek,* edited by P. J. Block and P. C. Molhuysen, 1201–3. Leiden: Sijthoff, 1933.

Wilks, Timothy. "Art Collecting at the English Court from the Death of Henry, Prince of Wales to the Death of Anne of Denmark." *Journal of the History of Collections* 9, no. 1 (1997): 31–48.

Willes, Margaret. *The Making of the English Gardener: Plants, Books and Inspiration, 1560–1660.* Padstow, Cornwall: TJ, 2011.

Williams, Jean. "Olympism and Pastoralism in British Sporting Literature." In *British Sporting Literature and Culture in the Long Eighteenth Century,* edited by Sharon Harrow, 33–54. New York: Routledge, 2016.

Williams, Raymond. *The Country and the City.* Oxford: Oxford University Press, 1975.

Williamson, Colin F. "The Life and Works of Sir Francis Kinaston." LittB diss., University of Oxford, 1957.

Williamson, Hugh Ross. *Four Stuart Portraits.* London: Evans Brothers, 1949.

Wilson, Charles. *England's Apprenticeship, 1603–1763.* New York: Longman, 1984.

Wilson, Guy M. *The Vauxhall Operatory: A Century of Inventions before the Scientific Revolution.* Leeds: Basiliscoe Press, 2009.

Winfield, Rif. *British Warships in the Age of Sail, 1603–1714: Design, Construction, Careers and Fates.* Barnsley: Seaforth, 2009.

Winkler, Amanda Eubanks. *Music, Dance, and Drama in Early Modern English Schools.* Cambridge: Cambridge University Press, 2020.

Winterbottom, Anna. *Hybrid Knowledge in the Early East India Company World.* Basingstoke: Palgrave Macmillan, 2016.

Wolfe, Audra J. *Freedom's Laboratory: The Cold War Struggle for the Soul of Science.* Baltimore: Johns Hopkins University Press, 2018.

Wood, Andy. "Custom, Identity and Resistance: English Free Miners and Their Law, c. 1550–1800." In *The Experience of Authority in Early Modern England,* edited by Paul Griffiths, Adam Fox, and Steve Hindle, 249–85. Houndmills: Macmillan, 1996.

Wood, Jeremy, "Gerbier, Sir Balthazar (1592–1663/1667)." *Oxford Dictionary of National Biography.* Oxford: Oxford University Press, 2004.

Woodcock, Brian A. "'The Scientific Method' as Myth and Ideal." *Science & Education* 23, no. 10 (2014): 2069–93.

Working, Lauren. *The Making of an Imperial Polity: Civility and America in the Jacobean Metropolis*. Cambridge: University Press, 2020.

Wrightson, Keith. *English Society, 1580–1680*. New Brunswick, NJ: Rutgers University Press, 1982.

———. "The Politics of the Parish in Early Modern England." In *The Experience of Authority in Early Modern England*, edited by Paul Griffiths, Adam Fox, and Steve Hindle, 10–46. Houndmills: Macmillan, 1996.

Yamamoto, Koji. *Taming Capitalism before Its Triumph: Public Service, Distrust, and "Projecting" in Early Modern England*. Oxford: Oxford University Press, 2018.

Zabel, Christine. "Introduction: The Search for Self-Interest and the Problems with Its Historicization." In *Historicizing Self-Interest in the Modern Atlantic World: A Plea for Ego?*, edited by Christine Zabel, 1–22. New York: Routledge, 2021.

Zilberstein, Anya. *A Temperate Empire: Making Climate Change in Early America*. Oxford: Oxford University Press, 2016.

Zilsel, Edgar. *The Social Origins of Modern Science*. Edited by Diederick Raven, Wolfgang Krohn, and Robert S. Cohen. Dordrecht: Kluwer Academic Publishers, 2000.

Zook, George Frederick. *The Company of Royal Adventurers Trading into Africa*. Lancaster, PA: New Era, 1919.

Page numbers in *italic* indicate figures.